THE
OXFORD HIS
OF AUSTRALIA

THE
OXFORD HISTORY
OF AUSTRALIA

General Editor Geoffrey Bolton

Volume 1 *Aboriginal Australia* Tim Murray
Volume 2 *1770–1860* Jan Kociumbas
Volume 3 *1860–1900* Beverley Kingston
Volume 5 *1942–1988* Geoffrey Bolton

THE OXFORD HISTORY OF AUSTRALIA

VOLUME 4
1901-1942
THE SUCCEEDING AGE

STUART MACINTYRE

MELBOURNE
OXFORD UNIVERSITY PRESS
OXFORD AUCKLAND NEW YORK

OXFORD UNIVERSITY PRESS AUSTRALIA

Oxford New York
Athens Auckland Bangkok Bombay
Calcutta Cape Town Dar es Salaam Delhi
Florence Hong Kong Istanbul Karachi
Kuala Lumpur Madras Madrid Melbourne
Mexico City Nairobi Paris Port Moresby
Singapore Taipei Tokyo Toronto
and associated companies in Berlin Ibadan

OXFORD is a trade mark of Oxford University Press

First published 1986
Paperback edition 1993
Reprinted 1997

National Library of Australia
Cataloguing-in-Publication data:

Macintyre, Stuart, 1947–
The Oxford history of Australia. Volume 4, 1901–42.

Bibliography.
Includes index.
ISBN 0 19 553518 9.

1. Australia — History — 1901–1945. I. Title.
II. Title: History of Australia. Volume 4, 1901–1942.
III. Title: Succeeding age.

994.04

Edited by Carla Taines
Designed by Guy Mirabella
Typeset by Asco Trade Typesetting Ltd, Hong Kong
Printed through OUP China
Published by Oxford University Press
253 Normanby Road, South Melbourne, Australia

CONTENTS

TABLES

PUBLISHER'S NOTE

The Oxford History of Australia covers the sweep of Australian history from the first human settlement down to the 1980s. It consists of five volumes, each written by a single author with an established reputation as a productive and lively-minded historian. Each volume covers a distinct period of Australian history: Aboriginal history; white settlement, 1788–1860; colonial growth and maturation, 1860–1900; the Australian Commonwealth in peace and war, 1901–41; the modern era, from 1942 to the present. Each volume is a work of historical narrative in its own right. It draws the most recent research into a coherent and realized whole.

Aboriginal Australia is treated in its entirety, from the dramatically recast appreciation of early prehistory to present-day controversies of place, identity and belief. Colonial Australia begins with the establishment of tiny settlements at different times and with different purposes on widely separated points on the Australian coastline. From these fragments of British society sprang the competing ambitions of their members and a distinctly new civilization emerged. As the colonists spread over the continent and imposed their material culture on its resources, so the old world notions of class, status and gender were reworked. The colonists came together at the beginning of the twentieth century and fashioned new institutions to express their goals of national self-sufficiency, yet they were tossed and buffeted by two wars and the dictates of the international economy. The final volume therefore reflects the continuity of Australia's economic and political dependence, and new patterns in the quest for social justice by women, the working class and ethnic minorities.

In tracing these themes, the Oxford History's authors have held firmly to the conviction that history needs to interpret the past as an intelligible whole. The volumes range widely in their use of source material. They are informed by specialist research and enlivened by vivid example. Above all, they are written as narrative history with a clear and dramatic thread. No common ideological orthodoxy has been imposed on the authors beyond a commitment to scholarly excellence in a form which will be read and enjoyed by many Australians.

PREFACE

This history relies, as all general histories must rely, on the work of others. Our understanding of many aspects of Australian history during the first four decades of the twentieth century remains incomplete, but insofar as an overview is possible, it rests on the pioneering work of earlier scholars and on the books, articles and unpublished theses that accumulated so rapidly. My indebtedness to this literature is declared in the endnotes: here it is appropriate to acknowledge the pleasure as well as the guidance that my reading has yielded.

Yet a general history must be something more than a mere accumulation of established knowledge, and the proposition that the general historian can work simply as a grand synthesizer, always dubious, is today even more difficult to sustain. Different forms of historical inquiry proliferate, employing particular techniques and embodying particular points of view. Women's history, labour history, Aboriginal, economic, social, urban, demographic, educational, administrative and military history—the list is far from exhaustive. It is not just that these sub-disciplines are practised as increasingly autonomous enterprises, their practitioners speaking to fellow enthusiasts all too often at the expense of a larger audience. More than this, the threatened fragmentation of the discipline calls into question the project of a holistic representation of the past. It is, for example, no longer possible to sustain the convention—so beloved of an earlier generation of Oxford historians—that history is made by the public endeavours of influential men. Nor will it suffice to present the Australian past as it was presented in more than one history of Australia—as a record of national achievement, a people moving towards self-realization. Organizing principles such as these cannot accommodate the specialisms which, at the very least, force the general historian to wrestle with the complexities and contingencies they have uncovered.

On the other hand, there was and is an Australian past. A people inhabited this land, they joined in material and social practices and ordered their affairs in associations and parties. Unless we grasp these relationships in their totality, we cannot properly understand any one of them. For this reason I am convinced that the fraying of old orthodoxies should be regarded less as an impediment to the work of the general historian than as its justification. Nor am I convinced that the present enthusiasm for history from below invalidates a concern for political processes. We may learn much from reconstructing the full range of social relationships and the texture of everyday life, but we will not understand the popular experience unless we recognize the effects of structures of power. Accordingly, I have paid considerable attention in these pages to the means whereby different classes and sections of society pursued their interests. I do not accept that this is to lapse into high politics, for I find that spatial metaphor as treacherous as its counterpart, history from below. Politics, properly understood, is not an end in itself but a response to social relationships. My approach has therefore been to concentrate in the early chapters on exploring those relationships and in the later ones on the endeavours to which they gave rise.

In carrying the project through, I have incurred some particular obligations. For information and advice, I thank Jeremy Beckett, Ken Buckley, Alison Churchward, Kevin Fewster, Margaret Hicks, Bill Latter, Jenny Lee, Paul de Serville and Jocelyn Treasure. For assistance with some of the research I am indebted to Stephanie Brown, Janette Ryan and Margaret Vickers. I have been helped also by the staff of various libraries and archives: the Baillieu Library, the Giblin

Library and the Archives of the University of Melbourne; the Menzies Library and the Archives of Business and Labour in the Australian National University; the Battye Library, the La Trobe Library, the Mitchell Library and the National Library of Australia. The Arts Faculty of the University of Melbourne and the Research School of Social Sciences in the Australian National University provided financial support. Geoffrey Bolton, Martha Macintyre and Rob Pascoe read drafts and made valuable suggestions. I am indebted to them all.

The subtitle of this volume comes from Nettie Palmer's *Fourteen Years. Extracts from a Private Journal, 1925–1939*: '... this is the succeeding age, and a difficult one; here we are, a part of mankind, and being forced to face the fact.'

Stuart Macintyre
University of Melbourne

NOTE ON MEASUREMENTS

This book employs contemporary units of measurement. Australians in the first half of the twentieth century reckoned in pounds, shillings and pence, and they were loath to give someone an inch lest they take a mile.

	currency	
12d (12 pence) =	1s (1 shilling)	
20s (20 shillings) =	£1 (1 pound) =	$2 (2 dollars)
21s =	1 guinea	
	weight	
	1 pound =	.453 kilograms
14 pounds =	1 stone	
8 stone =	1 hundredweight	
20 hundredweight =	1 ton =	1.02 tonnes
	length	
	1 inch =	25.4 millimetres
12 inches =	1 foot	
3 feet =	1 yard	
22 yards =	1 chain	
10 chains =	1 furlong	
8 furlongs =	1 mile =	1.61 kilometres
	area	
4840 square yards =	1 acre =	.405 hectares
640 acres =	1 square mile	
	capacity	
	1 pint =	.568 litres
8 pints =	1 gallon	
8 gallons =	1 bushel	

ABBREVIATIONS

ACTU	Australian Council of Trade Unions
AIF	Australian Imperial Force
ALP	Australian Labor Party
AWU	Australian Workers' Union
BHP	Broken Hill Proprietary Company
IWW	Industrial Workers of the World
ICI	Imperial Chemical Industries
RSSILA	Returned Soldiers' and Sailors' Imperial League of Australia
UAP	United Australia Party

INDIAN
OCEAN
Darw
Broome
N
TE
WESTERN
AUSTRALIA
Yalgoo
Kalgoorlie
Coolgardie
Boulder
Nullarbor Plain
Northam
Nangeenan
Perth
Blackboy Hill
Rottnest Is
Fremantle
Frankland R
Albany
King George S

North

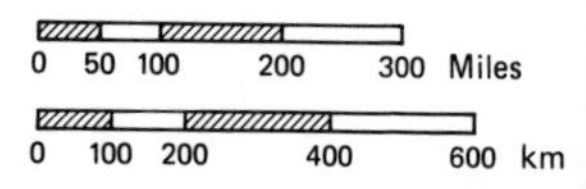
0 50 100 200 300 Miles
0 100 200 400 600 km

Gulf of Carpentaria
PACIFIC OCEAN
QUEENSLAND
Mungana
Cairns
Townsville
Charters Towers
Cloncurry
Rockhampton
Mt Morgan
Darling Downs
Ipswich
Brisbane
Warwick
Paroo R
Tibooburra
Mt Brown
Yancannia Station
Wilcannia
Darling R
Broken Hill
Grafton
New England District
Gunnedah
NEW SOUTH WALES
Hunter R
Maitland
Newcastle
Flinders Ranges
Whyalla
Port Pirie
Yorke Peninsula
Barossa Valley
Adelaide
Hahndorf
Mildura
Ouyen
Mallee
Sea Lake
Wimmera
Nhill
Minimay
Naracoorte
Penola
Mt Gambier
Booligal
Murrumbidgee R
Riverina
Murray R
Wyalong
Lithgow
Holsworthy
Sydney
Bulli
Port Kelba
Goulburn
Yass
Holbrook
Albury
Canberra
A.C.T.
Echuca
Kyabram
Bendigo
VICTORIA
Ararat
Mt Macedon
Creswick
Melbourne
Ballarat
Western District
Latrobe Valley
Frankston
Gippsland
Pt Nepean
Wonthaggi
Launceston
Mt Lyell
Hobart
TASMANIA

To my parents, and their parents,
whose place and time this was

PROLOGUE

THE MORNING OF Monday, 1 January 1901 dawned fine on the Western Australian goldfields. The churches conducted early morning services so that the ceremony to inaugurate the new Commonwealth could be held before the full heat of the day. The main Kalgoorlie procession was to assemble by the railway at 9.30 sharp. Only then would the commemorative medals be distributed among the schoolchildren, who would therefore march in strength and ensure a good turnout. The stratagem was unnecessary. A huge crowd of more than 20 000 was drawn to the celebrations, so far in excess of expectations that the railway officials were unable to prevent residents of nearby Boulder from spilling onto the carriage roofs as they travelled across the Golden Mile. An habitual drunkard who had awoken that morning in the Kalgoorlie lock-up and was anxious not to miss the occasion sent a message to a friend to bring clean shirt and collar and enough money to secure his release. The friend was a wag and replied with a note bearing only two lines from the poem 'Federation', recently composed by George Essex Evans:

> Awake! Arise! the wings of Dawn
> Are breaking at the gates of Day.

Not to be outdone, the convalescent replied:

For goodness sake, from prison pawn
Release me and the damage pay.

The events of the day had been planned carefully by the organizing committee. The Western Australian government had contributed £500, and this was swelled by a grant of £100 from the Kalgoorlie council and another £137 collected from residents. Generous prizes were offered for the best-decorated vehicles and buildings. The local Japanese community had built an enormous battleship mounted on three floats and carrying a crew of twenty; the Afghans were to turn out in native attire; and the Aborigines also, with weapons and warpaint. (There were second thoughts about including this last contingent, and they were removed by re-scheduling the annual 'blackfellow Christmas treat' to coincide with the street march.) With strict attention to protocol, the procession committee had worked out the order of the march. The chief marshal would be at the front and the Afghans and Japanese at the rear: that much was easy. The honour of following the chief marshal was given to the officers of the Reform League, who had led the agitation for federation on the goldfields. Then came the mayors of Kalgoorlie and Boulder, the magistrates, the Kalgoorlie town band and the fire brigade band, the town councils, the Australian Natives Association, the Boulder band, the local detachment of volunteers, the salvage corps, the Trades and Labour Council, the Chamber of Mines and the Mine Managers Association, the fire brigade, members of the licensed victuallers' association, the friendly societies, the Kalgoorlie Caledonians and, penultimately, the school-children. The main dignitaries and office bearers were to be carried in horse-drawn vehicles; the rest would march.

It was too much to expect such an unwieldy procession to pass off without a problem. Even at the assembly point the Japanese tailed off into the brothel area, and they were still putting the final touches on their man o' war at 10 a.m. when the parade was meant to begin. A marshal went back up the line to chivvy them along. According to one observer, he was unnerved by the residents of one house, the Terrace Geisha, who had gathered to watch the exhibit, and he barked out the order: 'The procession's away, bring along

Commonwealth Inauguration Day, Kalgoorlie, 1901: the Japanese man o' war turns into Hannan Street

the whore shop.' Or so the story went. In any event the parade did not get under way until 10.30. Even then, those in wagons and drags soon outpaced the marchers; the contingent of schoolboys straggled out of order singing snatches of unsuitable song, and when the pupils finally reached the terminal point, the distribution of buns and drinks degenerated into chaos amid the clamouring of several thousand children.

But it was a great day. In the decorated streets the appreciative crowd watched colourful floats, read the inscribed banners that the children carried, and matched cheers with the participants. The procession terminated at the schoolground where a hollow square was formed, the crowd sang 'God Save the Queen' and the 'Song of Australia', then gave three cheers for the new nation. That afternoon the annual Caledonian games attracted a record crowd of 10 000. And the evening was fine and cool for speeches and music at the Recreation Reserve. The mayor announced that henceforth Australia was a united nation and that this was the greatest and most glorious epoch his compatriots had ever seen. He was followed by John Kirwan, editor of the local newspaper and Kalgoorlie's first representative in the new federal parliament, who, as spokesman for the Reform League, was able to explain the meaning of federation for the goldfields. The creation of the Commonwealth, Kirwan proclaimed, was an affirmation of the 'white man's Australia', signifying that 'this great country should not be peopled by a mongrel or piebald race'; it was a triumph for the miners over the parochialists of Perth, it was a victory for democracy over vested interests; it had created a nation that would become a model for the world.

These sentiments commanded general consent. The recently established newspaper of the goldfields labour movement, the *Westralian Worker*, reminded its readers that 'genuine patriotism ... looks to the good of our fellow men and women', but agreed that the events of 1 January would become 'a happy memory cherished in the hearts of those who played a part in them'. Amid the rejoicing, there was only one dissident voice, that of 'Dryblower', the local poet:

> The garlands are all gathered in, the flags are stowed away,
> The streamers and the bannerettes are furled,
> And outwardly there's little need to remind us of the day
> When a unified Australia faced the world.
> In that memorable pageant there were functionaries bland
> Who bowed as rose the patriotic cheers,
> There were char-a-bancs and chariots for all who could command,
> For all—except forgotten pioneers[1]

1

SOME AUSTRALIANS

RICHARD GARDINER CASEY spent the first days of the new century on the Indian Ocean. He was returning to Australia from London where he had been in negotiations with British investors holding stock in a company of which he was a director. He had been on similar missions in the past and would return to London later in 1901 as well as on subsequent occasions. R.G. Casey was a businessman and his sphere of operations, the pastoral and mining industries, relied on British capital. He belonged to that group of Australians, thrusting and resourceful, who had risen to power with the expansion of the national economy and who were to consolidate their power in the years to come.

R.G. Casey was born in Tasmania in 1846, the eldest son of a doctor. At the age of seventeen he became a jackaroo on a Riverina sheep station and over the next twenty years rose to become the highly successful manager of large properties. His career was checked when he entered into junior partnership with the owner of some vast sheep properties in Queensland. The causes of Casey's difficulties were not uncommon: the partners had borrowed in Melbourne and London to finance the enterprise, and in an effort to reduce the debt they overstocked. Dry seasons, stock losses and falling wool prices caused increasing indebtedness, and Casey's efforts to renegotiate mortgages with the London investors proved unsuccessful. 'I'd rather deal with licensed pawnbrokers,' he declared. The Depression of the 1890s proved the final blow, and Casey was fortunate to escape from an enterprise encumbered with more than £600 000 of debt.

Casey exchanged roles to exact a speedy revenge. He set off to the Western Australian goldfields in 1894 with two associates, not to prospect but to seek out existing claims which could be bought out to provide business opportunities. The find that met their needs was the Golden Hole, a new discovery of rich surface deposits situated twelve miles south of Coolgardie. After buying the claim for £180 000, Casey's associates sealed up the mine and posted guards. They then went to London and floated their company with 700 000 £1 shares, of which they took half, a sixth went to the discoverers, and a third were sold to the public. Of the £230 000 thus raised, only £50 000 was directed to the development of the mine. With much boosting, the price of shares in the new company increased and the promoters realized further gains: then on April Fool's Day 1895, a Monday, a cable brought news to the London directors that there was little gold beneath the surface of the Golden Hole. Shares sold heavily before the news was released to the public at the end of the week. Casey himself was not a major shareholder in this dubious affair though he profited handsomely from his investment. His principal role on this and subsequent occasions was rather that of the adviser marrying his expertise to the capital of others to share in the killing.

Upon returning to Australia, Casey made Melbourne, the centre of finance, his base of operations. From investments, from fees paid for his services and from earnings as director of a number of important companies, he derived a large income, never less than several thousand pounds a year. He lived accordingly. He paid £5000 for a grand thirty-five-room house set on two acres of land in the fashionable suburb of South Yarra, less than two miles from his city offices. The house was extensively remodelled, fitted out with heavy furniture, decorated lavishly and bestrewn with valuable ornaments and bric-à-brac. A retinue of servants catered to his family's needs. He bought his cigars 500 at a time and his wine by the hundred dozen; he was in the habit of sending friends the precious sea delicacy, *bêche-de-mer*, half a hundred-

Richard Gardiner Casey: 'No doubt there will be difficulties, but difficulties are made to be overcome.'

weight at a time, with instructions on how to make it into soup. The city dinners he gave at the Melbourne Club featured such things as English partridge, Scottish grouse and salmon, a well-hung saddle of native lamb. As a result of such tastes, Casey put on weight so that by the early twentieth century he was 14 stone, nearly 4 stone heavier than when he had worked on the land. He dressed expensively and conservatively in three-piece suits and stiff-fronted linen shirts, buying many of his clothes from a Savile Row tailor during visits to London. He was a keen turfman who raced a number of horses and from 1907 to 1916 was chairman of the Victoria Racing Club. Apart from pictures of his horses, the paintings on his walls were English and indeed his lifestyle was little different from that of a wealthy Edwardian city gent.

Casey was a plain-spoken man of uncomplicated rich

tastes. His success was built on shrewd judgement, attention to detail and, when necessary, ruthlessness. Though he had sat in the Queensland Legislative Assembly, he had little taste for politics and in any case found that it was more convenient to employ lobbyists to protect his interests. In certain respects he assimilated into the ranks of the genteel: he belonged to the exclusive men's clubs in the cities where he did business, the Queensland Club in Brisbane, the Australian Club in Sydney and the Melbourne Club, of which he served for a term as president, he married the daughter of a wealthy Brisbane merchant and their three children, who were all born in the 1890s, were sent to the leading Anglican schools; the eldest son went on from the University of Melbourne to Cambridge. Yet Casey was not a social climber. A bluff, blunt man, he was accustomed to judging the worth of any individual 'on net balance' as he put it, and he preferred the company of self-made businessmen like himself. (Among his close associates were William Knox D'Arcy, a Rockhampton solicitor, and T.S. Hall, manager of the Queensland National Bank in Rockhampton, who bought shares in the Mount Morgan Gold Mining Company during the 1880s, and who both became millionaires. Casey also bought into Mount Morgan at the right time and served as director, and later chairman, of the company.) In similar fashion he took it for granted that Australia was part of the British Empire, but he was never overawed by the London financiers with whom he dealt. 'No doubt there will be difficulties,' he once remarked, 'but difficulties are made to be overcome.'

We can observe him at work in Australia and England during 1900 and 1901 reconstructing Goldsbrough Mort and Company, the pastoral company of which he later became chairman. Having financed its expansion in the 1880s by the issue of £2 million of debenture stock, the company was unable to pay the interest in the Depression of the 1890s. Before Casey joined the board in 1896 at the invitation of his friend Hall, the largest shareholder, there was a financial rearrangement largely at the expense of the Australian shareholders. Casey now brought his friend J.M. Niall, a Queensland pastoralist, into the Melbourne head office and the two men worked out a new reconstruction scheme which made further calls on the shareholders but which also substantially wrote down the value of the debenture stock. Casey's two visits to London in 1900 and 1901 involved protracted negotiations and much hard bargaining. At one point he broke off discussions with the trustees of the London debenture holders. 'I think this did good,' he wrote back to Melbourne, 'as it showed I did not mean to be unduly squeezed', and indeed the trustees were subsequently 'very civil and conciliatory'. Within a few years Casey was chairman of a highly profitable company, one which was paying dividends as high as 20 per cent on the eve of the First World War.[1]

Throughout the early years of the century the Neilson family struggled to earn a living from the sandy soil of the Mallee region in north-western Victoria. There were eight in the family when they came to Sea Lake in 1895. John, the father, had turned fifty, his wife Margaret was forty-five, and their children Jock, Maggie, Jessie, Annie, Bill and Frank descended in age from twenty-three to twelve. John had been born in Scotland, Margaret was the daughter of a Scottish immigrant, and an extended family of Neilsons, Shaws, MacFarlanes and McKinnons lived on the land around the Victorian and South Australian border. John and Margaret began at Penola and farmed holdings at Minimay and Nhill before moving to Sea Lake; like many others, they were pushing inland from the more fertile coastal region into newer areas of settlement.

The Neilsons lived on intermittent earnings, credit and hope. On each of their various selections they had to clear and fence the land, obtain seed to sow, fight the rabbits and then rely on rain to produce a harvestable crop. They were always poor and did not have enough capital to develop their land properly or ride out the poor seasons.

> The weather's hot and horrid and the old man's got the
> blight,
> The grain is small and shrivelled and the bags are very light,
> The rain may come tomorrow and knock the whole lot down
> And last night came a letter from the mortgagees in town.

The father and sons earned money whenever they could by cutting timber, harvesting, fencing or making roads. In good times a man could earn 30s a week as a bush-labourer, but in bad times he was lucky to find a job paying half that. John Neilson preferred contract work but he was a poor judge of a tender and sometimes earned nothing after he had met his costs. When the girls came of age, they also worked for the wives of more prosperous farmers or else in nearby pubs. In the late 1890s Maggie worked as sewing mistress in the Nhill state school for 12s 6d per week. As many of the family who were together at any particular time lived in a rough timber dwelling they built themselves with hessian lining and earth floors. Slush lamps were used for lighting. The kitchen was the biggest room and the principal living area. Cooking was done in billies or frypan over an open fire underneath a primitive chimney. There was an iron camp-oven which was placed in the coals to bake bread. Their diet consisted largely of meat—mutton in good times, kangaroo or rabbit in bad—and bread, supplemented by milk and butter when the cow could be milked and by such fruits and honey as they could find. The men wore thick, heavy boots and rough working trousers and flannel shirts; the women made do with a few long-wearing garments.

John Neilson was the optimist, prepared to take on any task and always managing somehow to scrape together the means to stay on the land. Since his block at Sea Lake was overrun with rabbits, he obtained from a storekeeper enough netting to fence two sides of the 120 acres he had cleared. As his son Jock recalled, this did not interfere with the bunnies, who merely went around. Then dad conceived the idea of completing the fence with 60 chains of split palings. He actually managed to prepare and erect about 40 chains of paling fencing by the Christmas of 1895, but by then all the crop had been eaten. The following years were worse, as 1895 was the first of a run of eight dry seasons, the most sustained and severe drought in Australia's recent history. During the worst of the drought, the Neilsons were forced to dispense with their horse and thus lost the only source of motive power on the farm. They acquired more blocks in the children's names but could not grow a successful crop: they were lucky when they harvested enough to provide seed for the next planting. Their lack of success was not caused by want of skill or effort for it was reckoned that hardly any original settlers remained in the eastern Mallee within years of it being opened up.

> Oh, 'twas a poor country, in Autumn it was bare,
> The only green was the cutting grass and the sheep found little there.
> Oh, the thin wheat and the brown oats were never two foot high,
> But down in the poor country no pauper was I.

Several of the family suffered from poor health, made worse by unremitting toil, the shortage of fresh food and good water, and a lack of medical treatment. Margaret Neilson contracted typhoid in 1897 and died shortly after her husband took her to Bendigo hospital, an eight-hour train journey made in very hot weather. Maggie died of tuberculosis in 1903, Jessie in 1907. Still John Neilson battled on.

His eldest son Jock did not possess the same temperament. He thought that he had inherited a tendency to melancholy from his mother's people, the McKinnons. 'Poor country depresses me more than it does most people', Jock once wrote. Throughout his life he suffered from digestive disorders; in 1897 he fell acutely ill with gastritis; for the next eight years he was debilitated by a nervous condition, and from his early thirties he was afflicted with a painful eye ailment. The death of his mother, and later his sisters, troubled him deeply and he developed a 'strange horror' of the locality where they had died. In 1903 he had a comfortable job harvesting on a farm 50 miles distant, when his father wrote asking him to come home and help with the family farm. 'I had a horror of going back to the loneliness and the baching', he recorded, but he thought it his duty to return and he did so. In fact, the first twenty-five years of Jock Neilson's working life revolved around the family farm, from the late 1880s through to the First World War when Sea Lake was finally abandoned. He never married. Throughout these years he tramped as far as 200 miles to find a variety of jobs: harvesting back at Penola, picking grapes up at Mildura, road making and cooking, cutting and grubbing mallee roots, shearing, harvesting, fencing. He once estimated that he had more than 200 employers in thirty years. Always the job was a temporary expedient, a means of supporting himself and bringing a little money back home. Always he would return to Sea Lake and pitch in.

Jock Neilson, then, was a rural labourer, tied loosely to a locality and tightly to a family network. He worked with hand and muscle. He was not particularly strong, standing

Jock Neilson: 'Down in the poor country no pauper was I'

5′8″ tall and weighing 10 stone, and did not possess his father's ability as a bush carpenter and mechanic. He was simply a diligent worker who could turn his hand to most farming or labouring jobs. He neither drank nor smoked, finding his pleasures in a small circle of family and close friends. His chief interest was poetry. He had little formal education—fifteen months at Penola and another ten at Minimay—and he taught himself to read with some books in the home. John Neilson wrote and read poetry, so his son had an early familiarity with Burns and Scott. Some of Jock's early efforts were published in the local press but it was not until 1901, when he was twenty-nine, that he began to find his voice: in that year he received encouragement from A.G. Stephens who was editor of the literary Red Page of the *Bulletin* and who, several years later, became Neilson's adviser, editor and agent.

Jock Neilson composed his verse out of doors as he worked. Often he would start by humming a tune and then turn to composing verse which would be kept in his head till it could be written down at the end of the day. The poem would be worked and reworked until Neilson felt satisfied with it—he might polish a short piece for years before he was finally finished. The result was brief poems in short stanzas of deceptive simplicity, exhibiting strict attention to metre and rhyme and the sound of words. Using a relatively small number of themes and images, he wrote of the world of nature, of seasonal change and daily life, of love and death. Through A.G. Stephens he was to achieve limited recognition in later life, and to meet other poets and writers during the 1920s and 1930s, but not even his closest admirer always fully understood what he wrote.

> The New Year came with heat and thirst and the little lakes were low,
> The blue cranes were my nearest friends and I mourned to see them go;
> I watched their wings so long until I only saw the sky,
> Down in that poor country no pauper was I.

A.G.S. said that the metre was no good, but it sounded all right to me when I said it over to myself. He also said that it was too much like a catalogue. The last time he sent it back I kept it—I never sent it to him again, but I did it up for the collected book.

Jock signed his early verse Shaw Neilson, using his second Christian name to distinguish himself from his father, and it was as the work of John Shaw Neilson that his collected poems appeared in 1934. With this identity he eventually became known as Australia's finest lyric poet.[2]

Things were going better for William Somerville at the turn of the century. After an earlier spell of tramping the country in search of work, he had settled down. He had a steady job as an engineer, he was recently married, he had bought a house. Here was a workingman who enjoyed a modest sufficiency. Somerville belonged to that section of Australian society who grasped the opportunity for self-improvement and who sought to make those benefits available to the mass of wage-earners.

He was born on 24 November 1869 in the little town of

Merewether on the northern coast of New South Wales. His father was a coal-miner on the Newcastle field and the family lived in a simple split-slab cottage. William attended the local public school until the age of fifteen and was then apprenticed as a blacksmith to a Newcastle foundry. At the end of five years, he mastered his trade and was admitted to membership of the Amalgamated Society of Engineers. The engineers were a craft élite of workmen skilled in the hand technology of working in metal and in demand in all branches of industry that used machines; nicknamed the 'tin gods', they had a union ticket described as the best passport all over the world for a tradesman.

It was William Somerville's misfortune to enter the trade at the very moment that these traditions were undermined by the great strikes of 1890 and the severe depression that followed. For several years he tramped the roads of New South Wales and Victoria in search of work; he had spells of employment in workshops and on sheep stations; he even tried his hand at gold-mining. His experience of unemployment was cushioned by those spells of work and by union benefits, so he did not suffer acute hunger or desperate hardship. Yet the experience burned deep. He discovered that the possession of a skill and a willingness to work did not in themselves guarantee a workingman's paradise, 'that in our sunny Australia in times of depression if a man has too stiff a back to beg then he can starve like a dog'. Somerville's generation of workingmen set out to change that fact of life.

In 1895 he followed work to Western Australia. He landed at Fremantle at six o'clock in the morning, hunted up the local secretary of the Amalgamated Society of Engineers at his breakfast, and had the choice of two jobs before starting time at 7.30. For the next decade he worked as an engine-smith for the government in harbour projects and marine repair. Before long he was secretary of his union branch and active in the infant labour movement. In Fremantle he met Agnes Spunner who taught in a local school and shared his political interests. Two years younger than William, Agnes had also come west in search of steady work. She grew up in Creswick, a declining gold-mining town in Victoria where her father was the local bootmaker, and she trained there as an assistant teacher before striking out on her own.

William Somerville: 'I tasted the bitterness of being unemployed'

They married and in 1900 bought a house in the nearby suburb of Buckland Hill. The house was new and set on a large block of freshly cleared land—more than half an acre—but in other respects it was uncannily representative of the skilled working-class norm. Built of weatherboard, roofed with corrugated iron, it consisted of four rooms with a hall down the middle, verandah at the front and skillion kitchen at the rear. As was the custom, Agnes lost her job upon marrying, but William's regular earnings of more than £3 a week kept them both in reasonable comfort and provided for the four children who were born over the next decade. They kept poultry and a pig for the table. Drawing water from a well, they established a large garden plot 14 yards long and 8 wide. In a journal William recorded his purchases of seed and fertilizer, noted the weight of the crops taken from the plot, and compared the costs of his produce with market prices. With his first child, he conducted similar experiments, systematically introducing him to educational projects and checking his progress against manuals of child psychology.

The Somervilles were closely involved in their neighbourhood. They helped to establish a society of Buckland Hill residents and held office on the parents' and citizens' committee of the local school. Agnes was a stalwart of the labour movement and served subsequently as campaign manager in Buckland Hill for Labor candidates in both state and federal elections. George Bernard Shaw's *Intelligent Woman's Guide to Socialism and Capitalism* became her bible. Her public activities were limited, however, by arduous domestic responsibilities: cooking for six on a wood stove, washing clothes with copper and mangle; cleaning with bucket and scrubbing brush; caring for four children; supporting her husband.

Her husband was both less gregarious and more able to pursue his interests. He was a straight-backed man, a pipe-smoking teetotaller whose greatest pleasure was found in books. 'Stop that knitting, read something,' he would instruct his daughter. He had the autodidact's faith in the power of knowledge. From his readings in history, economics and politics, he prepared addresses for trade unions, Labor groups and branches of the Australian Natives Association. He was never tempted by doctrines calling for a radical upheaval in society, even though he played a role in strikes, for he was convinced that the best means of advance was constructive, practical effort. Progress would be achieved when the wage-earners possessed sufficient understanding and organization to make the state serve their needs. It was the duty of the government to develop the country and ensure work for all at adequate wages. Nearly all the engineers in his union branch worked in government enterprises, most in the railway workshop, so he was a strong supporter of state industry. Somerville was also an advocate of industrial arbitration which, he believed, offered the best path to economic justice. The engineers were not so much concerned with the power of a court to determine their wages, for they could usually protect these through their own efforts, as they were at the prospect of a tribunal eliminating sub-standard conditions and improving the wages of less-skilled workers, thereby reducing the pressure on themselves. An Arbitration Court was established in Western Australia in 1900, composed of a judge from the Supreme Court, one representative elected by the employers and one elected by the workers. The court was strengthened in 1902. In 1905 Somerville was elected the workers' representative and he remained a member of the court until the Second World War.

Somerville would be appointed to other public institutions. He became a trustee of the state library, museum and art gallery; he was one of the first senators of the University of Western Australia when it was established shortly before the First World War, served temporarily as chancellor and vice-chancellor on different occasions, and received an honorary doctorate. He remained a champion of working-class interests, never retreating from the undertaking he had given to the Labor premier who appointed him that he would 'guard it as a university for the working man'.

As one of the professors remarked, his integrity and confidence in his own opinion were matched only by his obstinacy: he was prepared to intervene at any point in their academic deliberation to remind them of the educational entitlements of the children of the working class. On more than one occasion he wrote to government house to rebuke the king's representative for siding with 'that small section of the community who with insular offensive arrogance arrogate to themselves the title of society'. In the court he fought for the principle of a basic wage to which all workers are entitled, and he piloted apprenticeship schemes in a range of industries. He looked with a censorious eye on those fellow pioneers of the labour movement whose achievement of ministerial office caused them to abandon their principles or succumb to easy living. He continued to live in his modest house, to see his children pass from government schools to the free university, to write and speak out for the rights of all Australians to share in the national wealth. Somerville stands out as a striking example of the successful workingman, convinced that his success could and should be available to all.[3]

Deborah Turnbull came from the same background. Her father also was an engineer who had fallen on hard times in

Deborah Watt (right) at her sister Jennie's wedding, 1900

New South Wales before trying his luck in the west in 1897, where he too found work and built his weatherboard cottage at Buckland Hill within walking distance of the Somervilles. Seven of the eleven Turnbull children were still living at home at the turn of the century, and Deborah, the youngest of the girls, was seventeen. A photograph taken at her sister Jennie's wedding shows her in special-occasion finery, her fair hair spilling down her shoulders. She spent the first months of 1901 with this sister, who had lost her first baby, helping her in managing the hotel that she and her husband ran in the goldfields town of Hampton, some 30 miles from Kalgoorlie. There Deborah met Bill Watt, a man in his late twenties who was sinking a shaft at a nearby mine, and before long Bill wrote to Deborah's father seeking his consent to an engagement. It was given. With matters thus arranged, she spent the next twelve months at home. She returned to Boulder to be married, taking with her the trousseau that her mother and a sister had made—blankets, sheets, four sets of underwear, two dressing-gowns. The newly married couple rented a house in Boulder, a simple affair of two rooms made of canvas, to which Bill added a kitchen and wash-house. There Deborah had her first daughter.

Marriage took Deborah Watt away from the companionship and comfort of a family home in the metropolis. She did not complain about the tasks of washing, cleaning and cooking in such cramped and primitive accommodation—water cost 7s 6d a hundred gallons until the pipeline to the goldfields was opened in 1903—and she found joy in motherhood: 'I think that was the happiest time in my life to have a baby of my own.' But she did feel her new isolation. 'Mum and Dad came up and had Christmas with us, which I thought was wonderful,' and later one of her sisters looked after Bill and the baby so that an exhausted Deborah could have a spell at Buckland Hill. She was unhappy about her husband working underground, especially after he was burned in an accident with explosives, and determined that they should get away from the mines. So at her instigation the young couple went onto the land. The government of Western Australia was opening up the country along the railway line between the goldfields and the coast, paying settlers to clear the bush and then offering them the chance to buy their block. Bill built their new house, canvas on a bush-timber frame with an iron roof and clay floor, at the new farming settlement of Nangeenan. There they fell into the time-honoured routine of the pioneer couple, he grubbing out the salmon gums and mallee, she taking baby Deb to join him for lunch amid the wildflowers.

They had barely settled in, however, when Bill was seriously injured in a tree fall that crushed an arm and severed blood vessels. After a nightmare journey by foot, cart and train, he underwent operations to set the arm, first in Northam, later in Perth and finally at the hands of a specialist in Melbourne who pronounced the damage permanent: Bill would be unable to use his injured arm. The farm was abandoned, and Deborah worked first as a live-in domestic in Northam so that she could be near Bill, and later found shelter at Buckland Hill. Eventually Bill found employment as a mine watchman back at Boulder, while Deborah earned money as a laundress 'till we got on our feet again'. A second child was born in 1905, a third in 1908, but a fourth died in infancy, and ill health among the other children persuaded the Watts that the lack of sanitation, the dust and harsh summer heat could no longer be endured. They purchased a small orchard at Mundaring in the hills above Perth, where Deborah lived with her children (a fourth was born in 1912) while Bill earned money at Boulder. With these savings they bought their passage to New Zealand where Bill's brother-in-law was offering a dairy farm. When the farm failed to eventuate, the Watts's predicament was grim indeed, for they had lost the support of the family network. Marooned in Auckland, Deborah did not write of her tribulations until Bill finally found work as a watchman on the Auckland harbour. A sister sent her 22s. 'I cried, I was so thankful to get it and I was so homesick, but I soon bucked up. I was thankful no one was home to see me cry.' The Watts had suffered much during their eleven years of married life in this country—Bill's accidents left him partially disabled, a child had been lost to dysentery, a farm had been abandoned. They leaned heavily on the assistance of Deborah's family; they drew comfort from their children, for whose welfare they uprooted their household, and their concern for each other survived frequent separation. Above all, it was Deborah's resourcefulness and indomitable optimism that held them together.

Deborah Watt set down a narrative of her experiences in her eightieth year. She wrote for a niece and much of her narrative is taken up with geneaological detail of when and where her sisters and brothers married and established their own families. Weddings were highlights which remained fresh and clear in Deborah's memory—'Ena looked lovely in her white wedding frock, I did not go to the Church as I was helping at home with the Breakfast ... Bill could not get away to the Wedding, he was very sorry about that.' Few outside events impinge on the narrative—the opening of the goldfields water scheme, a cyclone, some Aborigines asking for food. Though the account of the farm accident and the nightmare journey to hospital is remembered in vivid detail, she simply assumes an ability to cope with all her tribulations. 'It doesn't matter what trouble one gets in, there is some good Samaritan to help one.' The ups and downs of life were accepted and she found her happiness in helping and being helped by those closest to her.[4]

Sometime round the turn of the century George Dutton was initiated. His mother was an Aboriginal living at Yancannia station in the extreme north-west of New South Wales. She had died when he was seven, while his father, a white stockman whose name he bore, had already moved on. George spent his boyhood with a stepfather, working and travelling through the arid country at the junction of New South Wales, Queensland and South Australia, which is known as the Corner. The whites had settled in the Corner over the last three decades of the century, some seeking gold or opals, others establishing sheep stations on the mulga and saltbush plains. These were huge properties of a million acres or more and there was scarcely one that did not shear at least 100 000 sheep. A station such as Yancannia was in truth a small village situated on a run of 1500 square miles and consisting of the manager's homestead, the overseer's cottage, huts for the hands, the kitchen, a carpenter's shop, a smithy, the saddler's workshop, a large store and great woolsheds. Such a station provided work for 100 or more. And on the edge of the village was the Aboriginal camp.

White settlement had upset the delicate relationship of the tribal group and the land. By grazing the cover, interfering with the water supplies, exterminating the game and killing those blacks who got in the way, the pastoralists made it impossible for the Aboriginals to support themselves in the way they had maintained for thousands of years. In any case, European foodstuffs such as flour, tea, tobacco and sugar attracted them to the stations. The station managers found it useful to keep the men as a cheap pool of auxiliary labour and to employ the women in the homestead where they also served as concubines or casual sexual partners for white men. But there was little attempt to regulate their lives and they were free to maintain a large part of their traditional practices. This freedom was all the greater for those of mixed descent such as George and his stepfather who, because they acquired the skills of the bush-worker, were not tied to a station but instead roamed from job to job with all the independence of their white counterparts:

> They call no biped lord or 'Sir',
> And touch their hats to no man.

The world that George Dutton grew up in during the 1890s was a world of stations, pubs and tiny townships, but it was also a world of bush-camps, collective customs and secret rituals.

George's stepfather was a Maljangaba man, his mother had been one of the Wonggumara, both Corner tribes with links to the inland Aboriginals of north-eastern South Australia and south-western Queensland. George was regarded as Bandjigali, one of the tribes of the Darling River country. When he was sixteen and working in Corner country, he was sent for. He arrived to find a large meeting and knew what it meant. 'Don't try these capers on me,' he said and ran off. The gathering reassured him that since he belonged to one of the Darling tribes, they would not subject him to their initiation rites which began with a circumcision ceremony known as the *milia*. Unconvinced, Dutton set off with his stepfather to nearby Mt Browne but was followed by a mob who kept 'making a grab at me'. Claiming that they were going rabbiting, he and another youth sneaked off from the camp. It was no good. Message was sent that his stepfather was ill and would like to see George go through with the *milia* while he was still alive. George finally submitted to this circumcision and the rites that followed.

His adolescent fears were not realized and he found that he enjoyed his new companionship. During the decade that followed, he attended similar ceremonies wherever and whenever he could, and was himself inducted in the higher *wiljaru* ceremony. He was less interested in the metaphysical aspects of his people's culture than he was in its drama and colour. In accordance with the customs, he married an Aboriginal woman with three children but the relationship did not last. In this period and on through the First World War, George Dutton roamed far and wide. At various times he was a drover, a stockman, a carter, a horsebreaker, a fencer, even a gold prospector. His country was roughly bounded by the Flinders Ranges in the west, the Queensland Channel country in the north, the Paroo River in the east and a southern boundary running through Wilcannia and Broken Hill (though he travelled to Adelaide on one occasion). He was 'flash'. His clothes were made to measure and he wore long-necked spurs. He was a skilled man and usually commanded the same pay and conditions as whites with whom he worked. George refused to eat his damper and mutton 'on the woodheap', which was the usual arrangement for the station Aboriginals. He stood up for his rights: one boss would not respect them so 'I told him to go and get fucked and left him the next day.'

Even during Dutton's young manhood in the early years of the century, the good times were running out. The huge pastoral stations were already in decline as a result of drought, overstocking and infestation by rabbits. Soil erosion and sandstorms depleted the pasture further and caused the number of sheep to tumble to less than a third of what had been carried in the 1880s. Accordingly, there was less need for Aboriginal communities on the stations and their numbers fell rapidly: in the Corner settlement of Tibooburra where there were 187 in 1882, only seventeen remained by the First World War. This decline was accelerated with the subdivision of the stations into smaller holdings with less

George Dutton in his last years

need for extra labour. At the same time, the rough-and-ready tolerance of the Aboriginals gave way to increasing discrimination. Their culture was destroyed, their ceremonies and customs fell into disuse. George attended the last *milia* in New South Wales on the eve of the war, and a decade later the ceremony had disappeared in Queensland and South Australia as well. George's people would be removed to reservations or drift to the outskirts of towns, the children would fall into uncontrollable giggling when the old man sang the old songs. For the time being, however, he lived the peripatetic life of the drover, mingling the customs of his people with the habits of his occupational group. He would not settle down or compromise his independence. He took pride in his endurance and knowledge of the country. He did not store up wealth, he consumed it and shared it.[5]

Did William Somerville ever look with casual curiosity on a part-Aboriginal stockman and stepson during his tramp for work in western New South Wales in the early 1890s? Did Deborah Watt maintain any contact with her parents' neighbours, the Somervilles, after she left Buckland Hill? When R.G. Casey disembarked at Fremantle on his way from London to the ill-fated Golden Hole in 1895, did he ever cross the path of the newly arrived engineer on the wharves? When John Neilson came down to 'Stony Town' and was invited to call on the wealthy patroness Louise Dyer, did his steps ever take him past Casey's imposing South Yarra mansion? The answer to all these questions is, in all likelihood, no. The lives of these five individuals were separated not just geographically but by firm economic and social boundaries.

Yet it is possible to offer some tentative comments about the differences and even to point to some common patterns. First, extremes of wealth and poverty were on a colossal scale. There would have been no more than 500 men as rich as Casey; he was several thousand times wealthier than the Neilsons. Living standards varied accordingly. Casey owned a mansion and retained servants to maintain it; the Somervilles had a more-or-less average dwelling; the Neilsons and the Watts lived in little more than shacks. George Dutton and William Somerville ate adequately, Casey's table groaned, while the failed selectors subsisted on meagre fare. The persistent ill health of the Neilsons was aggravated, at the very least, by their poor diet and inability to obtain proper medical treatment, while harsh conditions and primitive sanitation also caused a death in the Watt family. Furthermore, the disability of the male breadwinner greatly exacerbated the Watts's economic hardship. These differences are hardly surprising, for this was a society in which resources were allocated by income and wealth; and it is evident also that physical labour brought a lesser income than entrepreneurial activity. On the other hand, there was an element of choice. George Dutton preferred not to have accommodation of his own until he was forced in old age into a squalid humpy on the outskirts of Wilcannia. In his better days his home was where he put his hat down, since he preferred to stay on the move and to spend his earnings among his people.

A further theme running through the five vignettes is mobility, both geographical and social. It would seem that Australians were almost incessantly on the move. With Neilson and Dutton this was essentially movement within a defined area. The Neilsons first shifted 70 miles, then 50, then more than 100 in search of viable farmland. Jock Neilson's subsequent movements, at least until the last selection was abandoned, were forays dictated by seasonal availability of work around the home place. Dutton's pattern was not dissimilar, though his travels were more protracted since he was less tied to a particular base. Somerville also tramped more than 1000 miles during the hard times of the 1890s before he shipped west to become part of the urban working class. Deborah Watt had little control over her girlhood journey to Western Australia, but the subsequent moves combined a search for security with the maintenance of family ties until she abandoned Australia altogether. Casey's travels were on a different scale. After his boyhood in Tasmania, he lived and worked in all but one of the mainland states as he carved out his career, and his ultimate place of residence was determined by the fact that Melbourne was the financial capital. His ten trips to London indicate the fundamental importance of the imperial relationship.

There were also marked variations of fortune over the individual's lifetime. With young children, Deborah and Bill Watt were particularly vulnerable to blows of fate. Jock's father was one of those who plugged on regardless of the odds stacked against him; like Dad Rudd he was spurred on by the belief that you can't lose every time, that prosperous self-sufficiency was there to be won by his own efforts. Jock himself had no such illusions, while George Dutton did not even aspire to such a goal. Casey began with the important advantages of education and social status, tasted success and failure, then finally triumphed. Why was he so successful? He was an adventurer, certainly, and he was prepared to go to great lengths to win, but the scale of his activities far outstripped most Australians of his class. How many other Australians were able to turn the tables on British financiers? Finally, Somerville was not interested in acquiring more than a certain quantity of material possessions, nor did he seek to climb out of his class. How many others shared his outlook? What were the aspirations and life chances within the larger society?

The five individuals were woven into immediate and wider communities. But the patterns of incorporation varied. With Deborah Watt and Jock Neilson the primary relationship was the family. The distance separating Deborah Watt from her parents and siblings did not weaken their importance. As she had helped her married sister in time of need, so an unmarried sister helped her when she bore her own children; and in emergencies she returned to Buckland Hill. But her husband and most of all, her children, became the focus of her concern, and for their welfare she willingly accepted hardship and isolation. Jock Neilson's emotional attachment to his parents and siblings was also close and continued long after adulthood, even when his brothers and sisters were living elsewhere. The family was the economic unit and members pooled their labour or their earnings. Furthermore, they remained to some degree an extended family: equipment or resources were borrowed from kin, and at various times the Neilsons fell back on the support of their uncles and aunts. George Dutton's links were even wider. Through his mother and his stepfather he had defined relationships with a number of kinfolk, and his participation in the ceremonies of other Corner tribes extended his ties further. His second marriage in the 1920s was to some degree an involuntary relationship formed out of obligation to these customs. The Casey and Somerville families, on the other hand, were more restricted. While they maintained contact with parents and siblings, the primary attachment of each man was to his wife and children. They bequeathed the greater part of their patrimony to their children and to them they channelled their closest affection. It is evident that the women played a subordinate and more domestic role, but at critical junctures it was Deborah Watt and not Bill who made the decision. And what of those for whom there was no family? Many other forms of association have been encountered—Casey's clubs and business links, Somerville's trade union, Labor Party and community organisations, the circle of workmates or friends of Dutton and Neilson.

Then there are interests and beliefs. Casey has appeared as a rather philistine plutocrat, yet he was a quick learner and a wide reader. Somerville's intellectual cast was that of the autodidact, training himself in various branches of knowledge and placing an inestimable value on education. The principal events in Deborah Watt's memory were those that marked the family cycle, above all marriages and births, and the unadorned stoicism of her chronicle hints at an extraordinary resolution and self-sufficiency, stripped bare of external supports. Neither she nor the two men suggest any strong awareness of the inner mystery that was felt in different ways by both Dutton and Neilson. Neilson had rejected his inherited Calvinism but retained a sense of wonder which he tried to express in lyrical nature verse; Dutton was not fully immersed in the Aboriginal Dreamtime beliefs but he was caught up in songs and stories and certainly felt the enduring relevance of the legends. How common were these different propensities in the wider society?

It is necessary to emphasize that there was a wider society. While we have looked from different perspectives at the circumstances of individuals, they were all caught up in a common social structure and a common economy, now joined in a common polity. The two men who were most conscious of these linkages were Casey and Somerville, the one representing propertied conservatism and the other labour. Yet the connections were no less real for the others. Decisions made in the boardroom of Goldsbrough Mort determined the fate of pastoral workers like George Dutton. Legislation made in the Victorian parliament saddled the Neilsons with rules concerning land settlement that doomed them to failure. In the end we must turn to a consideration of such processes.

2

GETTING AND SPENDING

'IT IS CERTAINLY a very self-conscious nation that has just made its appearance in the centre of the Southern Seas,' wrote Alfred Deakin in the immediate aftermath of the Commonwealth inauguration celebrations.

Platform orators and the Press have combined to instruct it as to its present importance and future potentialities. The newspapers of late have comprised many retrospects and statistical comparisons as to our progress and relative resources in population and wealth, all calculated to minister to that self-esteem which is by no means wanting among us.[1]

Yet as he wrote, sheep and cattle were dying for want of grass and water. Australia was in the grip of a devastating drought.

The dry years began in 1895 and did not break until 1903. Not all parts of the country suffered for the duration of the drought: the coastal districts of southern New South Wales and Victoria were not greatly affected; Western Australia had good falls of rain in the later 1890s, so did Tasmania in 1898 and 1901. But right down the fertile crescent of eastern Australia—from Queensland, through New South Wales, Victoria and into South Australia, the grass had disappeared. A man who journeyed from Echuca, on the Victorian side of the Murray River, where the river boats had come to a stand-

The Mount Lyell smelter, Tasmania.
A pall of sulphurous smoke hangs over the valley

still, up central New South Wales as far as Booligal, described the scene as 'desolation and dust, dying stock and disheartened settlers'. On a cattle station in central Queensland, it was reported the kangaroos were too weak to hop and the kookaburras could not fly. A member of the scientific expedition that in 1901 travelled across central Australia from Adelaide to the Gulf of Carpentaria, remarked that even the lizards had disappeared—though not, he lamented, the flies.[2] By 1903 the number of sheep and cattle had been reduced to little more than half. The wheat crop planted in 1902 was a disaster; like the Neilsons, most farmers counted themselves lucky if they harvested enough grain to sow in the following year.

The drought threw into relief established methods of cultivation and their impact on the environment. Most of the Australian landscape bore marks of settlement by the turn of the century. Arable land had been fenced, cleared and put under crop; a large portion of the inland plains was given over to livestock; here and there were heaps of waste over-hung by a pall of smoke which, together with the absence of timber for miles round, indicated a mine. There was little appreciation of the need to husband natural resources. Destruction of the tree cover, the eating out of the original grass and shrub growth, the spoiling of river frontages and the erosion of channels around man-made water catchments—these were signs of ignorance and greed. Rabbits, which did enormous damage to the pasture, had spread upward to Queensland and in the 1890s had even managed to cross the arid Nullarbor to Western Australia. The series of dry seasons at the turn of the century augmented these destructive processes and, by robbing the thin topsoil of much of its remaining protective cover, caused lasting damage in the precarious environment of the inland plains. Thirty years later, when the Australian sheep flock finally regained pre-drought numbers, parts of George Dutton's Corner country had still not recovered.[3]

The drought also emphasized the importance of the rural industries for the national economy. While most Australians lived and worked in towns—two out of every five were in the state capitals—the prosperity of all was intimately bound up with pastoralism and agriculture. Approximately one-quarter of the workforce was directly engaged in farming, but many more earned their living from the carrying, processing and selling of farm produce or by catering to farmers' needs. Furthermore, wool, wheat, meat and butter were vital export commodities, accounting for the bulk of overseas earnings. During the first decade of the twentieth century exports brought between £50 and £80 million annually; when one considers that the fleece of a single sheep was worth between 5s and 10s, the significance of the drought loss of 50 million sheep needs no further emphasis.

Despite the setback, the opening years of the century saw striking economic advance. The value of Australian production increased almost twofold in the fourteen years from federation to the outbreak of the First World War. The contributions of the major sectors of the economy are suggested in table 2.1, while table 2.2 indicates the changing shape of the workforce. A survey of the principal industries will put a

Table 2.1: Gross domestic product, 1900/01 and 1913/14 (at 1910/11 prices, £m)[4]

Industry	1900/01	1913/14
Pastoral	16.0	42.9
Agriculture	13.0	22.7
Dairying, forestry, fisheries	8.9	14.6
Mining	19.8	18.5
Manufacturing	25.2	50.0
Construction	14.8	36.0
Distribution	29.6	59.0
Finance	4.6	5.7
Railways, other public undertakings and government services	16.1	32.8
Other services	32.5	41.5
Rents	20.9	28.4
Other	2.4	3.7
Total	203.8	355.8

little flesh on the bare statistical skeleton and reveal that both in the towns and the countryside this was indeed a period of expansion.

On the land, first of all, the breaking of the drought

Table 2.2: Workforce by industry, 1900/01 and 1913/14 ('000s)[5]

Industry	1900/01	1913/14
Rural	373.4	475.8
Mining	118.4	87.6
Manufacturing	233.1	394.2
Gas, electricity, water	8.6	16.6
Construction	164.6	233.7
Transport	91.5	163.1
Commerce	215.1	247.4
Community and business services	84.1	95.2
Finance and property	17.1	29.6
Other	237.2	246.5
Totals:		
Workforce	1 543.1	1 989.7
Population (excluding Aboriginals)	3 824.9	4 940.9

Agricultural improvement: a team of six horses pull a ten-furrow disc on a Victorian wheat farm

brought a run of good seasons. The pastoralists learned from past mistakes and retreated from the more arid regions, paid greater attention to water storage and invested money in artesian bores. By subdividing paddocks and erecting wire netting they could check the rabbit, by improving their pastures they could rebuild the flock. Yet there was no sharp break with traditional practice—as portrayed, for example, in C.E.W. Bean's *On the Wool Track* (1910), the pastoral landscape, homestead and woolshed symbolized an unchanging order—and most graziers owed their fat wool cheques to kind conditions and keen demand from overseas buyers.[6] Wheatfarmers made greater progress. They had come to appreciate the advantages of dry farming techniques and could see the need to let the land lie fallow over the summer and reap the benefits of alternately cropping and grazing. With greater success than the Neilsons, a new wave of settlers advanced into the Mallee regions of the Victorian–South Australian border, the Yorke Peninsula and the south-western corner of Western Australia. By using superphosphate in conjunction with new varieties of wheat, notably William

Farrer's celebrated Federation variety, farmers increased their yields up to and beyond 10 bushels to the acre; furthermore, the multi-furrowed plough, seed drill and new harvesting machines enabled them to work larger areas.[7] Other rural producers to advance were meat and dairy farmers, both using refrigeration to reach wider markets, fruit-growers in the irrigation areas of the south-east, and sugar-growers along the Queensland coast.[8] Timber remained important both as a source of domestic fuel and as an export.

The other great primary industry was mining. If the drought had underlined Australia's heavy reliance on wool and wheat, then it was particularly fortunate that mining production and especially gold cushioned the drop in export earnings. For fifty years Australians had benefited from the gold deposits in the south-eastern corner of the continent; further discoveries in Queensland and Western Australia during the 1880s and 1890s created new boom towns like Charters Towers and Kalgoorlie. By 1900, however, the rich alluvial deposits were mostly exhausted. The real test was to mine the deeper ore and separate the tiny specks of gold from the surrounding rock. By drawing on international expertise and pioneering new techniques, enough gold was recovered to sustain output up to the eve of the First World War. Production then fell away and did not recover until the 1930s. The silver mine established at Broken Hill in the 1880s presented challenges of equal magnitude, but the reward for solving them was even greater. Broken Hill had lost impetus during the 1890s because of the treacherous movement of the ground, declining yields and a drop in the price of silver. Only by lengthy and costly experiment was a process developed that would separate out the silver, lead and zinc. The full rewards for this perseverance, and for the decision taken in 1911 to expand smelting operations into the production of iron and steel, would be reaped after 1914. Then there were the copper mines in Tasmania, the Mount Morgan mine in Queensland which switched from gold to copper while R.G. Casey was chairman, and a wide scattering of lesser mines yielding these and other precious metals.[9] Finally, coal-mining expanded throughout the period to keep pace with growing demand.

Most everyday needs were satisfied locally. A fair-sized

country town of, say, 5000 inhabitants possessed its own flour-mill and bakery, brewery and cordial factory, coach-builder and blacksmith. These were small-scale enterprises employing simple technologies. The same conditions that allowed such industries to flourish, notably the limited size and segmented character of the national market, simultaneously restricted the Australian manufacturer. Lacking a heavy industrial base, deficient in more advanced technology and unable in most cases to achieve the economies of scale enjoyed by the overseas industralist, he could supply only a third of the country's manufactured needs. Thus there was more scope for the production of food, clothing and furniture than there was for specialized machinery. In processing primary products for export there were similar limitations: abattoirs and tanneries did good business because they worked with perishable materials that could not be sent abroad in their original condition, but only a fraction of the wool clip was scoured, spun or woven in Australia. And while a tenth of the workforce was employed in the construction industry, which in turn provided a market for local brickyards and saw-mills, such basic materials as galvanized iron for roofing and steel girders for major works still had to be imported.[10]

The rapid expansion of Australian manufacturing in the early part of the century owed much to the growth and diversification of the primary industries. As agriculture advanced, for example, so did the demand for agricultural machinery. Again, the erection of wire netting around the sheep paddocks enabled the Lysaght factory in Sydney to expand, electrify its machinery and produce 19 000 miles of netting in 1913. Lysaght outsold overseas manufacturers because its product was superior.[11] At the same time there was a quickening tempo in the older trades. Across a wide range of industries the small workshop gave way to a larger building, the craftsman, and his hand tools, was replaced by a line of operators using machines connected by belt and pulley to a mechanized power source, and the production process was broken down into a series of routinized operations. Echoing Adam Smith, a bootmaker remarked in 1912 that whereas one man used to cut and stitch perhaps a dozen pairs of boots in a day, there were now as many as fifty hands involved in the manufacture of a single boot. 'At the present time there are as many men engaged in making a pair of boots as were engaged in making a pin in the old days.'[12]

The degree of change should not be exaggerated. Of 3843 factories situated in Sydney, the biggest industrial centre in 1911, only 138 employed more than 100 hands. Barely more than half used any form of power other than the muscle of the worker, and the total contribution of all power plants in Sydney factories was only 60 000 horsepower. On the land, despite the limited contribution of the steam traction engine, the principal source of energy was still horsepower of the four-legged kind. There were more than a million horses in 1901—a wheatfarmer with 250 acres needed a team of at least six—and each one required 24 pounds of feed daily. Transport boasted greater mechanical progress, with the steam train hauling the primary products to city ports, and the electric tram linking the residential suburbs to the city centre; yet here too the horse remained ubiquitous. The motor-car was still a rarity: there were fewer than 5000 of them on Australian roads in 1910 and the bowser was yet to make its appearance.[13]

The great majority of Australians worked by hand and their productivity was correspondingly low. Whether it be lumping bags of wheat, cutting coal or timber, laying bricks or railway sleepers, a labouring job called for the expenditure of immense physical effort over a long working day. Yet no job was without its own particular skill, its own trick of performing the necessary movements with the greatest possible economy of stamina. Take an agricultural labourer earning perhaps 20s a week and his keep. In harvest time he stands under the hot sun in a paddock of stubble to sew bags of wheat; taking the ears of the bag, whose weight is equal to his own, he raises it on his hip a few inches from the ground, then drops it to let the grain settle; drawing twine from the bundle on his belt, he threads the needle and almost in one movement puts six or seven stitches into the bag, and ties the twine; and so on to the next bag, and the next, till sundown. Or take a teenage boy working the crushing battery of a gold-mine. Using a long-handled shovel, he has to feed the quartz into the battery so it can be broken down for subsequent separation of the gold. Too much quartz will prevent

A furniture workshop in Launceston, Tasmania. The band-saw on the right is one of the few mechanized aids. Off-cuts and shavings litter the passages and the men wear hats to provide cover from sawdust

the sufficient impact of the stamper, too little may cause a fracture of the shank as it strikes the solid metal below. For eight hours at a stretch he feeds two banks of five stampers, and in return is paid 3s 6d per shift. Again, take women in a tobacco factory. For nine hours a day and more they sit in line, forbidden to talk, and pinch the stems of tobacco leaves. The workroom is humid and the air is filled with dust; for pulling too long a stem or leaving too much stem on the leaf, a penalty is deducted from the 4d a pound that they are paid. Even in the more genteel atmosphere of the office, work was still highly labour intensive. Sitting at high desks on hard stools, male clerks kept the ledger books and wrote up the correspondence. Nearly all writing and copying was done by hand, for the first 'lady typewriters' were only beginning to appear. Finally, there was the daily round of domestic labour. The wood stove, the copper, mangle and flat iron, the bucket and scrubbing brush all imposed a burden of toil.[14]

Mechanization was greatest where labour was scarce or expensive.[15] Hence shearing machines were in general use in the woolshed by the First World War, and coal-cutting machines were becoming common in large pits. But where labour was plentiful, and especially where female and juvenile labour could be used, little money was spent on mechanization because the human hand was cheaper. There were attempts to stamp out the 'sweated' trades and force up the price of labour: indeed, the spread of trade union organization during these years from the skilled trades to broader and less-exclusive occupational groupings encouraged the employer's search for greater productivity. Already the new growth of the Australian economy was bringing complaints of a shortage of labour. In the last years before the war employers grumbled of the difficulty of filling vacancies, and governments redoubled their efforts to increase the population.

Between 1901 and 1914 the population increased by a little over a million, from 3 825 000 to 4 941 000. (Aboriginals were excluded from these official counts: they swell the totals by about 100 000.[16]) By modern standards the rate of natural increase was high—births outnumbered deaths by roughly three-quarters of a million—but not high enough for the nation builders and guardians of public morality who chorused their alarm in the New South Wales Royal Commission on the Decline of the Birth Rate. Indeed the marriage and birth rates had dipped in the Depression of the 1890s, but more powerful long-term influences were at work, changing the structure of the Australian population. A declining death rate meant that more Australians survived to reach old age (in 1911 one person in 23 was sixty-five years or older, whereas fifty years earlier only one in 100 had reached that age), and the pronounced drop in infant mortality (down from 110 per thousand in 1901 to 72 per thousand a decade later) altered child-bearing patterns. 'A woman who began her childbearing in 1911 would probably have four children or less. Her mother would have had five children, and her grandmother, completing her childbearing in 1891, would have had at least seven.'[17] In short, Australia was moving towards the low-fertility-low-mortality pattern characteristic of economically advanced societies. How then were the empty spaces to be filled? Immigration was the traditional method, but the number of immigrants to Australia in the first years of the new century was less than the number of emigrants. Knowledge of the drought, and, before that, the economic depression, deterred many prospective settlers from coming to Australia, and in any case there was an alternative destination beckoning the British migrant–North America. A passage across the Atlantic cost no more than £8, less than half the cost of the cheapest berth on a steamer bound for Australia. Only through extensive publicity of the southern 'land of opportunity' and provision of assisted passages did Australia attract significant numbers. Between 1910 and 1914 slightly fewer than 300 000 persons came, all of them white and the overwhelming majority British.[18]

> To escape these endless vaults of brick, and pitch
> a tent outback,
> If I get a chance I'll graft until my very sinews crack,
> Meanwhile may all the angels up in Paradise look down
> On a man of sin who died not, but was damned and sent to
> town.

The poet, the eugenicist and the immigration enthusiast were united in their preference for the country over the city. The concentration of the population in towns was blamed for the decline in the birth rate, and the president of the Immigration League urged that 'every lad or man who decides for Australia should come here with the determination not to linger in the city for a day longer than necessary'. Writing on the eve of federation, the *Bulletin* insisted that 'the first business of the Commonwealth, for many years to come, will be transferance of the population from city to country', away from those 'huge cancers' where health and strength were sapped.[19] The cities continued to grow, however, both absolutely and in relation to the rest of the population. These swollen centres seemed to defy all precedent. In their civic architecture, their parks, galleries, museums and libraries, they seemed not unlike one of the large manufacturing cities of northern England; Manchester, perhaps, or Sheffield. But Manchester and Sheffield had grown around their mills and workshops. Here in Australia the order was reversed: the urban centre was established, then followed the industry. It

Table 2.3: Population of the capital cities as a proportion of state populations, 1901 and 1911[20]

	1901		1911	
	Population ('000s)	As proportion of state (%)	Population ('000s)	As proportion of state (%)
Sydney	496	37	648	47
Melbourne	478	40	593	45
Adelaide	141	39	169	41
Brisbane	119	24	141	23
Perth	61	33	107	38
Hobart	35	20	40	21

was the very aggregation of people in the Australian city that gave rise to the economic activity, notably the service and construction industries and manufacturing trades catering to the urban market. Then whence the growth? The cities had grown during the nineteenth century partly through natural increase and partly because many newcomers to Australia put down roots immediately they disembarked. In this phase both the city and its hinterland developed together. By the turn of the century a new phase was beginning: the cities were drawing people off the land and away from the inland towns, they were expanding at the expense of their hinterland.[21]

Periodically the citizens of the metropolis were reminded of the country by the chaotic passage of livestock through the city centre. Flocks of sheep and herds of cattle would be driven up St George's Terrace, the principal thoroughfare of Perth, on their way to the slaughteryards at Subiaco; cattle and pigs went through the heart of Melbourne to the market at Parkville; shoppers and businessmen in Sydney had to make way for the livestock en route to Glebe.[22] Such nuisances were a reminder that the cities were, above all, commercial and administrative cities, conduits of trade and seats of government for the regions they commanded. But by this time they had taken on an impetus of their own. An urban world had been created whose physical form declared its distinctive identity. The city centre, first, was given over to civic and commercial activities. Imposing stone façades

Inner Sydney housing conditions, The Rocks area, 1900

announced government offices, business houses and banks; department stores, pubs and clubs kept the streets busy with human movement; parks and gardens marked the boundaries, and church steeples pierced the skyline. The creation of public space had occurred in Melbourne and Sydney during the 1880s while in Adelaide, Brisbane and Perth the factories and houses were still being combed out. But in every one of these cities a specialization of functions was well advanced in the surrounding suburbs. The noxious industries were concentrated in certain insalubrious neighbourhoods; the larger factories and the homes of the men and women who worked in them clustered in the inner suburbs; affluent residents commanded the most desirable localities, and outlying commuter suburbs served the rest.[23] The precise layout varied with topography. Adelaide and Melbourne, lying on plains and making little of their rivers, followed a regular grid pattern. Some remarked on the North American appearance of this arrangement. C.P. Trevelyan, who visited Australia with the Webbs in 1898, remarked that 'Melbourne is like an American city, with broad streets all at right angles', while Jock Neilson observed simply,

> No curve they follow in Stony Town;
> but the straight line and the square.[24]

The other capitals were softened by their hills, indented harbour or serpentine river.

Whatever the arrangement, it was now established practice in the sprawling suburbs to set detached houses on individual blocks of land. The saving in labour offset the cost of the land, and suburban householders were prepared to outlay the necessary quarter of their earnings on rent or mortgage payments. Building styles varied little from one city to another. Brisbane raised houses on stilts and favoured wood and iron, while Adelaide made extensive use of stone, but in general brick and tile were becoming the preferred materials. The standard double-fronted cottage, as chosen by the Somervilles, was giving way also to a less symmetrical style: the front room was pushed forward, thereby truncating the verandah, and perhaps permitting a bay window; and the kitchen and bathroom were brought into the main structure.[25] The era of the ornate inner-suburban terrace had passed. Denser housing was henceforth restricted to such industrial suburbs as Collingwood, Redfern or East Perth where cheap jerry-built rows were still built, twenty or more to the acre, offering three small rooms and a tap.

With growth and differentiation came a number of further changes. Public transport in the form of a network of tram and train services had become necessary. Trams were electrified and electric lighting replaced gas lighting in the streets, though many homes still lacked either service. The telephone had become a required adjunct of commercial life and was possessed also by the well-to-do householder. The main suburban roads were surfaced and planted with a forest of wooden telegraph poles supporting their tangle of wires. The ever-increasing water requirements of the cities could only be satisfied by reservoirs, to which the majority of houses were connected. But the most pressing need was for adequate means of disposing of sewage and refuse, both noisome menaces to health. Outbreaks of typhoid and dysentery, such as that which carried off a baby of Deborah Watt in 1910, were always likely in the summer months, and Sydney suffered visitations of bubonic plague in the early part of the century. While improvements in garbage disposal could be achieved piecemeal, sewerage demanded major public works. Adelaide, Sydney and Melbourne all began construction in the nineteenth century and managed to connect the majority of houses to water-borne sewerage systems by 1914, but Perth, Hobart and Brisbane remained mostly or wholly dependent on the outhouse and night-cart, as did the lesser urban centres.

The provision of these urban amenities called for great quantities of money and labour (they were in fact a major source of employment for the city worker) and gave the state capitals distinctly better living standards than were available elsewhere. Rural housing was of an inferior standard, using local materials and lacking the conveniences of the cities. Country towns also lagged behind. Rockhampton, for example, with a population of 15 000 at the turn of the century, was Queensland's second city after Brisbane. Yet it had to make do with gas lighting and household water tanks until the 1920s, the pan system until the 1940s, and right up to the Second World War it relied on primitive steam trams, known

A view of cleansing operations in inner Sydney during the bubonic plague, 1900

derisively as 'pie carts', for public transport.[26] Some country towns were already contracting. Echuca, on the Murray River, suffered a loss of trade as the railways sliced into its river traffic, and its population fell from 4800 in 1881 to 3500 in 1911. Albany, on the south-west corner of Australia, lost shipping to Fremantle when the metropolitan port was developed at the turn of the century, and it slumped from 3600 in 1901 to less than 2000 five years later.[27] These are perhaps exceptional examples since both towns relied on transport and suffered when the lines of communication were rerouted. Even more dramatic were the movements of population that followed the exhaustion of a goldfield when whole rows of houses were taken down and reassembled elsewhere as veritable portable towns. New centres were springing up in areas of agricultural settlement, and larger towns held their own in the well-established regions. On the land itself, as the statistics for the rural workforce in Table 2.2 suggest, the numbers were still increasing. Long lists of names on the war memorials of hamlets testify today to the fact that they supported much larger populations in the early part of the century. Already, however, the flour-miller and the implement maker were finding it difficult to compete with the city product. The city was likened to a 'giant octopus, stretching forth its suckers in every direction throughout the State, draining its life blood in a vain effort to satisfy its insatiable stomach'.[28]

To assess this claim and to obtain a more comprehensive picture of the economy, it is necessary to step back and consider Australia in its international setting. Australia's rapid nineteenth-century growth was made possible by its role in the world economy. Along with the other white-settler societies, it attracted labour and capital and in return supplied the homelands with ever-increasing quantities of foodstuffs and raw materials. In the Australian case it was Britain that provided the investment and the bulk of the immigrants, and Britain with which approximately half the trade was conducted. This made for a prosperity that was heavily dependent on London. On the eve of the First World War exports accounted for more than a fifth of Australian production. The economic relationship can be expressed more emphatically if we consider that the annual domestic product amounted to

£80 for every man, woman and child, and of this £80, we exported £18 in the form of wool, wheat, gold, meat and other such materials, in return for £17 of imports, mostly manufactures. Furthermore, British investment in Australia amounted to £75 per head. Part of this foreign investment was applied to primary industries producing for export, but more was used to finance railways, ports and urban facilities. For it was a common feature of the settler societies that population was concentrated in cities; farms and mines, while highly productive, absorbed only a small proportion of the workforce, which therefore spilled into other fields. The flow of capital into Australia made it possible to build up industries and endow the cities without sacrificing the high levels of consumption that Australians enjoyed.[29]

During the first decade of the twentieth century there was a marked decline in the inflow of foreign capital. British investors whose fingers had been burned in the financial crash of 1893 were loath to entrust further money to Australia and turned instead to alternative fields. The value of new Australian issues in London between 1900 and 1909 was £44 million, half the amount raised in the 1890s and a third of that attained in the buoyant 1880s.[30] Not until the eve of the war did the value of new investment exceed the cost of interest payments on existing loans. All the more remarkable, then, was the growth achieved in this period, as it was without substantial assistance from abroad. At the same time, Australia retained its role as a trading nation in the world economy, one still bound closely to Britain by currency and trade. Benefiting from the high prices brought by its export commodities, Australia did well in the Indian summer of free trade.

Australians enjoyed high living standards. The average Australian ate better, was better housed and lived longer than his British counterpart. Both consumed approximately the same amount of the staples, bread (half a loaf per day) and sugar (2 pounds per week), but the Australian was the greater carnivore, putting away twice as much meat as John Bull (roughly 4 pounds per week as against 2¼ pounds) and he had a much wider range of fresh fruit and vegetables. He lived in a house of five rooms, which he had almost a one in two chance of owning and which provided a room for every occupant. His cities contained slum neighbourhoods, but not the vast stretches of tenements and back-to-backs that disfigured Britain.[31] Furthermore, after the Depression and drought, his lot was improving.[32] Invaluable as he (and sometimes she) might be to the historian, however, the citizen conforming to the golden mean remains a statistical construct offering only an imperfect measure of welfare. One needs to know not just the aggregate consumption of a society but also how evenly the cake was divided—for if the R.G. Caseys and Jock Neilsons, who represented the extremes, outweighed those who earned a modest sufficiency like William Somerville, then per capita consumption statistics will tell us little. There was a high degree of inequality in the early twentieth century, whether higher or lower than today it is difficult to judge, but probably not as high in most other countries.[33] By the standards of the Old World, Australia had indeed been topped and tailed since the relative short supply of labour enabled all but the lowest income earners to participate in a broad-based consumer market.

This survey of material life may therefore conclude with a household budget calculated by the Commonwealth Statistician which is reasonably representative of the skilled manual worker or clerk earning approximately £3 per week. (In 1907 Mr Justice Higgins estimated that 42s were required to keep a man, woman and three children in 'frugal comfort'.) Our breadwinner also supported a wife and three or more children. He paid out 10s per week for his house, which is some way below the market level and suggests an inferior standard of accommodation or a substantial measure of outright ownership. Each week the family spent 22s 3d on food, allocated in this fashion:

meat	4s 11d
bread	2s 9d
vegetables and fruit	2s 8d
butter and cheese	2s 8d
milk	2s 7d
sugar	1s 5d
tea and coffee	11d
other food	4s 4d

This suggests a diet that would please neither the nutritionist nor the gourmet. The family was consuming cheap cuts of beef and mutton in large quantities, garnished no doubt with tomato sauce or pickles; bread, cheese and jam were staple fare, along with sugary puddings. Yet the expenditure on milk, vegetables and fruit suggests a reasonably balanced diet. Furthermore, it is worth noting that while the Australian family spent a third of its income on food, the European family spent more than half. The remaining weekly expenditure included:

other groceries	1s 10d
clothing	7s 1d
fares	1s 4d
drink, tobacco and amusements	1s 5d
fuel and lighting	5s
other items (including repairs, medical expenses, etc.)	5s 8d

Again, this is an abstemious standard: it provides for only an occasional outing and no more than a packet of tobacco with a couple of beers on pay-day. Yet the final items

insurance	1s 8d
contributions to benefit society	1s

emphasize that this is a fortunate household, for such self-help institutions were within reach of only a bare majority of families at best.[34]

These, then, were the patterns of material life in the early twentieth century. The next task is to consider the society they supported.

3

CLASS AND SOCIETY

FOLLOWING THE COLLAPSE of the long boom and the confrontation between labour and capital at the close of the nineteenth century, the social order hardened. The wage relationship became more general, occupational hierarchies more precise, class boundaries more clearly defined. The most important development was the mobilization of an urban working class with its own distinctive identity. There was, of course, a larger web of relationships and forms of mutuality, at work, at play and in worship, that knitted the society together. Some of its strands came under strain, and it was a task of government to repair and reinforce them. Yet the new circumstances also brought into play other, more structured institutions which helped to define the classes more clearly.

The working-class majority sold their labour. Most did so on a weekly, daily or even an hourly basis. The price they received, their wage, depended partly on the nature of the job and partly on the scarcity of the skill that was brought to bear. A qualified tradesman who had served an apprenticeship as a carpenter or a boilermaker could earn up to 60s a week in good times, while an unskilled factory hand was lucky to make 40s. A man in peak physical condition could earn more lumping bags of wheat or wielding a pick and shovel than could a less vigorous man.[1] Some privileged sections of the working class were piece-workers. An expert compositor in a print-shop was paid by results and might clear more than £10 a week.[2] Such a system of employment—it might more accurately be termed sub-contracting in many cases—had been widespread in the second half of the nineteenth century, and embraced shearers, miners, and many urban craftsmen who not only enjoyed a measure of control over the work process but were able to make good money. On the other hand, piece-work methods of payment condemned women making up clothing at home to a miserable pittance, in some cases no more than 10s for a long working week.[3] Female wage-earners in domestic service or workshops also earned less than men. Taken as a whole, the male worker earned twice as much as the female, and the skilled worker 30 per cent more than the unskilled.

Yet wage rates tell us just part of the story, for only a fortunate minority enjoyed regular employment. The carpenter, for example, was engaged to work on a site and had to look elsewhere on completion of the project; like all outdoor industries, his calling was always slack in winter. The wharfie was engaged to work on a particular vessel and when it was loaded had to return to the pick-up and hope to catch the foreman's eye once more in the scramble for engagement; his peak season lasted from late spring to autumn. In the abattoirs the same peak period saw a trebling of the workforce. Even the labourer in a clothing factory or food-processing works experienced the seasonal rhythm of plentiful and slack trade. E.J. Holloway, who worked as a bootmaker before taking up a trade union position, recalled that broken time was a fact of life: 'I, who did so much better than many in the same trade, have worked nine months on end in one year without getting a full week's pay.' A wage-earner with security was envied, and if a lad from a humble background was taken on by the railway, his people 'would look at it as we would a win in the lottery now'.[4]

Taken together, the abundance of unskilled labour and the irregularity of employment meant that the majority of bread-winners were lucky to earn more than £100 a year. Earnings would diminish as the worker aged and might cut out altogether at any time through sickness or infirmity. Even the trained artisan whose insurance and membership of the friendly society offered a degree of protection against such a calamity, was said to be 'always in danger of falling out of the ranks of the skilled and decently-paid labour into the abyss of poverty'.[5] The search for security was a powerful force for the working class's entry into politics.

The most vulnerable workers were the casual and unskilled who drifted from one job to another and lived either in cheap lodgings or run-down rented accommodation. The men could be found fetching and carrying, washing dishes in cheap restaurants, working in the most noxious and lowest-paid factory jobs. Their weekly wage might be 20s or less and at a time when the standard working week was forty-eight hours (from 8 a.m. to 5.30 p.m. Monday to Friday, with three-quarters of an hour for lunch, and from 8 a.m. to 12.15 p.m. on Saturday), their week might well exceed eighty hours.[6] The women, too, worked long hours, some taking work into the home, others going out as process workers in factories or daily domestics for the well-to-do. Even children were put into service long before they reached school-leaving age; they were to be seen collecting bottles and scrap metal, carrying messages round the business area and selling newspapers, matches and such items outside the pubs or the railway station. In the pastoral districts you might see as many as a hundred men camped together, 'travellers' who could expect to receive a feed at a sheep station and in return offered a plentiful supply of auxiliary labour.[7] It was a harsh and precarious existence where the very conditions of survival perpetuated deprivation. Battlers had to follow work but in their frequent moves they could not establish credit with shopkeepers or make the links with foremen and neighbours that cushioned the established worker. The children had to contribute to the household income as soon as possible and consequently could not learn a trade. The very drudgery of such an existence had to be relieved by periodic sprees. A man who grew up in railway construction camps recalls the scene:

> the day following pay-day—Sunday morning—the usual thing was to see the men lying about all over the place, suffering a recovery from the night before. During Sunday, which was always a wild day, the men continued to fight among themselves, all day and right away until dark. In fact I have seen them, many a time, fighting by candlelight.

Another recalls: 'That was the pattern all the time, and consequently they were looked down upon.'[8]

This distinction between the rough and the respectable was a fault-line running through the working class. Essentially a moral judgement, it nevertheless rested on a substratum of economic fact. Respectability was achieved by an act of will and demonstrated by ensuring that one's children washed regularly and had boots on their feet; it was made possible by regular employment.[9] An approximate measure of the proportion of such respectable workers is provided by the principal agencies of self-help, the friendly society (or the life assurance company) and the savings bank. The value of an average life policy was £300 and a bank deposit might amount to £100 or more, but such assets were held by a little less than one-third of the population.[10] This finding is corroborated by the wealth census of 1915, which found that some two-thirds of all wealth-holders held assets of less than £100, and by estimates of home ownership which suggest that a minority of Australians had bought their own home. (The rate of owner-occupancy was higher in the country, where houses were cheaper; in the cities it was less than 40 per cent.)[11]

To what extent did these various wage-earners cohere as a class? The gulf between the respectable artisan and the itinerant casual was undoubtedly wide, the sectional allegiances of different work groups strong. The labourer in a public works gang, for example, was bound closely to his foreman: 'a popular ganger always knows where to find 20 or 30 favourite men. Some men follow their favourite ganger from job to job.'[12] The small workshop where master and men worked alongside each other smothered a sense of separate identity among the wage-earners. The female domestic, the slushy who worked and slept in the restaurant kitchen or the milkman who bunked above the stable were all highly dependent upon their employer. It is hardly surprising that class consciousness had emerged first among the occupational communities of the pastoral and mining districts. Yet the changes that were under way in the cities fostered a similar tendency. The increasing spatial segregation was producing working-class suburbs. As family businesses turned into companies, as small workshops gave way to factories and as mechanization replaced craftsmen with factory hands, these labouring communities became more coherent. The spread of trade union membership was marked, from 5 per cent of all wage- and salary-earners in 1901 to 34 per cent in 1914.[13]

An assortment of occupational groups—salary-earners, shopkeepers, minor professionals, small landlords, the petit bourgeoisie—can be gathered together under the title of the anxious class. They numbered about 200 000 or one-fifth as many as the manual labourers, and their incomes typically fell somewhere between £200 and £500 per annum. Why the anxious class? They enjoyed a better standard of living than the manual workers, they held property, their status was indisputably higher. The reason is that they created their own unease.

The young salary-earner was compelled to forgo the simple pleasures of the worker with calloused hands. Recruited on a low initial salary and required in many cases to put up a substantial bond as a surety of good conduct, he progressed by slow degrees under the vigilant eye of the head clerk. A trainee in the Bank of New South Wales, for example, was taken on at somewhere between £25 and £50 a year and forbidden to marry until he reached a salary of £200.[14] In dress, in choice of friends and in lifestyle he was expected to conform to a suitable standard; the need for diligence and sobriety was impressed upon him until these qualities became part of his nature. His reward came later when as bank manager he held the pursestrings and commanded the respect of the community. The same was true of the accountant. In an age of mental calculation, and with normal human frailty, the checking of accounts and balancing of ledgers possessed an augmented significance—there was a palpable tension in the office when an audit was in progress. Since qualifications in accountancy could be obtained by night school or home tuition, it was the standby of the bright middle-class lad whose parents could not afford university fees. However, the low initial salary and the common practice of requiring a bond of £100 or more closed these careers to the majority.

Teaching was a more likely avenue of social mobility for the children of skilled artisans or small farmers. A country writer remarked in 1906 that 'here, as in the city, there is the same genteel aversion to manual work, especially where mothers with notions will have their sons qualify for a profession'.[15] At the turn of the century it was still possible for a youth of thirteen or fourteen to begin as a pupil-teacher, teaching by day and studying at night, and after passing various examinations to achieve reasonable seniority with an annual salary of £300. The country writer's attribution of ambition to the mother is questionable, and in any case he overlooked the fact that girls (like Agnes Spunner from Creswick) as well as boys became teachers. Along with nursing, education was the main profession open to a woman, offering a salary about two-thirds that of her male equivalent. But, as in banking, promotion came slowly at the behest of a remote central authority, and a rigid propriety was enforced by the inspector. The clergy possessed higher status but lower incomes. An average stipend was between £200 and £300 a year, and the many calls on the manse ensured that the minister 'will always be a poor and hard-worked man with the constant worry of being obliged to maintain the position in society for which his means are inadequate'.[16]

The position of the small businessman was rather different. Some shopkeepers lived on the very edge of survival, working long hours under the shadow of cheaper department stores and caught between the terms imposed by the large wholesaler on the one hand and the need to provide credit to a local clientele on the other. Others prospered. There was still room in this vigorous transitional economy for the small entrepreneur to make his mark, whether as a builder, an estate agent or a maker and repairer. Even then his status was marginal, as the common practice of marking the tradesman's entrance to the rear of the middle-class residence emphasized.

Australians liked to think that success was available to any determined individual and they concluded that their society was free of class barriers. 'Class distinctions, as they exist in Britain, are here practically unknown,' claimed the New South Wales government in an advertisement for migrants. A British observer came closer to the mark:

> It would probably be truer to say that in no country in the world are there such strong class distinctions in proportion to the actual amount of difference between 'the classes'. Betwixt the society worlds of Melbourne or Sydney and 'the masses' is fixed a social gulf that nothing but money can hope to bridge.

Hence the definitive dismissal of the successful man was, ''oo the 'hell does 'e think 'e is—Blimey, I knew 'is old man!'[17] In the cities it was becoming more difficult to enter that moneyed circle. The business enterprise was becoming bigger and more institutionalized. In banking, insurance, shipping, mining, large-scale manufacturing and trading concerns, the public company had replaced the partnership or private company, and mergers and trusts were thinning the ranks.[18] The brewing industry in Melbourne provides a clear illustration. In the 1880s there had been a dozen breweries in the city and inner eastern suburbs; by 1909 all were absorbed into Carlton and United Breweries which became a public company in 1913 and operated just three large plants. Similarly, the Colonial Sugar Refining Company bought out its competitors to achieve a national monopoly by 1907. In these big companies the management occupied a subordinate position and real power was exercised by the principal shareholders who sat on the board.[19] Plentiful and frequent dividends were the rule, and although the story was often told of the original Broken Hill shareholders who turned a few pounds into fortunes, or of Casey's friends, the Mount Morgan partners, who became millionaires, it was those who were already rich who generally benefited. And they formed a tight ring. At the outbreak of the war there were just over 15 000 people (representing 0.67 per cent of all wealth-holders) with more than £5000 and together they owned 34 per cent of all personal wealth.[20]

There was an alternative path to success. The professions commanded extremely high incomes: the head of the medical school at the University of Sydney claimed that some specialists were making £15 000 a year, and leaders at the Sydney or Melbourne bar commanded a comparable sum. To become a judge was to accept a real drop in income, as in the case of the chief justice of the South Australian Supreme Court, on £2000 a year, with a grand North Adelaide villa, a retinue of servants and a pastoral property worth £40 000, who still grumbled that he had sacrificed £130 000 in lost earnings since his appointment, or his equivalent in Queensland who would not go to the bench until the stipend was increased from £2500 to £3500. When the eminent Sydney barrister Sir Julian Salomons condemned the federal constitution because it gave equal representation in the Senate to the smaller states, he hastened to add that he bore no hostility to Tasmania: 'On the contrary, I'm very fond of Tasmania. I spent my last vacation there and liked it so much that I made up my mind that, if I had a good year at the Bar, I'd buy the island.'

Few in other occupations could make such a jest, but not far behind the lawyers and doctors came the successful architects and engineers. By 1909 John Monash had at last climbed to the top of engineering and his income in that year was £7000.[21] Only a handful, admittedly, could attain such eminence, and the run-of-the-mill general practitioner or suburban solicitor jogged along on annual earnings somewhere between £500 and £1000.[22] The reluctance to disclose the value of a practice—symbolized by the use of that genteel unit of currency, the guinea—and the sticky-fingered propensity to supplement professional fees by investment and business ventures make it difficult to establish just how many of the 20 000 practitioners of these professions climbed into the ranks of the rich. In any case the professions possessed an additional attraction: in a society with little regard for inherited privilege or prestige, they enjoyed an exaggerated influence. As custodian and counsellor, the lawyer played a pivotal role in his local community. He had always been prominent in the state legislatures and one malcontent in the Queensland parliament complained that 'there was a tendency in the profession to collar bigger and bigger slices of political power'. A quarter of the members of the first Commonwealth parliament, two-thirds of the first cabinet, were lawyers. The medical doctor took longer to slough off a reputation for self-advertising quackery, but by the turn of the century he too had attained a 'social influence' that one immigrant judged was 'in great contrast to the position of the profession in the United Kingdom, where it occupies quite a secondary position—a position which I, personally, would not tolerate for an instant'.[23] In most neighbourhoods it was the doctor who was the first to own a motor-car.

Law and medicine were pathways to success for those who lacked substantial capital. Some private means were necessary, however, to meet the fees (about £75 a year) and living expenses (no less than that sum) of several years at university and the cost of serving articles (which could come to £200). Then, as one who began practising in Adelaide during the

1890s remarked, 'The income of a young man who has just embarked on the law is, during the first year or two, very limited indeed.' John Latham, who would become chief justice of the High Court, had to teach in a country school for several years in order to finance his legal studies; he then earned just one guinea during his first six months at the Melbourne bar in 1905 and he could not marry until four years after his proposal was accepted. A young Sydney barrister invested wisely in a pair of bluchers so that he could walk to country towns, when the District Court was in session, on the chance of securing a brief.[24] Those who set their sights lower might still have to pay in order to serve an apprenticeship in a solicitor's office. Similarly with medicine, the graduate with ambitions to specialize had to support himself while learning his trade in the wards. Such barriers created a freemasonry. The son of a Naracoorte solicitor who commenced legal studies at the University of Adelaide shortly before the war found that he and a Western Australian were the only students who had not come from the three leading schools of that city. By self-regulation of entry and professional ethics, and by patronage and preferment, the professions ensured that only the exceptional outsider would gain admission to their ranks.[25]

Rural classes observed a different course. For the past half century the bush had been the arena for the competing ambitions of the pastoralist and the small settler. Pastoralism meant sharply polarized class relations, with the land locked up in the hands of a few thousand families and the remainder of the population constituting a largely masculine proletariat. In the Hunter Valley some hundred property owners held two-thirds of the land, and marked marriages and births by putting on entertainment for their 'people' in the woolshed.[26] Agricultural settlement, on the other hand, held out the vision of an independent familial yeomanry, putting the land under crop, creating their own homesteads and escaping from the labour market. Four decades of settlement legislation had failed to realize this dream. Many small blocks had been selected but their owners lacked the implements, the techniques and the access to markets on which success depended, and most found it difficult even to scratch a bare living from the soil. Above all, they discovered that the farm was not a haven from the cash economy. Agriculture required capital—to purchase, fence, clear and work a selection required a minimum of £400—and intensive labour. Small-scale farming imposed 'toil from daylight to dark—it meant wives working like slaves and children milking before dawn and after dark'.[27] A good season merely allowed the battling selector to clear his debts at the local store or reduce his indebtedness to the money-lender.

In this respect the cockies' experience resembled that of the wool kings. The powerful landed elite had borrowed heavily in the second half of the nineteenth century, partly to purchase freehold title to their runs and forestall selection, partly to increase carrying capacity. Their ambitions outran their resources and the twin blows of drought and falling wool prices at the close of the century hit them all the harder:

> Baa, baa, black sheep
> Have you any wool?
> Yes sir, oh yes sir! three bags full.
> One for the master, who grows so lean and lank;
> None for the mistress,
> But two for the Bank![28]

The early years of the twentieth century saw a recovery which enabled the pastoralists to resume their familiar pleasures: entertaining house guests, keeping up a town-house for the city season, educating the children at boarding-school and perhaps sending them to Oxford or Cambridge. But not all had survived the crisis intact. Banks and finance companies like Casey's Goldsbrough Mort foreclosed on some and installed managers or else kept the original owner on a short rein. Larger runs were broken up into several lots and in some cases subdivided further for closer settlement.[29]

Australians clung to an idealized rural identity. The bush provided them with a set of values, mateship, endeavour, fortitude, that they liked to think of as national characteristics and celebrated in ballad and literature. Yet while Henry Lawson and Banjo Patterson sparred in doggerel over the attractions and disadvantages of bush life, and while 'Steele Rudd' made humour from the hardships of the selector (Jock Neilson said that his father always reminded him of Steele Rudd's 'Dad' and both were avid readers of the events in *On Our*

Selection), both the writers and the bulk of their audience were indulging a vicarious pleasure. They were town-dwellers wedded to the attractions of the metropolis.[30] The extension of farm settlement proceeded from the natural increase of the rural population, augmented by some immigrant newcomers, and barely kept pace with the overall growth of the population. Why were so many reluctant to go up country? The farmer of the early twentieth century enjoyed a number of advantages that had been denied to his grandfather. He had greater government assistance, better access to his market, better machinery, improved knowledge to draw upon. The official view was that any stout-hearted labourer could succeed: 'There is no able-bodied farm hand . . . , twenty years of age, who is not able to save £25 a year if he wishes to do so'; after three or four years he could share-farm, rent or select Crown land, and make his own future. The fact was that farm hands were paid less and worked longer hours than urban labourers, and the gulf between the newcomer and his employer was just as real and just as wide as it was in the city. A regional examination of probate records for north-central Victoria discloses that at the turn of the century 40 per cent of all farmers left estates worth more than £1000, a level of wealth achieved by less than 1 per cent of the miners and labourers in the district.[31] Yet rural society *was* different. It had its own unique generational pattern of ownership which was to solidify, bind the country community together, and endure for the whole of the period.

At first sight the rural workforce seems to fall into distinct categories. First there was the wealthy farmer or grazier with land and equipment worth at least £1000, who employed a number of labourers; in the 1901 census there were 67 000 rural employers. Next was the small farmer with a few hundred acres whose family supplied the labour; there were 76 000 of these in 1901, with a further 55 000 recorded as unpaid assistants. Perhaps share-farmers should be included as a sub-category since they planted a fifth of the wheat acreage by 1914. (The normal arrangement was that the landowner provided the land, the seed and half the cost of bags and freight; the farmer contributed horses, equipment and labour, and the wheat cheque was shared, usually half and half.[32]) Finally there are the wage-earners, 131 000 of them, ranging from the shearer to the lowly dairyman or harvest-hand. But upon closer examination, both the large employer class and the rural proletariat shrink. First of all, a number of contract workers were themselves small farmers. This should not necessarily be taken as evidence of backsliding among the knights of labour: a shearer bought a few acres or a small business because he had no other security against sickness or old age, and he was no less ardent a unionist for that. The more significant special group consisted of sons of small farmers for whom there was no demand on the parental farm—or, as in the case of Jock Neilson, whose outside earnings were more important to the family economy at that moment than their unpaid labour. These were not rural proletarians; they were only temporary wage-earners who either returned in time to the family farm or moved into the areas of new settlement. Thus for many of the 131 000 rural employees, wage labour was merely a compensating mechanism for the effects of the life-cycle or demographic variation between households.[33] The family was the typical economic unit. Its flexibility and productive efficiency—Dave worked longer and harder for Dad than he would have done for an employer—was achieved by imposing on all members, male and female, a burden of unpaid labour.

On the farm the family served as both the unit of production and the unit of consumption. Husband, wife and children all had their allotted tasks, mingling their efforts to satisfy the material needs of the household. Such a division of labour could be found elsewhere, for example in the corner shop, but it had become uncommon. Among the middle and upper classes the father earned the income, the mother managed the household and the children remained dependent until their schooling was completed. Many working-class families aspired to the same goal, and the claim that adult males were entitled to a 'living' or family wage sprang from this ambition. For the time being, however, most labouring families depended upon more than one source of income with women making a crucial contribution and children pressed into service as soon as possible. These different family economies were accompanied by different familial relationships, both between husband and wife and parents and children, as well as different ways of thinking about the family. For better or for worse, it was the middle-class model that motivated the makers of law and opinion.

Set against the earlier patterns of European society, the middle-class Australian family was 'born modern'.[34] Husband and wife came together on the basis of the man's established occupation and set up house for themselves away from kith and kin; mistress of her own home, the mother could thus address herself to domestic responsibilities and devote her attention to children and breadwinner. In short, the novel figure of the housewife had emerged. This demarcation of the private or domestic sphere from the public or social world, and the corresponding differentiation of the roles of wife and husband, is so axiomatic that it is easy to overlook the historical forces that made it possible. It rested, fundamentally, on the decline in infant mortality and the reduction of the birth rate, both substantially achieved before the First World War. Smaller, stable families strengthened the emotional ties between parents and offspring, and transformed the position of the child. Childhood became a more protracted, more sharply marked-off stage that lasted until a boy went into long trousers at sixteen and a girl first put up her hair at eighteen. The demographic transition was reinforced by a variety of institutional pressures. Previously children had been expected to contribute to the household income as soon as they were able, but factory legislation and more stringent school attendance regulations closed off openings in the labour market. There were new measures against infanticide and child neglect, a new attitude to infant welfare and 'scientific motherhood'.[35] In these and other ways the state entered into the domestic sphere to limit the rights and responsibilities of parenthood. Yet at the same time the state was helping to construct the family in its modern form as 'the basis of national life'.[36] The result was that the state created a hybrid domain of the public and the private, fusing morality and nation building, which treated a whole range of topics from housing to juvenile deliquency under the heading of 'the social question'.

There were two obstacles to the realization of this ideal.

First, it was inapplicable to the circumstances of large sections of the population, either those who did not fall within the ambit of the nuclear family or those labouring, farming and Aboriginal families who were wedded to different habits. With eleven males for every ten females, with an illegitimacy rate of 6 per cent, with the death of a parent a common calamity and many men whose work took them far from home, some were simply unable to play their prescribed family role. In any case, the women and children of the bush district or working-class suburb were caught up in a work routine that left little room for close affective relations to develop. When the children were not lending a hand, they were outside in a world of their own. Judge Stretton recalled of his early days in the northern working-class suburbs of Melbourne that children were unmercifully thrashed by parents and teachers whenever they caused offence. Put simply, 'children were not the main feature of living in those days'.[37]

The second impediment to the domestic ideal was the presence of women in the workplace. Approximately one-third of all women worked and they accounted for more than 20 per cent of the national workforce at this time. Some worked by choice and some out of economic necessity, some earned wages only until they married and others made a lifetime career, but their increasing involvement in the factory and the office made nonsense of the ideology that women were unsuited to the profane world outside the home. The largest single category of female workers, the domestic servants, were accommodated most easily to this ideology since theirs was a traditional female calling. However, there was an increasing reluctance to go into service, not just because of the long hours and low pay—Sydney factory hands who had tried domestic service were adamant that it was 'much harder' and 'more laborious' than their work—but because of its restrictions on freedom and the subservience it imposed.[38] Younger women, especially, preferred the factory where there was more companionship and more independence. Even though the clothing and processing industries seldom offered more than 20s per week, and employment was highly seasonal, employers could usually pick the most industrious from a plentiful supply of applicants. In many cases they favoured juveniles, whose wages went as low as 5s, or single women (for 'the girl who keeps herself ... is the girl who gets the wages'[39]). Married women in need of an income formed a special pool of labour, the outworkers, who were especially susceptible to exploitation. Taking in sewing, they might work longer hours than a factory hand for less than half her earnings.[40]

Women experienced particular difficulties in improving their conditions of work. Attempts at industrial organization were handicapped by the vulnerability of most female employees; and male trade unionists, if not overtly hostile to women, were more concerned to protect their trade against the incursion of lower-paid female workers than to support claims for equal pay. No sooner did the line of demarcation between the masculine and feminine occupation shift than the status of that particular calling deteriorated. Hence the introduction of the typewriter into the office created the specifically female occupation of the shorthand-typiste whose salary was well below the male clerk's, and the influx of women shop-assistants into the city stores that flourished in this period was accompanied by deskilling and loss of prestige. The manager of one emporium reported in 1909 that boys no longer sought to train as shop-assistants because 'they think that to serve behind the counter is derogatory and work not fit for a strong healthy lad'. Where women did gain equality, for example as postmistresses and telegraphists, they were squeezed out of employment.[41]

With the treatment of the home as a haven from the workaday world came a new emphasis on public education. While previously the authorities had condoned a high rate of absenteeism among children of the working class, now they policed attendance more strictly in an effort to remove children from the labour market and put an end to what the South Australian minister called their 'aimless perambulation of the village or city streets'. For six years every child was expected to undergo a schooling that would instil skills appropriate to his or her station in life. Every country settlement had its wooden schoolhouse, every industrial suburb its imposing edifice towering over the surrounding workers' dwellings. The school bell regulated the waking hours of children as closely as the factory hooter prescribed the routine of their fathers. Inside the school the meticulous marking of the roll, the formal manner of address to the teacher, the habit of working in unison and on command all reinforced the need for order and obedience. The pedagogy was calculated to teach the elementary skills of literacy and numeracy: children read from texts, practised their penmanship and encountered literature only to parse it; in arithmetic the need to reach the correct answer outweighed comprehension; geography was rote learning of places, populations and products. Since the classroom might contain eighty or more students at different levels, much of the instruction was given by pupil-teachers under supervision from the teacher; the teacher in turn was watched by the inspector, and examination results were the currency of success.[42]

These government schools differed little from state to state. Nor, apart from their religious component, did the parish schools of the Roman Catholics, since public examinations made for a convergent syllabus.[43] As yet, neither system provided further schooling except for the ablest and most determined (though a new generation of administrators were about to extend public provision). Well-to-do parents who sought to provide their children with a secondary education or who aspired to the higher professional qualifications offered by the universities still relied on the independent schools. A host of private establishments, many consisting merely of a couple of rooms in the proprietor's home, offered instruction in genteel accomplishments. Most prestigious of all were the leading Protestant schools, and the vicar's wife Ada Cambridge wrote from experience that 'you cannot keep a son at a public school, giving him all the advantages of it, for much under £100 a year'. Yet even these schools were still acquiring the paraphernalia of the English public school. Melbourne's Scotch College, for example, had shared preeminence in that city with Melbourne Grammar for half a century; yet its total enrolment was less than 250, and only with the arrival of W.S. Littlejohn in 1904 were masters put into academic gowns and a prefect system imposed on the fractious senior boys.[44] Finally, there were the four universities—in Sydney, Melbourne, Adelaide and Hobart—which taught some 2500 undergraduates. For all their Gothic architecture, their lakes and sylvan settings and their cultivation of other Oxbridge conceits, the universities were already vocational, with a marked emphasis on medicine, law, applied sciences and those arts subjects that led to a career in education.

Higher education was a slender peak on a broad base. There were some 750 000 students enrolled in the primary schools in 1901 (of whom no more than 600 000 were to be found in class on any given day); one in twenty would go on to secondary school, and of this privileged minority, one in twenty again would proceed to university. The great majority emerged as they reached their teens, some fretful, some obedient, with a rudimentary capacity for reading, writing and arithmetic. It was, as its name suggests, an elementary education, leaving specific occupational skills, whether industrial or commercial, to be learnt on the job as a junior employee. In this rough-and-ready fashion, the educational system prepared and allocated the child to his or her station in life.

What of the hours outside work and the domestic routine? Clearly, the use of leisure varied according to individual preference, so that while one man went down to the pub, his neighbour stayed home and read. Much entertainment was self-made—playing the piano and singing in the front room, walking the city block on a Friday evening, visiting friends. Still, there were larger patterns of recreation and association, shaped by means, opportunity and convention, that reinforced the social hierarchy.

At one end of the scale there were the exclusive gentlemen's clubs offering luncheon and dinner, accommodation when needed, a library, games and conversation in congenial company. The Melbourne Club, to which R.G. Casey belonged, was instantly recognizable to an English visitor: 'It is almost as luxurious as Brook's, there are almost the same proportion of old fogies ... no Unionist peer could fail to recognise the ideal comfort of the Melbourne Club.'[45] Less obtrusive, though hardly less exclusive, were special-interest groups like the Wallaby Club, whose members shared a taste for walking and talking; the Savages, for those with an in-

A member relaxes with newspaper and refreshments on the verandah of the Melbourne Club

terest in the arts; or the Boobooks, where the talking took place after a good meal. Women had the Alexandra Club, which began as the Wattle; the Catalysts, for those with intellectual interests; and the Lyceum Club modelled on London precedent. But these women's clubs were later creations, formed after the turn of the century, and their acceptance was incomplete: on learning that his wife was to become foundation president of the Lyceum, Alfred Deakin told her that he had 'evidently failed to observe that your stockings had recently acquired a faint tinge of blue'.[46]

These are Melbourne examples, catering to the establishment of that city, but their equivalents were to be found in all urban centres. The common principle for admission into a club was that the candidate had to be nominated by several members and elected by general consent since 'every member must be on terms of perfect amity with every member'; the rule of the Melbourne Club was that 'one black ball in eight shall exclude' and some other clubs were even more rigorous. This allowed the established élite to keep out self-made interlopers. 'Strong men have died frustrated and chagrined', wrote a Perth journalist, 'for one reason and one reason alone. All their wealth and all their power could not buy their membership of the Weld Club.' Pastoralists and eminent professionals were the most conspicuous in the clubs: among the fourteen presidents of the Melbourne Club from 1901 to 1914 there were seven pastoralists, four lawyers, a surgeon, a newspaper manager and a banker. These were the principal figures of city society and prominent in company boardrooms, but they were not directly involved in industry and commerce, and there was a general tendency to look down on trade. The historian of the Wallaby Club, which contained a high proportion of the Victorian bench and bar, regretted that 'some of our earlier members were not of the right Wallabian strain—they were merely men of business'. In similar vein was the rule of Sydney's Union Club that any member caught discussing business affairs at the table must buy the others a bottle of port.[47]

At the other end of the scale, the tribal loyalty of the larrikin push was just as strong. Here is how a South Australian parliamentarian described such exotics in 1904: 'Tall and thin with pale animal faces, constantly smoking the strongest obtainable tobacco, dressed wholly in black with white linen, they take the pavement and pass obscene remarks and comments upon the passers by.' The meeting place of the push was the street corner; its members were working-class youth on a local and occupational basis. In Sydney, for example, there were the Rocks Push, the Iron House Mob at Woolloomooloo, and the Livers who mostly worked in the Glebe abbatoirs. Members were distinguished by their high-heeled boots, flared trousers, short jackets and white shirts, and rules were informal—but a transgressor paid a more painful penalty than the price of a bottle of port.[48] Between these two extremes lay a host of voluntary associations. A medium-sized town like Mildura offered choirs, amateur theatre groups, dances, charity bazaars and whist drives, Masonic and other lodges, a chess club, a Caledonian society and a host of sporting clubs including cricket, football, tennis, golf, rifle shooting, swimming and rowing.[49]

Such facilities were a source of pride, an affirmation of a vigorous and healthy community spirit. For in a society that was characterized by wage labour, work and leisure were clearly separated yet complementary, and idle time was consequently a cause for concern. The great majority of workers enjoyed a respite from work on Saturday afternoons as well as Sundays, and there were more than a dozen special holidays in the annual calendar. All too often these festivals began as a spree and degenerated into riotous excess, New Year's Eve being especially notorious as a night of public abandon. Against this custom of treating holidays as a release from the constraints of work, moral reformers waged unrelenting war. As the members of a royal commission considering the hours of work put it, the working class was expected to use its leisure moderately and not in a 'wild burst' of excitement and extravagance. The fear was that Australians would become 'so devoted to pleasure and to gambling as to be incapable of serious, self-sacrificing national work'. This new endeavour to restrain or tame popular pleasures operated in all corners of Australian society. Thus had Archbishop Moran in 1896 taken control of Sydney's St Patrick's Day celebrations, replacing a street march which had traditionally dispersed into hotels with a family sports day where he paraded wealthy Catholic dignitaries.[50] The campaigns against violent sports and gambling, the temperance crusade, even the fencing of parks and playing fields, can be understood as an endeavour to impose order on the atavistic impulse.

That objective was never wholly realized. To take the particular case of gambling, both games of chance and sporting contests exercised an irresistible attraction on the Australian public. Money was wagered in sweeps, lotteries and two-up schools, at boxing contests, on running and cycling, horse-races both on the flat and in harness, dogs and pigeons. The principal sporting events drew enormous crowds: more than 100 000 watched the 1911 Melbourne Cup and 25 000 were turned away in 1908 from the world heavyweight contest in Sydney between the Canadian champion and his Australian challenger.[51] Even the most popular spectator sports, however, operated within an orderly, regulated framework where control was vested in the hands of the respectable and well-to-do. An examination of the principal Western Australian enthusiasms suggests that horse-racing's controlling body was the most exclusive, so much so that the self-made businessmen turned to trotting as their domain. Cricket was in the hands of the aspirant gentry and only the Australian rules football clubs were open to the wage-earner. The same would hold for the northern rugby-playing states where rugby league was created in this period by small entrepreneurs with links to the Labor Party. Whereas rugby union remained an amateur game for gentlemen, rugby league clubs were located in working-class areas.[52] But here, as in Australian rules, it was the local businessmen—publicans, estate agents and the like—who filled most committee positions. Such pastimes were already part of a commercial leisure industry, catering for a public with money to spare and access to urban transport, and competing for their shillings with the dance halls and skating rinks, the old theatre halls and the new cinemas.

On Sundays a high proportion of the people were at worship, certainly one in three and in some places one in two. For despite gloomy prognostications of the 'Arctic chill of religious indifferentism' and warnings that 'Australians are becoming a non-churchgoing people' this was in fact an age of religious growth and renewal.[53] The building of churches, the training of clergy and the expansion of congregations more than kept pace with the increase in population. Leaving aside the tiny minority who adhered to non-Christian faiths, the worshippers were divided in three roughly equal proportions among the Church of England, the various Nonconformist denominations and Roman Catholicism.[54]

Of the Protestants, the Anglicans and the Presbyterians enjoyed greatest prestige. Both retained close links with their parent bodies, the established churches of England and Scotland; the leading families sat in their pews, and their schools catered for the prosperous and successful. But their worldly success was at the same time their weakness: since membership imposed fewer demands, adherence was commonly a matter of convention rather than active faith, and in both liturgy and theology most church members clung to the familiar ways. Even the more evangelical denominations like the Methodists and the Baptists drew primarily from the middle and lower-middle class, and found fewer adherents in inner-city, working-class suburbs or newer country areas. The Protestants' growing awareness of this failure did stimulate missionary work among the 'white heathen', but for the most part it took the form of a moral crusade to root out evil and impose righteousness.[55] The labouring poor were to be saved by enforcing temperance, policing the sabbath and eliminating vice.

In stark contrast, a disproportionate number of Catholics were to be found in the inner suburbs and rural backblocks. As the bishops said in 1885, their flock was numerous 'wherever the hours are long, the climate merciless, the labour unskilled, the comforts few and the remuneration small'.[56] But it went further than this. Overlaying the social differences between Catholics and exacerbating a religious division that dated back to the Reformation, there was the overwhelmingly Irish character of Australian Catholicism. The great majority of Church members were of Irish descent; their bishops and many of their priests were trained in Ireland; since Ireland was in bondage to England, this conjunction of faith, class and nationality fostered an intense cohesion among the Catholics in Australia. The orientation was ambiguous: Catholics simultaneously sought acceptance within the larger society and closed their ranks against it; on the one hand they rejoiced in a freedom of opportunity that was denied in their homeland, on the other they resented the advantages enjoyed by the entrenched Anglo-Scottish majority. So at an immense cost they built primary schools for their children and established their own orphanages, charities and other institutions. A priest observed the results:

> The raising of the material structure for the present, at least, absorbs all our energies. Brains echoing the ring of the trowel and hammer, or racked by the very practical problem, how to clear a Church debt off, have little thought to spare to ponder or evolve.[57]

Such circumstances fostered a 'religiosity which emphasised duty, obedience, loyalty, hard rules, the black of sin, the white of purity, with no areas of grey'. Under Irish episcopal authoritarianism, submission and conformity intensified Catholic segregation.[58]

Sectarianism was endemic. Catholic and Protestant children would meet on the street and taunt each other with rhymes; job advertisements advised 'No Catholics need apply'; mixed marriages could result in ostracism.[59] The fire was fuelled by public disputation, most notably in the protracted conflict in Sydney between Cardinal Moran and the Reverend William Dill Macky, minister of Scots Church and grand chaplain of the Loyal Orange Lodge. This began in the mid-1890s with Moran's jibes at Protestant missionaries, simmered for the next few years and came to a head in 1900 with the Coningham case (or as Protestants would have it, the O'Haran case). Coningham was a well-known cricketer who had toured England in 1893 with the Australian team, Father O'Haran a handsome priest (he gave photographic portraits to admirers) who exercised great influence as Moran's private secretary. Coningham brought a suit against O'Haran for adultery with his wife. The case attracted enormous attention and 5000 waited outside the court house during the first trial to hear that the jury was deadlocked; in the retrial Coningham's case collapsed and O'Haran was cleared. Dill Macky had opened a Fighting Fund for Coningham, he had helped

Mrs Coningham (for she was in collusion with her husband) identify Catholics on the jury list—he had even lent the plaintiff his revolver. Within a year Dill Macky created an Australian Protestant Defence Association to expose 'Roman Catholicism as an element of danger to civil and religious liberties of the people'. Packed meetings in the Sydney Town Hall were followed by a tour of rural areas where Dill Macky would talk of 'the Siege of Derry' and the faithful would sing 'Rule Britannia' and 'the Protestant Boys'. There were several skirmishes when Catholics tried to disrupt gatherings and at Wyalong the Riot Act was read. By 1903 the association had 135 branches and 22 000 members.[60]

In this, as in other matters of religious controversy, the zealots spoke of events that had occurred 12 000 miles away before the white settlement of Australia. With the possible exception of the progressive Australian Church that had been established in Melbourne in the 1880s and was already in decline, Australians created no indigenous religious sect. Their religion was derivative. The secular character of the state had been settled and the most innovative, creative minds were either opposed or indifferent to religion.[61] The overwhelming majority were born, married and departed life in church, and a fair proportion observed its outward forms. But real, intense religious enthusiasm was an exceptional phenomenon and often created unease. A young man working in a Ballarat gold-mine was converted in 1905 by an itinerant preacher. His immediate response was to go to his neighbours and confess that he had shot their billy goat. At work the next day his companion shied away from him while his other workmates swore at him and repeated the dirty jokes that he had once told them. And the reaction of the convert? 'I went out of my way to be kind to "Monkey", the Chinaman.'[62]

Amid a kaleidoscope of different beliefs and attitudes, it is perhaps possible to draw out some general modes of thought. As an offshoot of the most prodigious imperial power, Australia shared many of the assumptions of the parent country: its insular confidence in the superiority of white civilization, its restless pursuit of material growth and its search for order and stability. Yet the distance from the imperial centre and the vastness of the environment imparted to these inherited characteristics a particular twist. As the Commonwealth inauguration speeches at Kalgoorlie suggested, the sense of race was more anxious and correspondingly more aggressively xenophobic, not just towards Aborigines but towards the Asians and Pacific Islanders with whom the white majority came into contact. And even towards other Europeans. Some years after the Commonwealth inauguration, a Dalmatian youth, Anthony Splivalo, came to Kalgoorlie to join his elder brother and seek his fortune. In order to learn to speak English without a foreign accent, he boarded with an Australian family. One afternoon, while walking with the children, he was approached by a bigger boy who spoke to him in words the Dalmatian could not understand. 'As I bowed in apology, he scowled and hit me hard on the face with his heavy hand. The shock was great. My two companions looked on silent and helpless.'[63]

The mistrust of the unfamiliar, the quick resort to physical violence and the inarticulate irresolution of the bystanders were all part of the cultural fabric. Beyond this, the sheer novelty of the civilization weakened the strength of custom and tradition. There was a raw, unfinished quality about its artefacts that permeated social relations. Differences of rank and station that appeared natural in Britain had to be asserted in Australia and held in place by force.

The sheer physical effort involved in the tasks that most people performed seemed to press in upon their lives. Anthony Splivalo found it a 'dreary sort of world' lightened by simple pleasures. This did not exclude the creative arts. Australians were avid readers of newspapers, magazines and romantic fiction (the shilling novels of the New South Wales Railway Bookstall Company sold 4 million copies from 1900 to the end of the First World War[64]), and they delighted in songs and ballads that demanded to be spoken aloud. On the unlined walls of the simplest bush cottage were reproductions of pastoral landscapes, moral or heroic art cut out from illustrated periodicals. Seeking an immediate and direct emotional response from a limited stock of conventional images, the majority shied away from the abstract and the metaphysical. They preferred a realism cast in the demotic form, celebrating the land and its people. Jock Neilson was working at road construction during one of the spells when his eyes were particularly troublesome, so he asked a fellow navvy to take down some verse that was in his head. After a few lines came the scribe's muttered comment, 'Christ!'. Before it was finished he had found full voice: 'Yeah, but what is it? What's it *for*?' and finally, 'But this ain't poetry anyway. Jock, you must be going off your onion.'[65]

4

PATTERNS OF POLITICS

We regard the State not as some malign power hostile and foreign to ourselves, outside our control and no part of our organised existence, but we recognise in the State, we recognise in the Government merely a committee to which is delegated the powers of the community...

The speaker was William Holman, deputy leader of the Labor Party in New South Wales; the occasion a public debate with George Reid, leader of the federal opposition, held in Centenary Hall, Sydney, during 1904. Holman was defending the Labor Party's nationalization objective, Reid hunting the socialist tiger. Labor, he warned, would 'make the free citizens of the Commonwealth into members of one greater Government gang, governed by socialistic overseers'. Yet Reid was quick to allow that there was much the state could do for its citizens: he pointed to public works undertaken by the administration he had led in New South Wales from 1894 to 1899, and he spoke with pride of measures, such as his land tax, that were intended to 'make wealth pay a fair share towards the burdens of the community'. These, he said, were just some of the ways whereby the government could provide 'openings for fresh development of individual talent and ability'.[1]

Such conceptions of the state found rich and plentiful expression in Australia. Governments had served as a means of self-fulfilment throughout the second half of the nineteenth century, responding to the needs of those best able to articulate their economic interests. The qualification, *self*-fulfilment, is important. The colonial state exercised only the most minimal social welfare function in the sense that we now understand that term, and the idea of systematic public provision for the needy was only beginning to win acceptance, the old-age pension enacted in New South Wales in 1900 being the first such measure. When the state came to the assistance of the unemployed labourer, it did not give him a handout—it offered him employment on public works. When it broke up big estates to put families on farms, it expected those families to become self-supporting through commercial agriculture. When it built and operated a public undertaking, it did not seek to supplant the entrepreneur—rather, by providing what private enterprise could not or would not do, it hoped to stimulate the growth of the private sector and, above all, to help the producer to get his goods to the overseas market. In short, the state did not shelter its citizens from the rigours of the market: it helped them to help themselves. The six colonial governments thus played a crucial role in bringing to this country the two scanty resources, labour and capital. They provided railways, roads and harbour facilities for the woolgrower, land for the farmer, public utilities for the town-dweller. The Depression of the 1890s merely emphasized the extent to which prosperity depended upon these economic activities of the state.

That was one political paradigm. The other was the competitive nature of popular politics. Certainly it was the wealthy élite who dominated the legislatures and reaped the greatest gains from their decisions. Even so, the free-trade interests of the merchant differed from the protectionist interests of the manufacturer, and the townsman did not always see eye to eye with the pastoralist. Furthermore, the élite relied on electoral support and since a popular franchise had been won for the lower houses, it was forced to cater to the voters. While class lines were hardening, the polity still exhibited a criss-cross of competing interest groups and localized aspirations. A disgruntled observer of the political scene remarked that 'A ruling caste, though it looks well after its own, is able to ignore petty local interests.'[2] Australia had no such ruling caste and petty local interests were still uppermost at the turn of the century. The successful member of parliament was the 'roads and bridges' member, one who made sure that the railway line ran through his electorate and that employment on public works was freely available to his constituents. 'Hang reform! Blow ideals! I am nothing but a dancer on a departmental doormat', complained one such parliamentarian in 1901.[3]

The machinery of administration accommodated these imperatives. Government departments and public undertakings in each of the six colonies still lay within the direct control of the minister—public service boards had been created in New South Wales, Queensland, South Australia and Victoria during the closing years of the past century, but their authority remained incomplete.[4] This allowed extensive networks of patronage to flourish whereby political factions satisfied their clients and supporters. Similar practices prevailed at the level of local government, though on a greatly reduced scale since so many of their tasks—municipal transport, water supply and drainage—were hived off to government-controlled boards. Yet in a rough-and-ready fashion the system worked, and patronage was dispensed with due regard to the political and financial exigencies. Thus while staff appointments were an important part of ministerial prerogative, the turnover of administrations and the periodical retrenchment of the public services allowed the most glaring sinecures and the worst cases of incompetence to be purged. In similar fashion, the exercise of discretionary powers in a minister's administration of his department and his enforcement of statutes was balanced by frequent disclosure and investigation of abuses.

Probably the most spectacular example in the early years of the twentieth century occurred in the New South Wales Lands Department. For years past, land had been sold to the Crown for more than its value, and leased by the Crown for less than its value, both types of transaction being handled by land agents who took a large share of the proceeds. While some land agents had friends in the Lands Department, W.N. Willis, a member of the New South Wales Legislative Assembly, worked on a grander scale: he was an intimate friend of the lands minister, W.P. Crick. Their partnership, which came to light because of a disgruntled client, was examined by a royal commission, which calculated that Willis alone had received £44 000, Crick no less than £15 000 and other public figures lesser amounts. Crick evaded conviction—he explained at his trial in 1906 that the £27 000 in banknotes he had banked over the previous four years had been won at the races—but was declared ineligible to remain in parliament and struck off the roll of attorneys.[5] More successful was the aptly named premier of Victoria, Sir Thomas Bent, who bought tracts of land in Melbourne's southern suburbs and then doubled his outlay by putting a railway line through them.[6] It was Bent, too, who closed off a royal commission of inquiry into the Victorian police force in 1906 after it had gathered evidence of favouritism, victimization, corruption, failure to enforce the gambling laws and malpractice in the licensing section (even the chief commissioner was trustee of a pub in Carlton). And in the same year the Victorian chief secretary resigned from politics after the disclosure that he was a business partner of the madame of Melbourne's most fashionable brothel.[7] Scandals on this scale were not typical. They were, rather, instances where the practice of treating public office as a system of spoils was taken to excess, and the resultant clamour served as a safety-valve for civic morality. Yet even here it was possible for the principal figure in the New South Wales land scandal to reconcile public interest with private advantage. 'These land agents are making a lot of money', Crick was reported to have told an agent,

> and there is no reason why you should not make something out of it, too, or why I should not make something... But you must understand this distinctly, if you bring me anything that will in any way prejudice the public interests, or that can be questioned in any way, I will not have anything to do with it.[8]

On the same basis, the procession of graziers who paraded before the royal commission denied that they had any intention of bribery—they paid high sums to Mr Willis simply because he was an influential member of parliament who could get things done.

Development in all its aspects had been at the centre of politics for the past half century, ever since the colonies had attained self-government.[9] Governments passed laws and organized foreign loans and tariff duties, land settlement and mining, railways and telegraphs, immigration and education. These activities certainly raised important questions of principle: each of the colonial legislative assemblies contained a group of liberal members favouring the opening up of the land to settlement by farmers and measures to improve the lot of the urban wage-earners, and a group of conservatives defending the sanctity of property and freedom of contract. The liberals sought democratic reforms, the conservatives resisted them and held fast to their control of the upper houses. But such divisions remained fluid and majorities were unstable. A ministry had to build and secure its support by manoeuvre, coalition, promises and patronage. The successful administration was the one that allowed citizens to prosper and made the benefits of material progress available to a sufficiently wide clientele. By 1901 this system of factional politics was giving way to party politics. Voters were increasingly choosing between candidates on the basis of the platforms on which they stood, and the candidates themselves were chosen by political organizations representing distinct ideologies. The transition to party politics was more obvious in New South Wales and Victoria, where its origins had been apparent in the 1880s; it was hastened by the emergence of the Labor Party in Queensland and South Australia, while in the smaller states of Tasmania and Western Australia the change was only beginning. Even in New South Wales it was still possible for a parliamentarian of the old school to flourish. John Gillies, the member for Maitland from 1891 to 1911, was one such highly successful 'roads and bridges' politician. His base of support was local, for he was proprietor of the local newspaper, an alderman and member of various civic and voluntary associations in the area. For twenty years he faithfully gave his support to the party in power, whether it be Reid's free traders or Lyne's protectionists, and in return he was able to secure the favours his constituents sought.[10]

Under the old order the conduct of elections had a similar local emphasis. A group of parliamentarians would create a central committee and establish working relations with local organizations. They would hope to secure agreement on a satisfactory candidate to represent their outlook (since voting was first past the post, a multiplicity of candidates could be ruinous) and he would seek endorsement and assistance from a spectrum of organized interests. Neither voting nor enrolment was compulsory at the turn of the century. In Queensland only 21 per cent of the population were on the rolls, and just over half of the eligible voters exercised their right in the first federal election.[11] While manhood suffrage for the lower house had to all intents and purposes been won in each of the colonies (and universal suffrage in South Australia and Western Australia), some property-owners still enjoyed a plural vote, and residence requirements disenfranchised that significant section of the working class that led an itinerant lifestyle. Since turnouts were low and electorates small—it was a large constituency in which more than 2000 votes were cast—the loss or gain of a few hundred voters might make all the difference. These conditions made it all the more important to get your supporters onto the electoral roll and those of your opponent off it. The rolls were brought up to date at regular intervals and the onus rested on the citizen—hence the prudent candidate employed an agent who used a team of professional canvassers. Then he would need to hire halls, place advertisements, provide transport to get the voters to the booths and scrutineers to watch over them. It could be done with voluntary labour. The Labor candidate set out for the New South Wales electorate of Gunnedah in 1901 with £19 in his pocket and a single ticket, confident that he would return with a member's railway pass because the local miners had pledged their assistance. But for most it was an expensive business. 'Enthusiasm is a poor vote-catcher, and mere patriotism is valueless, but a strong banking account can win seats every time', commented one political organizer.[12]

Parties became more important as local concerns were tempered by national issues. The style of election campaigning changed also with the enfranchisement of women. Back in 1898 Billy Hughes had been opposed in his inner-Sydney electorate by a candidate who gave all voters a ticket entitling them to a free glass of beer when they took his how-to-vote card.[13] It was common in those days to hold meetings outside the pub. But with women's franchise, the old rough-and-tumble pub-balcony style of politics was no longer adequate. The increased size of Hughes's new federal electorate of West Sydney (it had more than 14 000 voters compared with just 1027 who had voted in the state electorate), and the subsequent introduction of compulsory enrolment, optional postal voting and the regulation of electoral expenses all strengthened the need for new methods in place of the knockabout ways of the past. 'The question of the postal vote is said to be of importance to middle-class women who shrink from going to the poll,' noted the governor-general after a conversation with his prime minister. This did not mean that electoral malpractice disappeared. 'Vote early and vote often' remained good advice. The Labor politician who declared that 'An honest ballot is the breath that fills the lungs of the Commonwealth', was the same politician who as Minister for Home Affairs attempted to hand-pick the returning officers in his home state of Tasmania.[14] Elections remained intensely public occasions, and the evening meeting or street-corner assembly was still a vital forum of politics. Indeed, the ritual aspects of electioneering, the give-and-take of the public platform and the colourful confusion of catch-phrases and personalities, often outweighed policy considerations. Like the young A.R. Chisholm, many were 'indifferent to politics' but 'went to political meetings, quite impartially and mainly to be amused by replies to interjections'.[15]

Up to 1901 the six colonies exercised their powers of self-government jealously, delegating only limited responsibilities to the machinery of local government and holding aloof from intercolonial co-operation. But in that year a third level of government was created. The six colonies, henceforth to be known as states, yielded to the new Commonwealth a strictly limited range of powers and responsibilities. A number of these covered such basic necessities of administrative efficiency as even Adam Smith might have found acceptable; most of the remainder were concerned with the circulation of commodities. Trade, commerce and intercourse between the states was to be absolutely free. The Commonwealth would assume responsibility for external affairs, defence, navigation, customs and immigration, and for postal and telegraphic services. It was given the power to impose taxes and customs duties (though for the first decade three-quarters of customs revenue was to be paid to the states), to issue currency, supervise weights and measures, copyrights and various ancillary aspects of trade, and to regulate interstate commerce, finance and insurance. It could make laws for marriage and divorce, invalid and old-age pensions, and the conciliation and arbitration of industrial disputes extending beyond the limits of any one state.

The authors did not intend any thoroughgoing unification and their Constitution provided a series of bulwarks against the central usurpation of power. All residual spheres of government including public works, transport, the development of agriculture and industry, education and law and order remained with the states. The powers of the new Commonwealth were separated between a bicameral legislature, an executive drawn from the legislature and acting in the name of the Queen and her governor-general, and a judiciary with the power to interpret the Constitution. In order to safeguard the interests of the smaller states, the second legislative chamber, the Senate, was to comprise an equal number of representatives from each state regardless of population and this chamber could reject (though not amend) taxation and appropriation bills. Finally, changes to this Constitution required the agreement first of both chambers and then of a majority of voters in a majority of states as well as an absolute majority of voters. Clearly, the fathers of federation had good reason for their belief that the scope of the Commonwealth was strictly limited. The new federal authority would simply inherit from the states their post office employees, customs officials and a sprinkling of other personnel. Next, the new legislature would frame laws dealing with those matters delegated to it—uniform immigration laws, for example—and set up an administrative machinery in those cases where none existed already. Finally, it would arrange a uniform tariff on goods entering Australia, from which it would derive its revenue. Thereafter, it was expected, the central government would need little guidance and its legislative transactions would be of a simple character: one commentator even suggested that federal parliament might exhaust its business in the foreseeable future.[16]

One aspect of the Constitution calls for particular attention. Section 51, which gave the Commonwealth parliament power 'to make laws for the peace, order and good government of the Commonwealth', encompassed special laws with respect to 'the people of any race, other than the aboriginal race in any State'. A later section added that in reckoning numbers for the purpose of distributing electorates, 'aboriginal natives shall not be counted'.[17] The provision was interpreted to mean that Aboriginals should not be included in the census, nor would they be entitled to Commonwealth pensions and benefits. This was the high tide of social Darwinism and it was widely expected that Aboriginals, as a primitive race, were doomed to extinction. Even though the Commonwealth would assume responsibility for Aboriginals in the Northern Territory when it took that administration over from South Australia in 1911, Aboriginals simply did not have citizenship as far as the new nation was concerned.

The obstacles to enlargement of the central government were formidable. First of all, interpretation of the Commonwealth Constitution was entrusted to a High Court and the initial appointments to this Court elevated three of the founding fathers—Sir Samuel Griffith, Edmund Barton and Richard O'Connor—who were hardly likely to digress from the compact they had helped to arrange. On the contrary, they regarded the Constitution with an almost religious reverence. The advent in 1906 of Isaac Isaacs and H.B. Higgins, two radicals who had criticized the draft Constitution as undemocratic and inflexible, merely produced a pattern of dissenting minority judgements. Initially, the High Court majority took the view that the Commonwealth and the states each had their own sovereignties and that neither should trespass on the sovereignty of the other. Each was independent and supreme in its own sphere. According to this doctrine of 'implied immunity of instrumentalities', Alfred Deakin as prime minister did not have to pay Victorian income tax but neither did the Victorian railwaymen fall within the jurisdiction of the Commonwealth Arbitration Court. Taken to its limit, such a theory of co-ordinate federalism would have paralysed the development of government, and the court drew back when it resolved that state railways had no immunity from a federal duty on steel rails. But there remained the doctrine of 'implied prohibitions'. Since the Commonwealth possessed a list of specific powers, the majority of the bench reasoned, the Constitution must be interpreted strictly so as not to trench on the reserved powers of the states. Using this reasoning, the court declared invalid a number of statutes whereby the federal government pursued its economic and social programme. The excise duty on which Higgins based his Harvester wage judgement was invalid; so too was anti-monopoly legislation and a law providing for registration that goods were made by union labour. And a host of other acts met a similar fate.[18] In 1903, the year of his appointment, Chief Justice Griffith had predicted:

> I think it will be some time before the profession and the public fully realise the extent or the power of criticism and determination that is vested in this Court with respect to the decrees of the State and Federal legislatures.[19]

Long before 1914 they were disabused of their ignorance.

Another means of enlarging the Commonwealth, by constitutional referendum, was made difficult by the need to carry both a majority of voters nationally as well as in four of the six states. Attempts by the Labor Party to secure wider economic powers by referendum failed in 1911 and again in 1913. Indeed, only two constitutional amendments were carried by this process during the first fourteen years of the Commonwealth—one altered the date on which a senator's term began and the other transferred the public debts of the states to the Commonwealth. Nor was the provision for the states to empower the Commonwealth to act on their behalf or to set up machinery for co-operative activity of much use as yet. It took only one premier to frustrate such arrangements, and there were usually several of them loath to give up their entitlements and all too willing to exploit the rich electoral lode of state rights. The egregious Sir Joseph Carruthers, premier of New South Wales from 1904 to 1907, was one such champion who took the doctrine of implied immunities so seriously that in 1907 he sent a party of officials to seize wire netting from the wharf rather than pay the Commonwealth duty on it. 'If even at the risk of bloodshed they had to overcome tyranny and pull tyrants down, it would be done', he assured the voters. William Holman, Labor attorney-general and later premier of the same state, was hardly less effusive in his denunciation of centralism:

> I have always been deeply dubious, and at this day, after ten years, I do not see any function worth speaking of carried out by the Federal Parliament that could not have been performed by a mere customs and military union among the states.[20]

So little could be expected there.

There was, however, another source of Commonwealth power—finance. When the states surrendered the power to collect duties on imports, they lost an important source of revenue. With the exception of New South Wales, which had favoured free trade, tariffs provided between a quarter and a third of all colonial revenues in the period before federation. Since the states retained the most expensive governmental responsibilities—public works, railways, education and police—they had inserted in the Commonwealth Constitution a stipulation, known as the 'Braddon clause' after its Tasmanian author, that for the first ten years the Commonwealth would pay them three-quarters of all revenue from customs and excise duties as well as any surplus left over after the Commonwealth met its expenditure needs. In the first year the Commonwealth raised £8.9 million from customs and excise out of a total revenue of £11.3 million, and in accordance with the constitutional provision it paid to the states £7.6 million (£6.7 million as their share of customs and excise and another £0.9 million as surplus). As these sums indicate, the states' expenditure still far outweighed that of the central government. But as the federal authority took on new responsibilities, notably old-age pensions and increased defence expenditure, so its financial needs increased. In 1908 a Surplus Revenue Act was passed to allow the Commonwealth to appropriate unspent money for future purposes. And in 1910 the Braddon clause expired. Already the states had become supplicants of the Commonwealth, and at premiers' conferences they pressed their need for a bountiful portion of federal revenue. They did not get it. Instead they were given an undertaking that for the next ten years the Commonwealth would provide annual grants to the states of 25s for each citizen. In fact this agreement would last for seventeen years (though the weaker states of Tasmania and Western Australia did receive special grants) and its value would be eroded greatly by inflation. The Commonwealth also extended its own revenue-raising activities into areas of direct taxation previously reserved for the states: it created a land tax in 1910, estate duties in 1914, income tax in 1915. Furthermore, in 1912 it began to compete with the states as a public borrower. Alfred Deakin had shown prescience when he observed in 1902 that while the states retained major responsibilities, they lacked commensurate sources of revenue and hence the Constitution had left them 'legally free but financially bound to the chariot wheels of the central Government'.[21] By 1914 their financial vassalage was becoming apparent.

The first federal elections were to be held in March 1901, three months after Commonwealth Inauguration Day. Accordingly, it fell upon Governor-General Lord Hopetoun to commission the first ministry without electoral guidance. The unfortunate Hopetoun, weakened by sea-sickness, landed in Sydney a fortnight before the inauguration and sent for the wrong man: Sir William Lyne was indeed premier of New South Wales, the oldest and most populous state, but he had opposed federation and in any case was unacceptable to the other colonial statesmen who now proposed a transfer to national politics. On the Christmas eve of 1900, Lyne had to report his inability to construct a ministry to the governor-general, who now redeemed his error and turned to Sir Edmund Barton.[22] Barton, the leader of the federal movement in New South Wales, had already reached agreement with Alfred Deakin, his colleague in the same cause in Victoria, and they were in touch with the principal figures in other states. Barton's ministry was constructed with careful regard to the need to represent all the states: Barton himself took external affairs; Deakin was attorney-general; bluff and impetuous Sir John Forrest, who had dominated Western Australian politics, became minister for defence; Sir George Turner, the unassuming premier of Victoria, took the treasury; irascible Charles Kingston, who as premier of South

The first Commonwealth ministry, from the album of the Duke and Duchess of York, who attended the opening of the Commonwealth parliament

Australia had harried the notables of the Adelaide Club, received trade and customs; Lyne was given home affairs; and when death and refusal to cross Bass Strait prevented the inclusion of the premiers of Queensland and Tasmania, they were replaced with men from the same states. In short, it was a combustible combination of the leading figures of colonial politics.

Some of the ministry were protectionists by conviction, others by convenience. Barton, who himself belonged to the latter category, had attempted late in 1900 to defuse the fiscal issue by calling for a 'compromise' tariff for revenue purposes, one that would meet the financial needs of the states without outraging free traders. But the exclusion from the ministry of Sir George Reid, the dominant figure in New South Wales politics and leader of the free-trade cause, made fiscal policy the major issue of the first federal election. In presenting a programme to the electors, Barton therefore dwelt on the necessity of the tariff in terms calculated to appeal to all the voters and offend none. His government would impose neither a high protection tariff such as existed in Victoria nor 'a tariff of free trade character' such as that of New South Wales. It would meet the revenue needs of the states but 'make the increase of taxation upon the people as light as possible'. It would protect local industries but not discourage trade. Finally, in case these assurances left some uncertainty among the listeners, he pledged that the 'tariff will be thoroughly liberal and at the same time of a purely Australian character'.[23] Other than that, Barton merely foreshadowed the establishment of the new machinery of government, immigration control and, looking further ahead, a Commonwealth old-age pension.

In responding to such an appeal, Reid was at a disadvantage. While Barton enjoyed the prestige of office, he was merely the putative leader of the opposition in a parliament that had not met. While free trade was a popular doctrine in his own state, elsewhere it was a conservative creed associated with the pastoral and mercantile élite. Furthermore, Barton, himself a man of highly conservative temperament, had gathered under his ministerial banner such a range of political opinion—from radicals like the Victorian lawyer H.B. Higgins to Tories like Forrest, and from ardent protectionists like the Melbourne manufacturer Samuel Mauger to an erstwhile free trader like the Tasmanian merchant Sir Philip Fysh—that there was little room left for Reid. One temptation he resisted: that, like other conservatives, of shrinking from popular politics and lamenting the changes whereby 'the making of Parliament and the control of legislation was fairly handed over to the impetuous, unreflecting and easily cajoled crowd'.[24] In the past such die-hards had been able to take refuge in the undemocratic upper houses of the state legislatures, but in the Commonwealth they faced the prospect of a democratic franchise for both houses. Reid accepted the inevitability of democracy, indeed he welcomed it. No stand-pat conservative, he courted the masses: as Beatrice Webb had observed two years earlier, 'He watches public opinion exactly as a stock-jobber watches the market.'[25] Touring the eastern states, this Falstaffian figure employed his rich talents for humour and coarse repartee to prick the vague pomposities of the prime minister. But what did the public expect from a Commonwealth parliament? That was a question to which Reid had still to find the answer, and in 1901 he offered little more than warnings of the perils of protection. If Reid made the tariff the crucial issue of the election, it was because he had not yet found an alternative platform.[26]

The election of 1901 was inconclusive. The Protectionists outpolled the Freetraders but not sufficiently to secure a majority, and in the House of Representatives they had 32 members to 27. The balance of power rested with the Labor Party. This result surprised the leaders of the established parties as well as many of the 24 Labor men (16 in the House of Representatives, 8 in the Senate) who met in Melbourne for the opening of parliament in May 1901. As yet they had no federal organization, each of the state parties having developed independently, with various titles and in different circumstances over the past decade. New South Wales and Queensland were the states of greatest strength, with an established record of independent electoral activity (Queensland had achieved the first Labor government in the world in 1899, albeit a minority one lasting less than a week). In Victoria and South Australia, on the other hand, Labor existed only under the wing of the Protectionist liberals, while in

Western Australia there was a purely trade union organization and in Tasmania no organization at all.[27] The state Labor parties had gathered on the eve of federation to prepare a common programme but could reach no agreement on the fiscal question, so they settled instead for a four-plank programme, including old-age pensions and the White Australia Policy, on the basis that 'the shorter the programme, the more likely they were to secure united action'.[28] Once in Melbourne, they elected their leaders, constituted themselves a parliamentary party and declared bluntly, 'We are for sale.'[29] Even that was an exaggeration. The much-vaunted strategy of 'Support in Return for Concessions' depended on cohesive voting in the chamber and this Labor could not yet achieve, having agreed that its members should enjoy a free vote on the fiscal issue.[30] Thus for the time being they trailed in the wake of the Protectionist ministry.

Labor's rise to power would be swift and dramatic. In 1904 it would take federal office, in 1910 win a parliamentary majority; and by the outbreak of the First World War there would have been a Labor government in every state. The politics of the early twentieth century were in large part a response to the emergence of this new force. But what did it represent? The basis of the Labor Party was the working class, and as the trade unions grew (from a membership of just under 100 000 in 1901 to more than half a million by 1914), the party grew. The affiliated unions provided the bulk of party funds, they usually sent a majority of delegates to the state conferences, and in many electorates they constituted a majority of the branch membership. The working class mobilized around clear economic aims—a living wage, the eight-hour day, security of employment, legal recognition of trade unions and preference in employment for their members—and looked to parliament as one means of their fulfilment. To these bread-and-butter concerns they added by 1905 further objectives, some indicating the means whereby their economic goals could be won, others more expressive of the social order to which they aspired: maintenance of a white Australia, a citizen defence force, compulsory arbitration, old-age pensions, progressive taxation, nationalization of monopolies. The patchwork quality of its platform and the pragmatism of its policies, the eclecticism of its doctrine and the sheer indifference to questions of theory were the qualities that struck the observers who came from Europe and North America to study the antipodean labour movement.[31] In learning from their comments, however, we should not lose sight of the principal characteristic that had drawn them to inspect the Australian Labor Party—its precocity. Continental socialists could trace a lineage of more than half a century for their workers' parties and yet they remained on the margin of politics; in Australia Labor had achieved office while still in its adolescence.

In at least three aspects the character of the Labor Party remained unclear. First, what was its constituency? Was it to be the party of the manual working class only or was it to project a wider appeal in the search for electoral success? With class boundaries blurred in the rural areas by the overlap of wage-earners and small farmers, initially Labor achieved greater electoral success in the backblocks than it did in the cities. Hence from the beginning there was an impulse for a broad-based populism that emphasized the common interests of the 'battlers', and that impulse strengthened as the party grew. The report of the second federal conference of the Labor Party in 1905 claimed that it represented not just the wage-earners but

> the civil servants and downtrodden clerks . . . ; the sugar growers of the north coast, the small farmers 'by oppression's ruffian gluttony driven' from the arable lands; the business men struggling in the grip of the usurer. . . In short, every interest in Australia was represented except the interest of the parasitic classes.[32]

Second, there was the question of aims that crystallized at the same conference in a debate over the socialist objective. The political mobilization of the working class was carried forward by the ethic of solidarity, expressed in references to brotherhood and unity; and its enthusiasts looked forward to a radical break with existing society for the construction of a co-operative commonwealth. As the leader of the Queensland Labor Party put it:

> To me this movement is a religion. It is the religion of Humanity. It is an endeavour to uplift and elevate. We want every man and woman to be freed from the system of wage slavery and in my opinion the Socialistic principles embodied in our platform are a step towards that end.[33]

But the socialists remained in a minority. At the 1905 conference the Queenslanders proposed that the party should take as its objective 'the securing of the results of their industry to all producers by the collective ownership of the means of production, distribution and exchange'. In reply the secretary of the rural-based Australian Workers' Union (AWU) declared flatly that the Labor Party 'is not a Socialistic but a trade union movement', and, on the suggestion of the federal leader, the conference settled for the objective of 'the collective ownership of monopolies'. For as William Holman put it, 'It is not easy to go as sheep among wolves and they should consider what would be their fate in regard to eight or ten wavering seats.'[34]

This tension between the pragmatism of the politicians and the principles of the party membership provided a third area of uncertainty. While parliamentarians sought the greatest possible freedom of manoeuvre, the party imposed a close discipline. A member of the party executive expressed a common feeling when he said that 'Once you allow the politician to "boss the show", he will give away everything because he believes himself indispensable to the show, and in fact he ends up becoming the show.'[35] Hence when the workers entered politics they brought with them their symbolic rituals of working-class life—the ballot, the pledge and the caucus—as expressions of a tribal solidarity. The selection of candidates by rank-and-file ballot, the insistence that members of parliament be bound by pledge to carry out the platform, the control of parliamentary leaders by caucus and the provision for caucus election of ministers were all intended as safeguards against the temptations of opportunism. J.C. Watson, the first federal leader, would be an early victim of these rules, protesting bitterly against the instruction from the 1905 conference that ministers would be elected by caucus and that there should be no electoral alliances. 'Coalition is corruption said softly', insisted the Sydney *Worker*.[36] Thus there were powerful countervailing forces operating within the Labor Party, and they would produce serious conflicts as its power increased. But Labor would not achieve a parliamentary majority in the federal parliament until 1910, and for the time being it remained one party among three.

The immediate beneficiaries of the three-party system were the Protectionists. As the middle party they stood between the two irreconcilable elements, the Freetraders and the Labor Party. Either they held office themselves (1901–04, 1905–08) or they determined who did (Labor, 1904, 1908–09; Freetrade–Protectionist coalition, 1904–05). They were the fulcrum of national politics and the architects of the emergent Commonwealth.

Much of the business of the first parliament was given over to forging the machinery of government: the creation of the High Court, the working out of the electoral system, the organization of the public service and the beginning of the search for a site for the national capital. Party divisions were still fluid, much of the legislation uncontentious and the leisurely style of debate—the guillotine had yet to be included in the standing orders—suited the easy-going style of Prime Minister 'Toby' Barton. Three issues raised serious disagreement. First, there was the White Australia Policy, disguised as a fifty-word dictation test in any European language, where the government narrowly defeated a Labor amendment that Asians and Africans be specifically excluded. Second, there was the tariff question on which debate lasted a full year. Here the government proposed moderate duties as would meet its financial needs and those of the states, only to see piecemeal reductions by an alliance of Freetrade and Labor members in the House of Representatives and Freetraders who controlled the Senate. At last the compromise tariff was accepted. Third, there was the government's bill to establish an Arbitration Court, a matter close to the heart of Labor and the radical wing of the Protectionists. The government was not prepared to accept the successful Labor amendment that the court's jurisdiction extend to railway workers who were employees of the states, and allowed the bill to lapse in 1903. The disagreement brought the government down in the following year. Barton had retired to the High Court and been replaced by Deakin. An election confirmed the Protectionists' reliance on Labor support (in the House of Representatives there were now 25 Protectionists, 24 Freetraders, 25 Labor and one independent member) and Labor insisted on amending the Arbitration Bill as before. Deakin treated the motion as a vote of confidence and in April 1904 he resigned.

The first Federal Labor Cabinet, 1904. Lord Northcote, the Governor-General, occupies the armchair; on his right sits J.C. Watson, the Prime Minister, flanked by H.B. Higgins, the Attorney-General. On the left of the Governor-General is a watchful W.M. Hughes, Minister for External Affairs. Andrew Fisher stands, top left

Such a decisive response surprised the Labor members, for most of whom, as indeed for most of his colleagues, Deakin remained an enigma. A man capable of enormous personal charm, 'Affable Alfred' kept a protective reserve that masked his nervous sensitivity. A visionary intellectual and genuine reformer of impeccable rectitude, he was always surprised to find blood on his hands. Earlier in 1904 he had declared that it was 'absolutely impossible' to continue with three parties. 'What kind of a game of cricket', he asked, 'could they play if they had three elevens instead of two, with one playing sometimes with one side, sometimes with the other, and sometimes for itself?' The uncertainty must end. 'It was absolutely essential that as soon as possible the three parties should somehow be resolved into two.'[37] His resignation was designed to force such a resolution. By demonstrating to both the rival elevens that they could not occupy the batting-crease single-handed, he would remind them of his centrality. First he put in Labor. 'To say that we were surprised at finding ourselves in office describes our feelings very mildly', said one of the new Labor ministers. Four months later Deakin allowed Labor to fall, again on the Arbitration Bill, and in a rare lapse he described the same minister as an 'ill-bred urchin whom one sees dragged from a tart-shop, kicking and screaming as he goes'.[38] Next he allowed Reid to take a turn. Even during the short lifetime of the Labor government, he had been negotiating terms for a coalition with Reid on the basis of a tariff truce; now, adroitly, he stood aside to permit Reid to take office with some lesser Protectionists in his ministry. This administration did little more than see the Arbitration Act onto the statute book and hand the tariff question to a royal commission, where it was safely out of harm's way, and then parliament went into recess at the end of 1904. Before it reassembled, Deakin gave Reid public notice to quit in a speech to his Ballarat constituents in June 1905. For by this time Deakin's manoeuvres had brought what he wanted, a promise of support from the Labor Party.

The second Deakin ministry lasted from June 1905 to November 1908. Despite the losses his party suffered in the election of 1906 (17 Protectionists and 26 Labor men were returned to the opposition's 32), Deakin retained office with Labor support. This was the Indian summer of Protectionist liberalism. As Deakin put it in an election address in 1906, his Liberal Party—for this identification had replaced the old Protectionist label—was at one with Labor in its commitment to social justice and willingness 'to seek these ends by a free use of the agencies of the state'.[39] Old-age pensions, anti-monopoly legislation, higher tariffs and accompanying legislation intended to guarantee domestic living standards were the principal means; and though the High Court broke the statutory link between protection and wages, Labor's conversion to the doctrine of protection was complete. Shortly after the passage of the Customs Tariff Act of 1908, in fact, Labor decided that it had nothing more to expect from the Liberals and turned them out. Even Reid had bowed to the inevitable and acknowledged that free trade was a lost cause. The fiscal issue was dead and protection would henceforth be a fact of Australian political life.

That being so, the original basis of party divisions no longer obtained and new alignments were necessary. Reid had chosen to base his politics on a conservative, though not reactionary, opposition to the extension of state activity. As early as 1905 he hit upon the image of the socialist predator, found that audiences responded to it, and made it his business in the 1906 election to 'stalk the socialist tiger'. Labor, on the other hand, looked for further instalments of social reform by extending federal powers, strengthening the Arbitration Court and regulating living standards. While the erstwhile Freetraders—now styling themselves Anti-Socialists—spoke for property and Labor for the working class, Deakin's Liberal Party tried to span the classes with diminishing success. Even during the years of achievement, its progressive orientation had caused the defection of conservative Protectionists who now sat as an opposition 'Corner'. On the left there were similar difficulties. Deakin, with his liberal outlook, could not accommodate the demands of organized labour, nor was he prepared to enlarge further the activity of the Commonwealth. No longer the innovative force of the centre, the Liberals were squeezed from left and right; and Deakin warned during the 1906 election that there was a danger of 'being crushed between these two conflicting powers'.[40] In the previous election their share of the vote had fallen from 44 per cent to 30 per cent; in 1906 it fell further to

21 per cent. Meanwhile Labor increased its support to 37 per cent and the Anti-Socialists theirs to 38 per cent.[41] Outside Victoria the party organization of the Liberals was almost defunct, and even in this protectionist heartland, employers and manufacturers were drifting to the Anti-Socialist cause.[42] So when Deakin's government was dismissed in November 1908, that danger of being squeezed out of existence seemed all the more urgent.

His response was to join forces with the other non-Labor elements. Reid, for whom Deakin had an unbridgeable antipathy, had resigned his leadership to Joseph Cook, and Cook was prepared to serve under Deakin; since Deakin was making a clean break with Labor, the arrangement gathered in the Corner group also, leaving only a handful of intransigent Liberals to protest their sense of betrayal in bitter recriminations. The Fusion accomplished, parliament reassembled and Deakin put out the Labor administration that had been formed on his dismissal. The parliamentary session of 1909 was marked by angry exchanges between former allies now facing each other across the chamber, and on the evening of 22 June the Speaker was carried stricken from the House. 'Dreadful, dreadful' were his last words. Nonetheless, for the first time Deakin took office with a clear majority behind him, and when this uneasy combination of former enemies was consolidated into a party, it took the name of the Liberal Party. It was left to George Reid, speaking from the backbenches, to point to the real meaning of the Fusion. Henceforth, he said, there was a broad line of cleavage separating the two sides of the House. 'The question that separates us is whether the development of Australia on lines of private enterprise is the right method of development, or whether the industrial development of Australia along lines of state control is the proper one.'[43] Closer examination will suggest that this oversimplified both the Labor and non-Labor positions: private enterprise was quite prepared to call on assistance from the state and Labor's schemes of state regulation were wholly consistent with private enterprise. Of his underlying assumption, that there were now two parties based on distinct class interests, there can be little dispute.

The Fusion ministry lasted less than a year before it was swept from office at the 1910 election. Labor secured the first

Propaganda in the 1911 constitutional referendum when the Labor government sought power to control prices

electoral majority in federal politics, winning 41 seats to 31 for the Fusion, and held power for the next three years under the leadership of Andrew Fisher. While its record of social legislation included several major achievements—maternity allowances, a land tax, amendments to the Arbitration Act, the creation of the Commonwealth Bank—it followed the broad lines of national development laid down by Deakin. Admittedly the government sought and failed to obtain the constitutional power over industry and employment without which it could do little to alter inequalities of wealth and power. Even so, the moderation of the parliamentary Labor Party was confirmed. That government was defeated at a general election in 1913 and replaced by one led by Joseph Cook, who had succeeded Deakin as the leader of the non-Labor forces.

The ambit of federal government remained narrow. The Commonwealth provided a framework within which its citizens traded with the rest of the world and decided who might enter it; it provided encouragement for local enterprise, regulated the terms of employment for part of the workforce and made some provision for those whose working lives were over. But in the matters that chiefly touched everyday life, the states remained more important. They controlled land tenure and housing, education and health, public amenities and public morality. Furthermore, political parties, like the polity, were still only loosely federal. In organization and emphasis they focused on the state arena and looked to that level of government to realise their principal aims.

The early years of the Commonwealth saw the consolidation in each state of a two-party system. This was encouraged by the growth of the Labor Party and its abandonment of the liberal alliance. By 1911 Labor had nearly half the seats in the six state lower houses, and held power in New South Wales, South Australia and Western Australia. Faced with the Labor challenge, the non-working-class groups had a strong incentive to sink their differences and develop their own forms of party organization. They did so with remarkable success, and in some cases outstripped Labor: thus as early as 1904 the Liberal and Reform Association of New South Wales claimed a membership of 100 000, and in South Australia the Liberal Union boasted 343 branches and 24 000 members by 1912. The central organization became more important in the formulation of policy, selection of candidates, financing of campaigns and maintenance of parliamentary discipline—indeed it was the success of central finance committees in channelling funds from business houses that made it possible for the non-Labor parties to charge such ridiculously low membership fees, as low as a shilling a year in some cases, and thus to enrol so many members.[44]

Another factor that allowed the non-Labor parties to amalgamate was the disappearance from state politics of the fiscal issue which up to 1901 had generally divided mercantile and financial interests from manufacturers, and city businessmen from rural producers. Such sectional differences still touched areas of state activity after 1901, but they were increasingly overshadowed by the task of ordering relationships in a more demanding, class-bound society. Equally, the various pressure groups that had flourished in the factionalized political setting of the nineteenth century did not disappear, but they no longer operated independently. Instead of endorsing and financing individual candidates, pressure groups now focused on the party organization. Alignments became more systematic: Protestant religious campaigners and temperance advocates gravitated to the non-Labor party, Catholics and publicans to the Labor Party.

Shortly after federation, two men in the Victorian country town of Kyabram fell to discussing the alarming extravagance of the state government. The new Commonwealth seemed merely to have added another layer of regulation and interference to burden the taxpayer. The two men called a meeting and the meeting passed resolutions for cuts in public expenditure, public salaries, the civil service and parliament. Under the slogan 'Retrenchment and Reform', these resolutions were taken up by a National Citizens' Reform League and propagated across the state. The Kyabram movement then spread into New South Wales, where it took the form of a People's Reform League, and to Tasmania. It appeared for a time that the movement for small government was irresistible and that the fear of Kyabram would sweep away the established world of politics. The demand for retrenchment was confirmed in the Victorian election of 1902 when candidates endorsed by the Citizens' Reform League dominated the polls; subsequently, public salaries were reduced and the Legislative Assembly was trimmed from 95 members to 65. In New South Wales a referendum to reduce the size of parliament was carried overwhelmingly and the Legislative Assembly was cut from 125 members to 90.[45]

The actual genesis of the Kyabram movement was not as simple as the legend suggests. One of its founders had links with the food-processing firm of Swallow and Ariell, while

Kyabram campaign button, 1902

the treasurer was a wirepuller in Melbourne financial circles.[46] So while the movement grew among the fruit trees of Kyabram, it arrived in the cities canned, labelled and ready for sale. What, then, were its aims? The Kyabram movement undoubtedly tapped the long-standing anti-city prejudices of the farmers who felt that they, the producers of wealth, were supporting an army of parasitic public servants. As it spread to the middle-class belts of Melbourne and Sydney, it emphasized the prodigality of public works and denounced the wastefulness of minimum wages and day-labour in public works. Both the urban and rural groups shared an aversion to the familiar patterns of patronage and political spoils. The objective of the New South Wales league was 'to promote economy of government, to oppose unwise social legislation, and to oppose Government interference with private enterprise'; and it sought representatives 'who had no political past to answer for', for 'they did not require old politicians, but able businessmen who would give a straight-out vote'. In practice, however, the reform leagues were soon assimilated into the conservative parties and cuts in parliamentary numbers did little to curb the activities of the state. For all his huffing and puffing, Carruthers's first budget reduced the spending of the New South Wales government by just

£50 000 out of more than £11 million.[47] How could it be otherwise? Public authorities had played a crucial role in the development and support of primary industry. Kyabram itself was a product of a state irrigation scheme undertaken late in the previous century. And even as the Kyabram movement flourished, the state was taking on new responsibilities.

5

MAKING A COMMONWEALTH

THE TITLE COMMONWEALTH of Australia was proposed back in 1891, quite early in the deliberations that preceded the federation. Some of the delegates puzzled over the meaning of this strange term, Commonwealth. Was it not a reference to Cromwell's protectorate? If so, was it appropriate for Her Majesty's loyal Australian subjects? Deakin, as it happened, rejoiced as a good liberal in the memory of that parliamentary triumph, but he hastened to explain to the delegates at the Sydney convention that the word had an older lineage. It meant 'the common good of the people'. The South Australian Tom Playford explained its etymology, 'common weal', and gave as a modern reading the common good or common well-being of the people of this continent, to which his crusty fellow-colonist Sir John Downer retorted: 'It means common goods now.'[1] Both senses ran together in the public life of the early twentieth century. That the business of government was to pursue the general welfare of all citizens was taken as axiomatic. More particularly, the task of nation building demanded that every citizen have the opportunity to achieve that material prosperity on which civic virtue depended.

Clearly, Australian circumstances were very different from those in which the expressive notion of a commonwealth had drawn its force centuries earlier: the bonds of community and continuity were broken, the customs of mutual obligation replaced by possessive individualism. When the celebrants of Commonwealth Inauguration Day proclaimed their allegiance to nationhood, they were creating an object of loyalty;

> The fair new nation cometh, drawn
> By six proud states so fair and tall.
> She was their child; now, strange to tell,
> She is the mother of them all![2]

But their enthusiasm was no less real for that. Similarly, disagreements about how competing interests should be balanced did not detract from the belief that it was possible to strike such a balance. The colonial governments acted as landowners and landlords, immigration agents and educators, borrowers, investors and employers, and builders and operators of transport and communication facilities. In all these activities the needs of the individual were commonly interpreted as being consistent with the good of the community. If small-holders sought land to farm, then in making it available the state was promoting the growth of agriculture; if woolgrowers wanted a railway to take their bales to the coast, then in building it the state was providing employment for the navvy as well as ensuring a growth in export income; and if manufacturers asked for protection from imports, then in erecting tariff barriers the state was again creating employment for the city labourer who in turn would augment the local market. Their self-fulfilment was at the same time serving the common good. Granted, there were disputes between pastoralists and agriculturalists, exporters and domestic producers; all the same, the availability of foreign investment up to 1890 allowed the colonial treasurer to imitate the superintendent at a Sunday-school picnic and ensure that there were prizes for everyone. As for the allocation of the best prizes, of course, the race went to the swift.

With the collapse of export earnings and the loss of foreign investment in the financial crisis of the 1890s, that happy arrangement could no longer continue. The colonies simply could not find the resources to maintain their bountiful provision of public goods. Nor could the free operation of the labour market satisfy the expectations of the working class once employers sought to restore profitability by wage cuts and retrenchments. And even though the employers had won the decisive trial of strength in 1890, the hardening of class feeling and the entry of the working class into politics posed a new challenge. Under these circumstances the state took on new and more directly interventionist tasks.

In the first place it interposed itself between belligerent employers and workers to regulate conditions of employment and provide machinery for the settlement of disputes. There were factory acts to limit hours and ensure safe working conditions, and further legislation to provide for workers' compensation. Above all, there were tribunals of industrial arbitration and wage determination. Arbitration courts were established in New South Wales, Western Australia, Queensland and the Commonwealth in the 1890s and early years of this century—Tasmania and, after some hesitation, South Australia followed Victoria in preferring wages boards. Whereas the courts had a comprehensive jurisdiction to settle disputes and make common rules applicable throughout an industry, wages boards operated on a piecemeal basis, one to each trade, bringing together employers and employees under an outside chairman simply to lay down rates of pay and conditions. In practice the two models converged and by 1914 the majority of wage-earners worked on legally enforceable terms that had been determined by tribunal.[3]

The sponsors of arbitration rejected brute force for sweet reason:

> It was felt that the settlement of labour disputes by the strike and the lock-out—in other words, by a tug of war—was antiquated and barbarous, and meant that the industry was to be controlled by force and cunning merely. The desire for something juster, fairer and more peaceful than this was strong. Men turned to the Government and asked for a remedy.[4]

But their motives went further than a concern for justice between the parties to a dispute. Deakin, Kingston, Higgins and Isaacs—advocates of industrial peace in their respective colonies and architects of the Commonwealth Arbitration Court—were lawyers who stood at a remove from industry. As liberals they occupied the precarious middle ground of politics and the rising class turmoil threatened their vision of an orderly, prosperous society. Beyond the combatants, capital and labour, they discerned a larger and more fundamental unit, the community. If conflict between employers and workers paralysed industrial life, then they insisted that the government was entitled to act on behalf of the community and restore peace. They found little difficulty in justifying this extension of the law, little difficulty in adjusting their liberalism to collectivist needs. 'Can it seriously be questioned', asked Bernhard Wise in introducing the New South Wales Industrial Arbitration Act of 1901, 'that the community at large has a vital interest in the prevention of strikes, and to do it, may even interfere with a man's "freedom"?' More commonly they presented compulsory arbitration as simple pragmatism—thus Deakin likened the Arbitration Court to an engineer carrying around an oil can with which to cool any overheated bearings in the industrial mechanism.[5] Employers were more wary. While some saw advantages in industrial peace, the majority were reluctant to surrender their precious 'freedom of contract' and hence resisted arbitration. Unions, on the other hand, generally were enthusiastic if only because in their position of weakness at the turn of the century, they saw arbitration offering the opportunity to safeguard their organization and make good their losses. Arbitration certainly assisted in the rapid increase of union membership (as, for that matter, it encouraged the formation of employers' organizations), but as they gained strength and discovered how wage tribunals could block the improvements to which they believed they were entitled, more militant unions revised their attitude. By then it was difficult to shake the attachment of the political wing of the labour movement. From the 1890s compulsory arbitration was part of the Labor Party programme and there it remained.

Initially the tribunals made their awards according to the prevailing economic circumstances. Where conditions permitted, they might agree to an application for a wage increase, providing it did not outstrip the capacity of the industry to pay, since 'imprudent' wage increases would only lead to unemployment.[6] Had this basis of wage determination continued, the significance of arbitration would have been limited. But the Deakin administration of 1905–08 went a step further with its doctrine of the New Protection. 'The "old" Protection', announced the Commonwealth government, 'contented itself with making good wages possible. The "new" Protection seeks to make them actual.'[7] Hitherto protectionists had appealed for working-class support on the grounds that the tariff safeguarded Australian employment and wage standards; now they were proposing to tie the one to the other. At the same time as it increased the tariff duty on agricultural machinery in 1906, the government created an excise duty on local products which would be waived if the Australian manufacturer paid his workers 'fair and reasonable' wages. It fell to H.B. Higgins as the newly appointed president of the Commonwealth Arbitration Court sitting in the Harvester case of the following year to determine the meaning of a fair and reasonable wage.

Higgins decided that a fair and reasonable wage must be based on need. Neither the market value of labour nor the profitability of the industry was an acceptable criterion, for the task of the court presupposed some higher standard than the higgling of the market; and if an enterprise could not pay its workers a living wage, then it would be better abandoned. The minimum wage should be that amount that would enable a worker to live as a 'human being in a civilized community' and to keep himself and his family in frugal comfort. Higgins then considered some household budgets and settled eventually for a wage of 7s a day—the very sum that had been regarded as a minimum standard before the depression of the 1890s.[8] The declaration of an inviolable minimum wage calculated to meet human and family needs seemed to meet workers' expectations. But the day after the Harvester judgement was handed down, the Chamber of Manufactures met to consider its implications and within a week announced that one of its members would appeal. By a majority of three to two, the High Court declared the Excise Tariff Act unconstitutional on the grounds that while purporting to exercise the Commonwealth's taxation power, it was in fact regulating the conditions of industry in a manner that lay outside its competence. And Labor's subsequent attempts to extend Commonwealth power failed. In short, the New Protection could not be institutionalized. While Higgins continued to use the Harvester standard as the basis of awards to workers who came before his court, he saw but a small proportion of the whole, and the state courts, which dealt with the majority, did not follow him. Wage standards, which had fallen in the depression, did not make a sustained recovery until the very eve of the First World War, and then only because of market conditions. The Harvester standard was not generally attained in awards until after the war.[9]

If Higgins's decision was a myth, it was an extremely powerful one. Within five years of his judgement Labor governments in three states had legislated for the judicial determination of a basic wage, and by the 1920s the practice of basing a minimum wage on the cost of living would be generally accepted. This served as the bedrock of wage determination in Australian industry, on which was erected a structure of increments or 'margins' for skill, a practice that narrowed income differentials among wage-earners. As measured by its effect on pay packets, wage fixation and compulsory arbitration were of greatest benefit to the weaker, unskilled workers who had been largely at the mercy of their employers. They offered less to the skilled and better-organized workers, and took away more by the institutionalization of industrial relations.[10] For as Australia emerged from the depression, the nature of industry was changing: small workshops were giving way to factories, hand-tools to power-driven machinery, and these more capital-intensive enterprises required a more specialized, permanent workforce. By smoothing out fluctuations in the labour market, arbitration provided a framework within which a workforce could be assembled and organized. In this way the institutionalization of industrial relations looked forward to new patterns of employment. Nevertheless, Higgins's proclamation of this 'new province for law and order' and his obvious sympathies for labour did much to popularize arbitration. And despite the removal of the statutory nexus, protection became firmly established as the basis of Australian living standards.

The search for balance is a persistent feature of the regulatory activities of Australian government. Just as the colonial administrations had sought to provide prizes for all, so the federal government strove to maintain the fragile equilibrium between the states, between producers and consumers, and between urban and rural interests.[11] The importance attached to economic growth and, above all, the alarm caused by the disparity between the swollen cities and the empty hinterland reinforced a determination to settle the land. Within tight budgetary limits—for overseas investors were still reluctant to risk capital in Australia—governments sought to encourage closer agricultural settlement, and with some success. They assisted immigrants (two-thirds of the 300 000 newcomers were assisted) with the intention that they should go onto the land. They built additional railway lines and embarked on further irrigation schemes, provided technical assistance and credit, imposed new land taxes aimed at breaking up the big estates and made small blocks more easily available. With all this Labor was in full agreement: the radical demand for access to the land went back to the middle of the last century, and broad acres still held out the promise of release from servitude. For the time being, Labor was also an electoral beneficiary, and in the two states where Labor enjoyed its greatest initial success, New South Wales and Queensland, the majority of Labor parliamentarians represented rural electorates. But as Jock Neilson had cause to appreciate, the possession of a selection did not in itself bring economic independence. A selector needed capital to make a go of farming, and even if he survived the vicissitudes of climate and market, he was unusually successful if he could look the storekeeper or bank manager squarely in the eye. Rural credit and co-operative marketing schemes did not alter his reliance on the processor and financier. Furthermore, the products of the areas of closer settlement—notably fruit, sugar and dairy items—were especially susceptible to exploitative forms of economic control. The sugar companies, for example, broke up their large plantations and sold or leased them to small farmers during this period because they found farmers more productive than hired labour; but the central mill system still held the growers in bondage.[12] Herein lay the seeds of the populist ideology of resentment against the exploiting middlemen, big businessmen and parasitic financiers. Yet the same ideology could easily tilt towards hostility to the unions.

One Labor response was to establish public enterprises in

the hope that they would at least provide a degree of competition for the private operators. New South Wales, Western Australia and Queensland developed a wide variety of such state ventures: insurance, coal-mines, quarries, brick and pipe works, timber yards and sawmills, bakeries, butchers' shops, even hotels and tobacconists. The Commonwealth entered the field after Labor won power in 1910, with woollen mills, clothing factories and a dockyard. Sometimes described as exercises in state socialism, the actual significance of these enterprises was far more modest. Their retail activities made little impact and for the most part were meant merely to supply material to government departments more cheaply than private suppliers, and to act as model employers providing optimum pay and conditions for employees. The Labor premier of Queensland emphasized that such undertakings 'were not inaugurated as the commencement of a state-wide scheme of nationalization of industry', merely as 'a check on profiteering and a regulation on commodity prices'.[13]

In one area, perhaps, the Labor Party went a little further—banking. For if Labor had an ingrained suspicion of middlemen who cornered the market, it reserved a special hostility for those who controlled the 'money power', and a national bank came high on its list of objectives. Up to this time the financial system had been controlled by a group of private banks—some Australian, others British—pastoral finance companies and insurance companies. The states had established their own savings banks, but relied on the private banks to handle their transactions and London banks to raise their loans. Currency issue and all other aspects of monetary policy rested in private hands. The very idea of a central bank of issue, one that could compel the private banks to deposit a proportion of their funds, was described by the bankers' journal in 1905 as 'too absurd, for no well-informed or sensible person would deposit money in an institution where it would be liable to confiscation or detention'.[14] Nevertheless, they accepted readily both the introduction of the Commonwealth note issue in 1910 and the establishment of the Commonwealth Bank in 1911. So long as the financial system was protected against an irresponsible use of the printing press, and providing the new bank followed 'sound and prudent'

An official publicist's depiction of the entitlements of citizenship from cradle to grave. Note how the needs of women, children and the aged fuse in the workingman's 'living' family wage

practice—as its charter guaranteed and the appointment of an official of the Bank of New South Wales as its first governor confirmed—they could rest easy. The chief opposition, in fact, came from the states, who complained that the savings branch of the Commonwealth Bank took depositors away from their own banks.[15]

Up to 1914, in short, there was little disturbance of the social and economic order. Rather than altering class relationships, liberal and Labor reformers regulated them. Chiefly by the devices of protection and arbitration, they sought to alleviate the lot of the working class while strengthening the economic fabric. Moreover, they did so within a framework of self-fulfilment. The living wage, the safeguards for Australian industry and the land settlement schemes offered incentives to those who were prepared to help themselves, and for those who would not or could not do so provision was much more limited. In New South Wales (1900), Victoria (1901) and then the whole Commonwealth (1908) there was an old-age pension—10s a week for a single person, £1 for a married couple—available to those who satisfied a stiff means test. In 1910 the Commonwealth added an invalid pension for those permanently incapacitated for work, and in 1912 a maternity allowance of £5 paid to the mother on the birth of her child.[16] These were special instances where the state accepted a responsibility for those outside the workforce, but the insurance industry and friendly societies resisted further public provision and in general it was assumed that all family needs could be satisfied within the labour market. So while governments were sometimes prepared to create public works for unemployed men, direct relief was reserved for women and children of unemployed families and dispensed with a niggardly hand by charitable agencies.

Another group of reformers approached social problems from a different perspective. Combining middle-class philanthropy with a strong concern for the national welfare, these influential administrators and energetic professional men and women sought to eradicate conditions that they associated with social degeneration. Like the older generation of liberal reformers, these progressives were convinced that *laissez faire* was no longer an adequate basis for public policy; they too aimed to avert disorder and class conflict. They went beyond the liberals, however, in their reappraisal. Industrial civilization, they insisted, was no longer susceptible to piecemeal reform. Order and progress required not just new laws but new institutions to guide and shape all aspects of social development. The progressives were administrators and publicists rather than politicians, and they liked to think that they transcended politics with their altruistic professional expertise. Environmental solutions appealed to their understanding of the social process and although their schemes for slum clearance gave rise to more investigation than action, they achieved a measure of success in creating playgrounds and kindergartens with which to save working-class children from 'the street and gutter influence'. Medical inspection in schools and more rigorous enforcement of juvenile protection legislation were other initiatives aimed at the physical and moral welfare of the coming generation, who were thought to be more amenable to improvement than their parents.[17]

The distinctive ethos of meliorism, nation building and efficiency is better illustrated in the field of education. There was general agreement that the state schools were unsatisfactory, that they were neither training their pupils in necessary skills nor inculcating the values that the new nation required. A pedagogical doctrine, known as the New Education, emerged in the early twentieth century. Like the New Protection, it linked the needs of the individual with those of the state so that personal opportunity and national efficiency went hand in hand. The inspector-general of Queensland schools saw the relationship like this:

> The State would reap the benefit that would follow from the unearthing, rearing and developing into full flower and fruit the latent seeds of genius which would otherwise perish unfulfilled or be born to blush unseen in poverty and obscurity.

Or, more simply, 'education is the chief industry of the State, and produces its most valuable asset'.[18] Royal commissions or reappraisals in each state resulted in the appointment of young, vigorous directors of education, notably Frank Tate in Victoria, Peter Board in New South Wales and Alfred Williams in South Australia. They remodelled the curricula to reduce the old rote learning and introduced a new emphasis on self-expression in drawing, nature study, health and manual activities. Fees were abolished, attendance enforced more rigidly, training of teachers made more systematic. Against considerable opposition from the private schools, which Tate criticized as 'locked against the mass of the people', the states entered the field of post-primary education of their abler students.[19] Even the universities came under assault for their exclusive and rarefied character. Opening the University of Queensland, the premier boasted that 'Oxford was established by a King: the University of Queensland is established by the People.' 'University is no longer regarded as the luxury of the rich', advised the royal commission set up to establish a University of Western Australia, and the 'University of this state ... must have particularly in view to help the sons and daughters of the working man.' William Somerville would quote these words to his colleagues on the university senate and protest when they were ignored.[20]

Moral reform shaded imperceptibly into outright repression, and nowhere was this more apparent than in official policy towards the Aboriginals. Philanthropy, pseudoscientific racial theories, a concern to preserve cheap pastoral labour and unadorned contempt all dictated that the Aboriginals should be removed from white society and more thoroughly supervised. Queensland first, in 1897, and then all the other states with the exception of Tasmania (where they were thought to be extinct), adopted comprehensive measures for the segregation of Aboriginals on government reserves. Here they could not marry, consume alcohol, accept employment, manage their own assets or even leave without the permission of white officials. In the pastoral areas the reserves became enclaves where the Aboriginal family produced labour for the stations and to which they were returned when no longer needed. Older reserves in Victoria and New South Wales, where Aboriginals had achieved a degree of self-sufficiency, were reorganized: their farms were broken up and sold; Aboriginals of mixed descent were pushed out with the intention that they should disperse into the larger society; viable communities were turned into ration depots.[21]

Women played a prominent part in movements of social reform, partly because these were among the few avenues of activity open to them and partly because the issues bore directly on their circumstances. 'Women may well share in housekeeping the State without neglecting their own homes', claimed the Women's (Non-Party) Political Association of South Australia at its inaugural meeting in 1909. A crucial role was played by a new generation of educated middle-class ladies. Females had gained admission to university in the 1870s, and by the early twentieth century they were an accepted part of the undergraduate scene. The overwhelming majority, of course, came from well-to-do homes and were treated by their brothers and their brothers' friends with an uneasy condescension. Thus we find Enid Derham, one of a remarkable group of girls from Melbourne's Presbyterian Ladies' College to go on to the University of Melbourne, reporting the Arts results in 1903: 'I was the only one to get a first, Sissie Lothian and another girl got a second, and two other girls and the only man got thirds.'[22] But what was the female graduate to do? Even those higher professions that did not actually exclude women placed formidable obstacles in their way, so that schoolteaching remained the usual alternative to matrimony. Enid Derham was one of the very few to obtain a university teaching post. As in so many other areas, it seemed that women stormed the fortress only to discover an inner citadel.

Most feminists concentrated their efforts on the franchise, believing that the struggle for full citizenship held the key to their emancipation. They did not so much dispute the notion that men and women were different by nature as invert it: the vote was to be the means whereby women extended their private sphere concerns into the public sphere to purify and regenerate national life. Thus in Queensland the legislation to enfranchise women simultaneously disenfranchised wife-beaters, habitual drunkards and men with unsatisfied maintenance orders.[23] More generally, the winning of the vote (in Commonwealth elections from 1903 and in all states by 1909) coincided with measures such as the maternity allowance, designed to strengthen the family. A multitude of laws and regulations defined the age of sexual consent, marriage, divorce and the protection of 'normal' as well as 'neglected' children. The very concept of the living wage was intended to enable a man to support a wife and children. What, however, if there was no male breadwinner? A man of his times, Higgins was reluctant to recognise the importance of women's labour: 'fortunately for society, the greater number of breadwinners are men. The women are not all dragged from the homes to work while the men loaf at home.' He assumed that working women were either single or else part of a larger family unit and making only a supplementary contribution to the family income, and in his determinations he erected a barrier between men's work and women's work. Female workers were awarded a wage sufficient to keep a single person; male workers were awarded a wage sufficient to keep a man, a wife and three children, and only in certain marginal occupations where women worked alongside men (fruit-picking, for example) were they awarded the same wage—and then only to prevent the men from being replaced. The same preconceptions guided the decisions of state tribunals. Here was a graphic example of how the concern for the family enshrined the inequality of women.[24]

Rather than protecting women from exploitation, the effect of these endeavours was to fix the inferiority of their occupational status. Much the same outcome occurred during the same period in one area that was reserved for women—prostitution. Throughout the nineteenth century, a significant number of females (just how many it is difficult to say, but a figure of 10 000 in 1901 is probably not excessive) resorted to the sale of sexual intercourse either because of economic necessity or because this seemed a less oppressive way of earning a living than domestic or factory employment. By the turn of the century there was a concerted attempt to stamp out the practice. Many brothels were raided and closed down, street-walkers were driven out of business. The campaign did not eradicate prostitution—nor was it meant to, despite the strictures of moral reformers. The police and the magistracy tacitly allowed brothels to operate within defined neighbourhoods, provided they were conducted by entrepreneurs who maintained orderly houses and were generous with their kickbacks. The outcome was the enmeshment of prostitution with organized crime and the reduction of the prostitutes themselves to the status of a proletariat.[25]

The drive for moral reform spilled into other areas as well. The puritanical parson who denounced all unholy pleasures was already a stock figure when John Norton, the proprietor of the scandal-mongering *Truth* newspaper, christened him with the name by which he would henceforth be known—the wowser.[26] Wowserism took on a special force during this period through a combination of specific circumstances. First, it took advantage of the extension of state activity to translate its strictures into legal prohibitions: 'Laws were the expression of the sentiment of the people, and were absolutely necessary if moral suasion were to be made effective', declared a leading Melbourne publicist. Second, it exploited opportunities offered by the emergent two-party system to attach itself to the anti-Labor conservative party, drawing particularly on its influential women's organizations. Third, it capitalized on the upsurge of anti-Catholicism. Hence the Methodists of New South Wales characterized the 1907 state election as a battle between the decency and responsibility offered by Sir Joseph Carruthers' ministry and Labor's 'Rum, Romanism, Socialism and Gambling'.[27]

The achievements of the wowsers were impressive. In New South Wales alone they influenced the Obscene and Indecent Publications Act, the Juvenile Smoking Suppression

Act, the Royal Commission on the Decline of the Birth-Rate, a raising of the age of consent, regulations against mixed bathing, the Gaming and Betting Act, sabbatarian measures and de-licensing of many pubs. In Melbourne also they turned many of the outer suburbs into publess deserts and even managed to close down John Wren's celebrated Johnston Street betting shop.[28] But they were not omnipotent. In three successive temperance referenda in New South Wales, 1907, 1910 and 1913, the 'no licence' vote peaked at 39 per cent while the vote for continuance of current licensing regulations grew into a solid majority. Again, after ill-managed prosecutions brought by the Commonwealth Department of Customs in 1901 against importers of Balzac's *Droll Stories* and other books, federal censorship lapsed for a quarter of a century.[29] Furthermore, the political alignment of the wowsers strengthened a conviction among Labor voters that the killjoys were concerned only with working-class pleasures. The influence of wowserism was episodic, and its successes in the period reflected the desire to impose harmony and stability on a restless society.

Such efforts to bring peace and order to the new Commonwealth were seldom completely successful. Whether it was an inner-city child wagging school or Aboriginals insisting that their children should not be locked out of the local school, there was evidence aplenty of obdurate independence. The larger project of balancing the classes was barely sustained. The nation's new institutions were imperfect tools, the underlying divergences of outlook and interest too powerful. The erosion and eventual capitulation of Deakin's centre party seemed to symbolize a polarization of capital and labour, raising the threat of what he described as 'the overwhelming of the "classes" by the "masses"'.[30] It was far from clear that New Protection and arbitration could control such powerful forces, for even as the economy emerged from depression and drought into renewed prosperity, there were obvious signs of labour's impatience with arbitration and the inability of the state to impose its discipline on industrial capital.

The first in a wave of major industrial disputes was

The New South Wales police guard the BHP mineworks, Broken Hill, 1909

brought on by the employer. Caught in a cost–price squeeze, the Broken Hill Proprietary Company (BHP) demanded in 1908 that its workers in that inland town accept wage cuts. Both sides had ample opportunity to marshal their forces: the unions transferred their funds to prevent confiscation by the conservative New South Wales government, and brought in the socialist organizer Tom Mann, who arrived from Melbourne in a white suit and red tie; the company secured police reinforcements and the general manager, G.D. Delprat, noted in his diary that he had taken 'all necessary precautions to stave off trouble at the mine—rifles and pistols—food supplies—blankets'. On New Year's Eve, 1908, the pickets and police took up their positions and the fires went out. After several days of mounting tension, the police swooped to arrest Mann and a number of other unionists; he was released on bail providing he took no further part in the strike, a restriction he circumvented by addressing a meeting of several thousand from the South Australian side of the state border less than 40 miles away. Meanwhile, the workers sought an award from the Commonwealth Arbitration Court (arguing that the court had jurisdiction since the smelters at Port Pirie in South Australia were involved) and Justice Higgins handed down a favourable decision in March. 'Delly! Delly! Higgins is winning!' the men taunted the general manager. But the company appealed and had the Port Pirie award set aside; the Broken Hill award it avoided simply by keeping the mine closed. So, after twenty weeks, the strike collapsed. Mann's trial for sedition and unlawful assembly was transferred to Albury, 500 miles away, where a supposedly unsympathetic jury acquitted him. Others were tried, convicted and imprisoned. Mann was not alone in concluding that 'this experience of the admittedly most perfect Arbitration Court in existence, with a [federal] Labor Government in power, dampened any enthusiasm I might have felt for such an institution.'[31]

Later in the same year the coal-miners of New South Wales were embroiled in conflict. Once again the unionists had made financial preparation for the siege by allowing two independent collieries to work and sharing the profits with the owners. For their part, an owner declared, 'the feeling among the proprietors was that the only way to settle our differences was to fight to the finish'.[32] The men of the northern district struck early in November for an eight-hour day and better pay, and they called on the transport workers to support them. But the key union, the Waterside Workers' Federation, was persuaded not to do so by its masterful secretary, the federal parliamentarian Billy Hughes.

A difference of opinion within the labour movement had been widening for some time, and this strike brought about an open breach. As an industrial negotiator Hughes stood for pragmatism; his successes were won by setting limited objectives, concentrating on a particular group of employers and mobilizing public opinion to force them to the negotiating table. Wherever possible he avoided a walk-out: 'I am now, and I have always been, in favour of the settlement of industrial disputes by peaceful means', he declared at an early point in the coal strike. His aversion to strikes and his particular hostility to all-out confrontation between workers and employers—'it is but a nightmare which if by any chance translated into actuality would mean social suicide'—were shared by his parliamentary colleagues. But a growing number of trade unionists, including the miners' leader, Peter Bowling, were disenchanted with the methods of the ballot box and arbitration. All the efforts of the Labor Party seemed merely to have produced a group of well-paid politicians whose careerist tendencies were mocked in the Wobbly song of the period:

> I know the Arbitration Act
> As a sailor knows his 'riggins'
> So if you want a small advance
> I'll talk to Justice 'Iggins
> Bang me into Parliament
>   Bump me any way
> Bang me into Parliament
>   On next election day.

Against that advice, militants such as the Wobblies, which was the nickname given to enthusiasts of the Industrial Workers of the World (IWW), urged solidarity among workers, the bypassing of parliament and direct action at the point of production—'Arbitrate on the Job,' their slogan put it. Neither the IWW nor the socialist groups attracted a large membership, but their doctrines caught the impatience of many workers in a period of full employment and rising prices.[33]

The coal strike did much to reinforce this impatience. It quickly spread to all districts and even with coal coming into the country from Japan and India, was beginning to paralyse industry and transport as the Christmas of 1909 approached. At this point the New South Wales premier, C.G. Wade, intervened. A lawyer who had often appeared for the coal-owners before winning ministerial office, Wade amended the Industrial Disputes Act in 1908 to introduce penal sanctions against recalcitrant parties to a dispute. At an early point in the coal strike he had the miners' leaders charged with conspiracy, now he pushed new legislation through parliament. The Industrial Disputes Amendment Bill was not distributed to members until the debate began—with an explanation from the premier that the whole affair had been worked up by 'the glib tongue of the loud-voiced agitator'—and its passage was completed in the same sitting, which ended at 8.30 in the morning of the next day. Known afterwards as the 'Coercion Act', it made the instigation, assistance or continuation of a stoppage in an essential industry an offence punishable by twelve months' imprisonment. Summonses were taken out immediately against the union delegates and they were given the full sentence. Bowling received an additional eighteen months for conspiracy and was taken to Goulburn jail in leg-irons. The miners of the western district were the first to capitulate, followed by the southern district and finally, in March 1910, by the northern district.[34]

Nevertheless, Wade's heavy-handed actions played no small part in the election of a federal Labor government in April 1910 and a state Labor government six months later. Hughes told the Labor voters: 'You are like rabbits facing the gun ... Turn it round against the vend and elsewhere, and you will find nothing so effective as the gun of the law!'[35] But could the artillery of the state be turned round? There was already an action commenced by the Commonwealth against the Coal Vend, which was an agreement among the coal-owners to fix prices and divide up the market, to which the interstate shippers were parties. The Coal Vend was perhaps the most notorious example of a widespread phenomenon in Australian industry: one writer found evidence of trusts and restrictive practices in sugar, tobacco, bricks, meat, beer, bread, even jam, as well as coal and shipping.[36] The fear of monopoly was strong and not simply on economic grounds. That a handful of businessmen could divide up the national market in such a fashion to prevent proper competition and extort excessive profits was a moral affront to small producer, worker and consumer alike. When the Labor Party federal conference of 1905 took as its objective not the 'collective ownership of the means of production, distribution and exchange' but rather 'the collective ownership of monopolies', it thought of the monopolist as the true capitalist, Mr Fat Man as he was portrayed in cartoons. His profits were not earned by genuine productive effort, they were the illicit spoils of restraint of trade. The Deakin government responded to this feeling first with anti-trust legislation in the form of the Australian Industries Preservation Bill aimed primarily at foreign cartels dumping on the

Australian market and strengthened in 1906 after Labor pressure to prohibit local combines as well. But when the first Fisher government had invoked the Act against the Coal Vend, one of the shipping firms simply refused to provide information, and in this it was protected by the High Court. The new federal Labor government managed to obtain the records of the Vend and used them to commence an action in the High Court against no less than forty defendants. After a hearing lasting over a year, Mr Justice Isaacs fined all of the defendants the maximum of £500 and ordered the combination to break up. But on appeal the Full Bench reversed that decision. 'Cut-throat competition', they decided, 'is not now regarded by a large proportion of mankind as necessarily beneficial to the public.' The Australian Industries Preservation Act remained a dead-letter for the next fifty years and monopolistic practices flourished unchecked.[37]

The Broken Hill and the coal-miners' disputes were but two major exchanges in the struggle between labour and capital. The unions lost them because they were dealing with powerful foes who could draw on their financial reserves, assemble non-union labour and call on the direct assistance of the state. As with BHP and the Coal Vend, so with employers generally. They forged new and more broadly representative organizations at the turn of the century to advance their mutual interests in the key states of New South Wales and Victoria. As part of their work, these employers' federations raised political funds, lobbied, ran candidates and generally threw their weight behind the non-Labor parties at the state and federal levels. The creation of the Commonwealth, and in particular the legislative programme associated with New Protection, was clearly a stimulus to action. 'The object of our legislative work, to use a homely phrase, is "to try and temper the wind to the shorn lamb"', confided the president of the Victorian Employers' Federation in 1910, 'and when we fail in this object in Parliament, we sometimes have to go to the High Court.' They did so with manifest success. Four of the eight major cases in which the High Court invalidated reform initiatives were brought or financed by the Central Council of Employers, and in three others the appellants were particular companies.[38] Thus protected, the employers were usually victorious in the major

Charles Hoskins, the Lithgow ironmaster, stands in the entrance of his works after strikers smashed the windows

confrontations. Protracted strikes such as those of the Harvester implement works in Sunshine, Victoria, and the ironworks at Lithgow, New South Wales, ended in the workers' defeat. A particularly serious set-back was the Brisbane general strike, which began when the manager of the tramway company sacked employees for wearing their union badge. The Queensland Labor Federation called out 20 000 members of forty-three affiliated unions and brought the city to a stand-still; the Queensland government responded by recruiting 3000 special constables from the public service and rural areas. On 9 February 1912, 'Black Friday', the police attacked a procession of women who defended them-

The Brisbane general strike, 1912: a government show of strength

selves as best they could with hat-pins against batons. Once again, the High Court overturned Higgins's ruling in the Arbitration Court that the badges be permitted. The workers capitulated after five weeks.[39]

But if the bosses won most of the battles, it was by no means clear that they were winning the war. In myriad smaller skirmishes across the wider front, the wage-earners were extending union organization and winning wage increases, helped by buoyant conditions and the keen demand for labour. Furthermore, they could win set-piece encounters. E.G. Theodore and William McCormack led a successful strike of the Queensland sugar workers in 1911, just four years after they had launched their general Amalgamated Workers' Association. They would see that union become part of an even larger body, covering the great majority of rural workers, the AWU, and both men would become Labor premiers.[40]

While the Colonial Sugar Refining company lost out to the union, it was more than a match for the Commonwealth government. Hard on the heels of the strike, Fisher and Hughes appointed a royal commission to inquire into the sugar industry, a major beneficiary of tariff protection and notorious for its restrictive practices. The general manager of the company simply refused to answer questions or provide documents sought by the commission, and again the High Court upheld him.[41] With commendable restraint, the chairman of that royal commission observed that 'The difficulty of controlling the forces of capitalism in the interest of society is a theme upon which it would be easy to elaborate.' On the one hand there were workers impatient with Labor's moderation and contemptuous of the attempts to regulate capitalism—he described them as syndicalists; on the other there were businessmen oblivious of their social responsibilities. 'Both, though in very different ways, constitute a menace to the stability of the social order, and to its development by evolutionary process.'[42] In 1914 that menace seemed very real. When the state brought capital and labour into the framework of the New Protection, it gave them a privileged institutional status. The mechanisms for safeguarding industry and resolving issues of employment became fixtures to which other arrangements had to be accommodated. But even as these devices were forged, the mobilization of wage-earners and employers was still proceeding. As more workers organized themselves into unions and pressed for improvements in pay and working conditions, as employers resisted them and pursued their own strategies in the enlarged economic unit, the new institutions came under greater pressure. Above all, the conflict between capital and labour at the point of production threatened their reconciliation in the machinery of the state.

6

AUSTRALIANS IN THEIR WORLD

THE COMMONWEALTH gave expression to a burgeoning national consciousness. From the late 1880s powerful forces had created a heightened sense of identity: there was, in the popular writing of the period, a celebration of the indigenous and an anticipation of the 'nationality that is creeping to the verge of being'; in the arts, a revaluation of the Australian landscape and a conscious exploration of local forms, while the membership of the Australian Natives Association grew to 20 000 by the turn of the century and its nativism extended to the proposal that Australia adopt an Order of the Wattle Blossom in place of imperial honours and three 'cooees' as a substitute for three cheers.[1] Augmenting these impulses were the aspirations and anxieties of a white settler society in an increasingly troubled world. The strategic threat presented by the scramble for empire caused Australians to cling to their motherland, while the decline of Pax Britannica and special fears in the immediate Pacific region made for greater self-reliance. The loss of British investment in the 1890s forced a reconsideration of that economic relationship, and the desire to build local industries suggested that Australia should become something more than a supplier of raw materials.

The strength of the new nationalism was therefore undeniable but its meaning remained ambiguous. Were the Australian people to chart their own course, alone, as William Lane had prophesied? 'Behind us lies the Past, with its crashing empires, its falling thrones, its dotard races; before us lies the future into which Australia is plunging, this Australia of ours that burns with the feverish energy of youth.' Or were they instead to hold fast to the 'crimson thread of kinship'? This was how a representative of the Melbourne Chamber of Commerce, in London for the passage of the Commonwealth Constitution Act, reassured his audience:

> We sometimes talk of a new nation under the Southern Cross. This is scarcely correct; we are not a new nation—we don't want to be—but in reality we are only part of a nation, a large and growing part ... [and] desire nothing better than to be a part of the great people from whom we have sprung.[2]

These positions were certainly unequivocal, but they lay at the extreme edges of a wide spectrum of opinion in which the quickening sense of separate identity fused with imperial loyalty. Many thought of themselves as 'Independent Australian Britons' and saw no difficulty in building the new nation as a component of the Empire. Others, more impatient with Old World ties, found that their endeavours to forge a distinct national identity merely emphasized dependence.

Nowhere was this more apparent than in the immigration laws passed by the Commonwealth parliament in its first year. While the importance of immigration control as a motive for federation should not be exaggerated—if only because there were already effective colonial laws—the prevailing racial sentiment demanded affirmation. Labor had made clear that it would insist on immigration restriction as the price of its support for the Protectionists: 'Our chief plank is, of course, a White Australia. There's no compromise about that.'[3] The Protectionists themselves had pledged not just the prohibition of non-white immigration but the repatriation of Pacific Island labourers from the Queensland sugar plantations (a promise they redeemed at considerable expense to the Australian consumer). The fifty-word dictation test in any European language adopted for immigration control was to meet the objection of Joseph Chamberlain, the British colonial secretary, that overt discrimination on grounds of race or colour would be offensive and painful to Her Majesty who was, after all, the Empress of India and whose government was hoping to negotiate a treaty with Japan.[4] Here, already, was a divergence of interest between Britain and Australia. The Freetrade opposition exploited it, declaring that the government sought to do in a crooked and indirect way what ought to be done straightforwardly and honestly; Labor members took the bait the Freetraders dangled before them. Their prejudice was the most extreme, and in 1905 they adopted as their objective 'the cultivation of an Australian sentiment based on the maintenance of racial purity'. The fear of economic competition, declared their leader on this earlier occasion, was one reason for preventing Asian immigration, but the essential danger was racial contamination: 'The question is whether we would desire that our sisters or our brothers be married into any of these races.' Deakin, however, was able to reassure members that the legislative effect was the same—'Unity of race is an absolute essential to the unity of Australia'—and parliament acceded reluctantly to the imperial request.[5]

Henceforth the White Australia Policy was the accepted basis of population policy, allowing no more than a handful of non-Europeans into the country with temporary permits for specific purposes. Australian racism was a mass hysteria fed on ignorance and fear. When local communities had dealings with particular non-Australians, toleration was possible; but when they regarded them collectively in their teeming millions, they fell back on emotive stereotypes. A Chinese market gardener who sought permission for his wife to remain in Australia found widespread support from the people with whom he mixed, ministers, shopkeepers and trade unionists. A Japanese doctor in the pearling centre of Broome came under attack from an Australian rival—some residents had fallen so low, the local man alleged, as to allow their womenkind to be attended by this Japanese doctor—yet was defended in a petition signed by more than a hundred Europeans.[6] These were isolated particularities in an ocean of generalized xenophobia. As the naval supremacy of Britain came under challenge, so Australians' sense of geographical isolation increased and their racism became more strident. But even here, in their emphasis on the purity of the Anglo-Saxon stock, Australians were dissolving their distinctive national identity. In practical terms, White Australia meant a British Australia policy.

Australia was bound to Britain by ties of law, economics and sentiment. The head of state was the British monarch and her vice-regal representative retained prerogatives that were no longer available to her in dealings with the British parliament. The Commonwealth had no power to declare war or peace, no diplomatic status in foreign countries, and its external relations outside the Pacific region were conducted almost exclusively through representations to London. These arrangements were as natural to the founding fathers as the national anthem, 'God Save the Queen'.

However, two particular points of dispute arose where the constitutional claims of the Commonwealth touched British economic interests. The first concerned section 98 which claimed for the Commonwealth the right to regulate shipping in Australian waters. The British shippers, who controlled Australia's overseas trade, believed that the Commonwealth could not upset their existing rights, since these were laid down in British law, but they objected to the prospect of the Commonwealth regulating their vessels when they worked cargo along the coast. When the British objected to the first Australian maritime legislation, it was withdrawn and an imperial shipping conference convened. There the Australian delegate begged the shipowners to 'trust Australia a little bit'. 'We don't', one replied. Not until 1912 did the Labor government pass a law to protect the conditions of Australian seamen, and its operation was delayed until 1921.[7] The other point of contention was the right of appeal from the Australian High Court to the Privy Council, which Chamberlain as Colonial Secretary thought essential to the security of British investors; he admitted that he was strengthened in his stand by 'banks and other financial and commercial institutions having large investments in Australia'. Even before the draft Constitution went to London in 1899 for its passage through the imperial legislature, he had worked surreptitiously through colonial governors, judges and the premier of New South Wales to secure concessions.

Then in 1900 he summoned Australian delegates to England where he tried to browbeat them into compliance. Holding firm, they won the right to control appeals from the High Court on cases between the Commonwealth and the states, but on other matters imperial law would prevail.[8]

Commenting on the inability of Australians to protect their limited sovereignty, a contemporary writer drew attention to

> the pressure, subtle but unmistakable, with which the mortgagee knows how to check the independence of his victim. A whisper from the City, and Australian patriots sorrowfully weigh the prospects of the imminent loan or the impending conversion against the behests of the national conscience.[9]

The strength of this pressure is undeniable. Britain bought half of Australia's exports and provided more than half its imports in the early twentieth century; three-quarters of the shipping clearing Australian ports was British; the City of London held more than £300 million of Australian investments, the overwhelming bulk of loans was still raised there and of course Australian currency was tied to sterling. Economically, Australia was largely reliant on Britain. Yet it was less reliant than it had been even a generation earlier and as the economic recovery gathered pace, the degree of dependence declined further. As table 6.1 suggests, Australia was finding new markets for its products and new suppliers for its needs. Other European countries were taking more of the wool and wheat, while Germany and the United States were providing machinery, mining equipment, chemicals and other manufactures that had once come from Britain, and this despite a 5 per cent imperial tariff advantage.[10] More worrying still in the long term for Britain was the ability of Australian manufacturing industries to finance from local sources much of the expansion during this period.

For all that, the ties with the home country were ubiquitous. They began at Government House, which in each of the state capitals served as the measuring-rod of social acceptability. 'These society people are very British', remarked Sir Ronald Munro Ferguson shortly after he arrived as the new governor-general in 1914, and he added that their 'exclusiveness is unmitigated'. A visiting Englishman noted how

Table 6.1: Foreign trade, 1901 and 1913 (£m)[11]

	1901		1913	
	Imports (£m)	Exports (£m)	Imports (£m)	Exports (£m)
United Kingdom	25.2 (59.4%)	35.2 (50.7%)	47.6 (59.7%)	34.8 (44.2%)
Rest of Empire	4.8 (11.3%)	12.0 (24.1%)	9.9 (12.4%)	9.5 (12.1%)
United States	5.9 (13.9%)	3.4 (6.8%)	9.5 (11.9%)	2.6 (3.3%)
Germany	2.8 (6.6%)	2.6 (5.2%)	5.0 (6.3%)	6.9 (8.8%)
Other foreign countries	3.8 (9.0%)	6.5 (13.1%)	7.7 (9.7%)	24.8 (31.5%)
Total	42.4	59.7	79.7	78.8

it was the business of the aide-de-camp, invariably English himself, to discover 'who is who in Australia'. By such means the manners and fashions of Mayfair reached Toorak or Vaucluse:

If the high handshake is fashionable in England, it must become fashionable in Australia. If it is the custom to take your partner's arm in the West End of London, it has to be the custom, a little later, in certain quarters of Melbourne and Sydney.

Similarly, a visitor to the Melbourne Law Courts found them 'astoundingly like England'. The judges wore wigs and red robes, the barristers wigs and gowns, the courts were the same shape, the law the same, 'everything the same'.[12] For though the profession was now predominantly Australian-trained, its members read the same journals and followed the same precedents. The same was true of the medical profession which, moreover, emphasized the connection with membership of the British Medical Association. The churches imported their bishops and the leading schools their headmasters. The Australian press took its foreign reports from British news services. Australian schoolchildren saluted the Union Jack and learned about Robin Hood rather than Ned Kelly.[13]

For a great many talented Australians, singers and surgeons, scientists and writers, ambition pointed to London. Even if they would eventually return, there were just not the facilities and opportunities here for them to develop fully their talents, nor to win the recognition they craved. Others, like R.G. Casey, went on business or simply satisfied a desire to revisit the old country. Reactions followed a similar bitter-sweet pattern. The colonials remarked on the weather, the reserve of the people, the sheer size of the imperial capital. Since Australian education was so derivative, there was an odd sense of literary familiarity about the metropolis, so that the visitor found 'nearly every street and square we passed had a familiar name. It was all novel and strange—and yet remarkably homelike and familiar.' One member of the AIF described London to his parents:

> Take the city of Sydney and place it where North Carlton is, and Melbourne city where it is, and Adelaide city where South Melbourne is, with a few more on yet, and then you may have an idea of London.

Drawn to that metropolis to arrange the reconstruction of Goldsbrough Mort, Casey found that he was a small fish in a very large pond:

> You cannot understand the immense amount of time occupied in fixing these details in London. The people are all so busy and the distances apart so considerable, and the amount of time wasted in waiting so great, that whole days pass without apparent result.

A senior Commonwealth public servant, visiting London in 1907, thought that 'after our free life, our practical absence of snobbery, and our climate, . . . existence here would be practically impossible'; the Chief Justice of the High Court, in London to sit on the Judicial Committee of the Privy Council, found the ignorance of Australian affairs was exceeded only by the lack of interest. Mabel Brookes, accompanying her husband Norman while he competed in the Wimbledon tennis tournament, could not understand why rich Australians chose to live on the fringes of London society where in Australia they would be at the centre. In her experience you had to beat the English before they would become your friends. Some colonials succeeded and others failed, but there was a common tendency to remember the rebuffs and dwell on the ways the mother country fell short of their expectations—or perhaps confirmed their latent prejudices. The poet Henry Lawson complained of the snobbery, his publisher George Robertson of the dirt, the journalist Keith Murdoch of the 'squalor, cold and hunger and depravity'. Even the first Wallabies, the rugby players who toured England and Wales in 1908, were disappointed. Their captain, a medical man, had prepared his charges with lessons in table manners and lectures on venereal disease, but he had to admit that their patriotism wilted before the constant carping of their hosts and that 'most of them developed a dislike for everything English'.[14]

In return, well-bred English men and women toured the colonies and set down their reactions in that tone of easy condescension that never failed to infuriate an Australian. 'You know what "having a chip on your shoulder" means?' asked one. 'Well, may I suggest in kindliness, get yours off.' The wife of a governor wrote to her mother that her eight-year-old daughter would remain in her care rather than attend school 'as I can bear her demonian ways better than I can put up with an Australian accent'. Another newcomer compiled a comprehensive denunciation of 'God's Own Country', its heat, its wine and the virtue of its women, and concluded by advising his fellow-Harrovians who contemplated emigration to leave their money in a British bank for a minimum of two years. 'If the two years has not been sufficient to prove the unsuitability of Australia, by all means cable back for your money—you deserve to lose it.' 'Upon the whole,' one of his local readers replied,

> Australia may well pray to be saved, not only from imported convicts, rabbits, stoats, weasels, sparrows, thistles, and snails, but also from the visitations of globe-trotters, especially globe-trotters with a high and mighty tone, who peer at us from a pedestal when here, and patronise us loftily in their books after they have left us.[15]

Yet what was this but the perennial antagonism between a metropolis and its province, the classical affirmation of a colonial relationship?

Throughout the second half of the nineteenth century that relationship had mostly been taken for granted. Notwithstanding the republican sentiment of some radicals, Australian colonists had not needed to affirm their imperial connections because they were obvious. Only when the links began to fray was there a conscious endeavour to strengthen them, and the Boer War, the creation of Empire Day celebrations and some attempts to weld the Empire into a more coherent unit were part of this process.

Australian participation in the South African war would seem to demonstrate a reflexive loyalty to Britain. In 1899 each of the colonies was able to select contingents from hosts of volunteers and altogether 16000 crossed the Indian Ocean to help put down the Boers. A crowd of a quarter of a million saw the New South Wales volunteers depart. The actual merits of the dispute between Britain and the Boer republics hardly mattered. 'What do we care if she is right or wrong?' asked a Victorian parliamentarian, 'Our mother is attacked.' When a member of the Western Australian legislature asked the reason for the conflict, Premier Forrest told him that 'We do not want to know.' So overwhelming was the display of imperial loyalty that Deakin could write, 'We are now able to see that the supposed agitation for independence was of the most superficial character and of the narrowest dimensions.'[16] In fact the initial response was more tepid than this. The colonies, when Chamberlain asked them to make a spontaneous offer of support, were at first noticeably cautious. Who would pay? The early meetings to express support for the Empire were thinly attended. It was when patriotic organizations were formed to orchestrate public opinion and when the men actually set off to win glory, that the full force of jingoism was felt. Only then were critics harassed, German communities attacked, and the relief of Mafeking greeted with wild abandon. The change in mood caught out twenty-eight of the hundred New South Wales Lancers who had been exercising at Aldershot when the war began and chose to return to job and family rather than disembark at Cape Town on the voyage home; it must have seemed to these unlucky men, who were met with white feathers and public vilification, that Sydney had turned topsy-turvy during their short absence. Even then, the war mood was ephemeral. It began to dissipate with reports of British officers riding roughshod over Australian high spirits, and

Volunteers for the Boer War leave Sydney, 1899. The man in the right foreground lifts his hat as the troops march by

The pupils of this South Australian primary school show signs of restlessness as the visiting speaker warms to his subject on Empire Day

waned further during the drawn-out operations of 1901 and 1902 to wipe out Boer resistance. The soldiers returned to crowds much smaller than had seen them off.[17]

New circumstances called for new measures. To inculcate a deeper reverence for the Empire it would henceforth be necessary to mobilize and mould public opinion. From 1905 the custom began of marking the late Queen Victoria's birthday, 24 May, as Empire Day. Each year there would be public addresses and press articles, parades and patriotic homilies of the deeds that had won the Empire and of the need to uphold its ideals. Empire Day was directed especially at the schools. Lessons were put to one side for the day to allow politicians to go from school to school and deliver what the *Bulletin* called unkindly 'the same shop-soiled oration, carefully stowed away in camphor from Empire Day to Empire Day'.[18] As with the Boer War, however, enthusiasm was not as clear-cut as these practices suggest. Empire Day had begun in Canada in the 1890s and was urged by the Earl of Meath at the Imperial Conference of 1902 as a link to unify all the far-flung dominions. In Australia it was taken up by branches of the Empire League, whose members had been so active in anti-Boer agitation, and they put the proposal to celebrate Empire Day in Australia to a premiers' conference in 1903, explaining that 'a national holiday, circling the earth with Britain's drumbeat ... would be helpful, and increasingly so as the years rolled on, in holding the race together'.[19] But they were rebuffed. The conservative provenance of the proposal damned it in the judgement of Barton, who was at this time head of a federal government dependent upon Labor support. Thus the Empire League had to wait until 1905, when the conservative George Reid was prime minister and Japan's defeat of Russia in the war of that year had awakened Australian anxieties. Furthermore, Reid linked the theme of Empire to the anti-socialist crusade he launched at that time. His exploitation of Empire Day as a 'surety against a precipitate nationalism and socialism' was enough to discredit it in radical nationalist circles, so that it was denounced in the *Bulletin* and Labor press as St Jingo's Day, All Fools' Day or Hempire Day, an occasion for gilded toadies and prosy bores to celebrate the glories of a country that was not their own. Furthermore, the Catholics resented the Anglophile initiative, so from 1911 their schools adopted 24 May as Australia Day and organized their own national festivities.[20] An attempt to promote imperial unity had led to domestic discord.

The same hazard threatened another institution created at this time, the Rhodes Scholarship. In 1902 Cecil Rhodes, the adventurer whose imperial enthusiasms had triggered the war in South Africa, died and he willed much of his considerable fortune to a scheme that would take young men from the colonies, the United States and Germany to Oxford. Rhodes sought 'the best men for the world's fight' and he believed that by exposing them to all that was best in the homeland, they would return to leaven the public life of their places of origin. There were six Australian Rhodes scholars each year, one from each state, and they were chosen by selection committees operating under detailed instructions laid down by the Rhodes Trust in accordance with the founder's intentions. Rhodes sought scholars who could best exemplify the imperial ideal of 'manhood, truth, courage and devotion to duty'. While they would possess literary and scholastic attainments, they should 'not be merely bookworms'; hence they should have a 'fondness of, and success in many outdoor sports' and exhibit 'moral force of character and instincts to lead'. The Rhodes scholars would constitute an influential élite in Australian public life. (And also in New Zealand: one of the Watt boys went to Oxford as a Rhodes Scholar between the wars.) Not all of them returned from Oxford and not as many went into politics as the founder would have hoped (though Australia's only Communist member of parliament was one of them), but especially in the professions and the universities they were well represented and well able to pass on their reverence for the old country.[21] On the whole, the Rhodes scholars found Oxford a more congenial representation of English values than did the visitors to London or the north. In the period before Morris built his factories at Cowley, they did not see the ravages of industry, and in their dealings with college servants they found little of the class antagonism that poisoned social relationships elsewhere. If the English undergraduates sometimes seemed aloof and patronizing—in Max Beerbohm's aphorism, the Germans loved Oxford too little, the colonials too much—most Australians did not take offence. The term colonial was still accepted 'as a matter of course and carrying no hint of obloquy'.[22]

But the annual selection of Rhodes scholars was a straightout competition between the denominational education systems. In Melbourne it was the practice for the undergraduates of the three university colleges, Anglican, Presbyterian and Methodist, each to elect a corporate candidate. To their mutual consternation, the honour fell in 1904 to an outsider; worse, to an outsider whose manly outdoor sport consisted of keeping score at cricket matches. So incensed were the colleges that they sank their rivalry to hold joint protest meetings and send a remonstrance to the selection committee, the premier, the Rhodes Trust and the press.[23]

The Rhodes scholarships were a mixed success and so was a more confidential device created by the same circle of imperial enthusiasts who had gone out to South Africa and there conceived a vision of unity. They could see how new trading patterns were tugging the dominions out of Britain's orbit and they could feel the discordant effects of separatist impulse, but their administrative experience in Africa convinced them that the Empire was a living reality. The battle to give it closer cohesion was fought and lost at Westminster, and the defeat of the Conservatives by the Liberals in 1906 effectively destroyed these imperialists' chances of official success, yet their influence reached further. If there was to be some form of imperial federation and if the imperial economy was to be safeguarded by trading agreements, as they hoped, then in every corner of the Empire there had to be statesmen imbued with imperial instincts who could prepare public opinion. To this end they created a Round Table movement, the name echoing their ideal of Arthurian chivalry, and they established a journal with the same name. Their organizer came to Australia in 1910 to form local groups of the Round Table. He was disingenuous about the aims of the journal, presenting it simply as a medium of information and discussion, and circumspect in his choice of members. By talking to likely prospects he was able to settle on a suitable figure and with his help draw up a list of men with what he called the 'leading spirit' who were then inducted into the aims of the Round Table.[24] In Melbourne the chairman was Professor Harrison Moore, the influential constitutional lawyer, and the group included Herbert Brookes, businessman, son-in-law and confidant of Deakin; Walter Murdoch, another friend of Deakin and already a prominent writer; F.W. Eggleston, a rising politician; John MacFarland, the vice-chancellor of the university, and Sir George Knibbs, the Commonwealth Statistician.[25] Sydney included Henry Braddon, the manager of a leading pastoral company, influential professors and a brace of Labor men. Similar groups were established in other state capitals. With ready access to the press, they were able to keep the Empire in the forefront of public attention. But with what success?

The English, Irish, Scots and Welsh were but some of the European nationalities to throw off large contingents of settler-colonists during that continent's era of world supremacy. If they did not choose to go to their own colonial territories, most non-English speaking migrants preferred the United States, where opportunities were greater and the British flag did not fly, but Australia had pockets of most European nationalities. There were distinct clusters of German farmers in South Australia and Queensland; Scandinavians were more scattered, while Mediterraneans tended to settle in coastal areas, though their influx was only beginning. Precise numbers are difficult to establish because the census distinguished only those born in the homeland and not their Australian-born children: if descendants are included, it is possible they made up more than 5 per cent of the population. But counting them opens up further questions. When do Europeans who settle and establish families in Australia cease to be European? Should the German-speaking community in the Barossa Valley, with its own churches, schools and press, be regarded as Australian? That question had hardly occurred to them or their neighbours over the previous fifty years because unnaturalized residents suffered few civil disabilities, but it took on a fresh urgency with the rise of national sentiment. Whereas in 1891 only a third of the German-born South Australians were naturalized, two-thirds had taken that step by 1911.[26] Further, what of migrants who did not create such stable ethnic communities or, more particularly, did not even stay? Among Italians, Yugoslavs and Greeks it was common for a group of young men to cross the ocean and take on the more demanding labouring jobs that are always available on a frontier. Often they had no intention of settling, planning to take back their savings and stories to their places of origin; but if conditions were favourable, then settlement could follow either by bringing out females or marrying locally. Before 1914 few Mediterraneans put down such roots in Australia.

One reason for reluctance was that they were victims of the same prejudice as was shown to Asians. Both in the labour market and in popular attitudes there was a hierarchy of racial antagonism: the 30 000 Chinese were the most fiercely resented, followed by the 15 000 other Asians, the 10 000 Pacific Islanders and then the 10 000 southern Europeans.[27] In an age that defined racial characteristics as freely as psychological labels are now employed, the Anglo-Saxon race (some with Irish ancestry preferred to say Anglo-Celtic) was superior.

Australia's dealings with the rest of the world, apart from Britain, were primarily dealings with an Anglo-Saxon civilization. That especially advanced social laboratory, New Zealand, offered influential precedents in land settlement methods, progressive taxation and arbitration. The relationship with New Zealand was close at the end of the nineteenth century. One New Zealander in twenty was Australian-born, much of its finance and external trade was in Australian hands, and since Sydney could be reached by saloon passage for as little as £2 10s, there was a steady traffic across the Tasman of visitors and men seeking work. But the Commonwealth tariff and an emergent New Zealand nationalism caused the neighbours to drift apart after the smaller country decided in the 1890s not to federate.[28] From the United States Australians adopted mining and agricultural methods, techniques of irrigation and advances in locomotive design. Many social and political ideas came from the same source including the land reform message of Henry George, the radicalism of Edward Bellamy, American models of anti-trust legislation and industrial unionism. Bryce's *The American Commonwealth* was a storehouse of precepts at the federal conventions of the 1890s.[29] To the South African Rand were sent miners, mine managers and techniques developed in Victoria, Broken Hill and Western Australia and, in addition, Australian miners' took an insistence on strict segregation of white and non-white labour. Even this country's contacts with the ancient civilizations of India and China were mediated through the Raj and the British concessions.

Only in the Pacific did Australia strike out alone. Traders plied the islands of the south-west Pacific, miners had ransacked them for gold, planters had taken land for copra and sugar fields, blackbirders had persuaded or coerced the men to work in Australia, and missionaries insisted on saving their souls. Here Australia had out-run the Union Jack and then, as other colonial powers began to lay claim to the area, it had pressed the Colonial Office to declare British sovereignty over the south-east portion of the island of New

Guinea and remaining smaller islands. While the New Guinea trade amounted to less than £50 000 in 1901, Australia's stake in the region was significant: the investments of the Colonial Sugar Refining company and other Australian enterprises in Fiji were worth more than £3 million, and the copra trade yielded £400 000 to the Burns Philp trading company in 1914.[30]

'The race is to the swift and the strong, and the weakly are knocked out and walked over.'[31] This battle cry from the principal partner in Burns Philp characterized the Australian determination to prevail in its corner of the Pacific. Seeing France, Germany and the United States all establishing sovereignties and extending the operations of their merchant fleets, Australians pressed Britain to act more vigorously on their behalf. But the British were feeling the pinch of naval competition and had little desire to quarrel over these distant and insignificant specks on the map. They were only too pleased to hand British New Guinea, which Australia renamed Papua, over to the new Commonwealth, and they were not going to take on new responsibilities. A glance at that map convinced them of the absurdity of Australia's claim that the New Hebrides were essential to Australian security—the Antipodean 'Channel Islands' as the Commonwealth described them in 1903—so they settled for a joint protectorate with France. 'Mr Marsh will compose a soothing but serious and unyielding reply leading to the conclusion of the correspondence.' Thus wrote Winston Churchill, the under-secretary for the colonies, in response to Australian protests.[32] More than any other factor, the Australian chagrin and growing realization that British support could not be relied on aroused the desire for greater self-reliance.

Under the terms of an 1887 Agreement, the Royal Navy maintained an auxiliary Pacific fleet and the Australian and New Zealand colonies contributed £126 000 annually towards its cost. That agreement was revised at an Imperial Conference in 1902 where the Australian contribution was raised to £200 000 and the British undertook to provide a larger squadron; however, it was specified that Britain might deploy the ships outside Australian territorial waters if it chose. The Australian government tried to put the best construction on the new condition. Forrest, as minister for defence, affirmed the doctrine of imperial unity whereby 'Our aim and object should be to make the Royal Navy the Empire's Navy', and Barton explained that 'The principle surely should always be this: Touch one of us and you touch us all.'[33] In reality there was little alternative. The Commonwealth had inherited from the states a motley collection of antiquated gunboats, and the Braddon clause meant that there was little chance it could afford to build an effective force of its own, so Australia was forced to buy its protection on the cheap. Deakin, who succeeded Barton as prime minister in 1903, was nevertheless convinced of the need to develop an Australian navy and the Russo–Japanese War of 1904–05 did much to convince the Labor Party, on which he relied. The moral drawn in Australia from that short engagement on 27 May 1905 in the straits of Tsushima when the Russian fleet was destroyed was that 'the yellow man had taught the white man a lesson that Australians can neglect only at their peril'.[34] The difficulty was to convince the British Admiralty, which for several years fobbed Deakin off with assurances that Japan was an ally and his fears were groundless. One Australian response was to invite the American fleet to visit Australia on its world voyage of 1908, when huge crowds welcomed them—Admiral Sperry estimated the throng on the cliffs and shores as his white-painted battleships entered Sydney Harbour at half a million. Deakin presented this visit as evidence of an '"entente cordiale" spreading among all white men who realize the Yellow Peril to Caucasian civilization, creeds and politics', and he hoped in addition that Britain would respond possessively with a fleet of its own. He pursued the strategy in the following year with an invitation to President Roosevelt, but nothing came of the proposal and the British still held aloof. In 1909 Australia began constructing a fleet of its own.[35]

Compulsory military training was another aspect of the same increased defence effort. The scheme was established by Deakin in 1910 and enlarged by Labor in 1911—compulsory training was a bipartisan policy. For many Labor supporters a citizens' militia was preferable to a caste of professional soldiers and Hughes particularly was active in the National Defence League that had campaigned for conscription since

1904: 'Our people should be taught obedience which is a primary virtue and an essential of citizenship in a free state.'[36] Boys and young men were compelled to spend between 64 and 90 hours a year drilling and training. The militarist strain of Australian nationalism was not new. During the Boer War there had been anticipations of the baptism of blood that would consecrate the new nation:

> A nation is never a nation
> Worthy of pride or place
> Till the mothers have sent their first born
> To look death in the field in the face.[37]

As military preparations increased, sending defence expenditure from £1 million in 1908/09 to £4.3 million by 1913/14, when it made up a third of all Commonwealth spending, such sentiment became more fiercely Australian:

> The wealth you have won has been wasted on trips to the English Rome
> On costly costumes from Paris, and titles and gewgaws from 'home'.
> Shall a knighthood frighten Asia when she comes with the hate of hell?
> Will the motor-launch race the torpedo, or the motor-car outrace the shell?[38]

It was in keeping with this sentiment that Australian uniforms dispensed with gold lace. Responding to past British slights against 'windy-minded underbred spouters' and 'transoceanic mediocrities', many Australians were no longer prepared to identify their own strategic interests with Britain's. The First Lord of the Admiralty announced an emergency programme in 1909 to meet the German challenge. The press, imperial pressure groups and conservative politicians demanded that Australia follow New Zealand's example and offer to donate the cost of a dreadnought to the home country, but the Labor government refused and a voluntary fund fell far short of its target. Unable to obtain an assurance at the 1911 Imperial Conference that his government would be consulted by Britain during increasingly tense national negotiations, Australian Prime Minister Fisher was quoted as saying, 'Don't talk of Empire. We are not an Empire.' He denied the statement.[39] In the end, Australia did remain part of the Empire. Both the Australian army and navy were developed in accordance with the recommendations of a British field marshal and a British admiral.

An election was in progress when the world war finally broke. 'With almost miraculous celerity,' declared Hughes, 'the din of party strife has died down, the warring factions have joined hands, and the gravest crisis of our history is faced by a united people.' There was even talk of postponing the election and reconvening the last parliament, but in the end the election went ahead with both parties as one on the question that mattered most. On behalf of the Labor Party, Fisher declared that Australia would help Britain to the last man and last shilling. For the Liberals, Joseph Cook promised that 'if it is to be war, if the Armageddon is to come, you and I shall be in it.'[40] In the event it was Labor that won a substantial majority and assumed responsibility for the nation at war.

7

WAR

ON THE MORNING of 5 August 1914, Australia found itself at war. That much was determined by its membership of the British Empire and while it was up to the federal government to decide how much support the mother country would receive, there was never any doubt that Australia would come to her assistance. Even before Sir Edward Grey's ultimatum to Germany, Australia had put its navy at the disposal of the British Admiralty and offered to send a contingent of 20 000 soldiers to any destination required by the War Office. These decisions were made by the outgoing ministry and confirmed by the Labor ministers who were sworn into office after the election on 5 September. No legislation was necessary, indeed parliament did not meet until October when it simply voted £100 000 to Belgium and provided the executive with further powers under a War Precautions Act. It was in fact the commander of the contingent who coined its name, the Australian Imperial Force (AIF), and who resisted the British suggestion that Australia send separate brigades such as could readily be incorporated into British divisions. Rather, Australia would provide a complete division of infantry and a brigade of light horse, and since they were to be volunteers, the Australians would have their own code of discipline as well as their own organization and a commander who would be responsible to the home government.

Blackboy Hill training camp, 1914. The recruits display different degrees of military polish

The limit of 20 000 was imposed not by any difficulty in raising volunteers but the logistical task of equipping, training and transporting the troops to Europe by the end of the year—for few expected the war to continue much into 1915. So keen was the desire to volunteer, when enlistment began in August, that applicants jostled each other in the queues that stretched before the recruiting tables and strong men choked back their disappointment when rejected as unfit for service. The pay, 5s a day (and another shilling after leaving Australia), was undoubtedly an attraction, yet there were some who sold up businesses to enlist; and while the prospect of travel and adventure lured many a Ginger Mick into uniform almost as by casual impulse ('wot for? Gawstrooth! 'e was no patriot'), there were ardent patriots who travelled long distances to reach the nearest enlistment centre. One Queensland grazier rode 460 miles to the nearest railhead to offer himself in Adelaide for the light horse; finding that quota filled, he sailed to Hobart only to discover again there were no vacancies, whereupon he took the boat to Sydney and there, after a journey of 2000 miles, was enlisted as a private.[1]

The training of these recruits was a makeshift affair, hardly

Seeing them off: HMAS Persic *at Princes Pier, Melbourne*

an advance on the abbreviated instruction that many had received during their compulsory military service and merely preliminary to the thorough preparation they were expected to undergo in England. The men were quartered at camps on the outskirts of the cities, enabling friends and relatives to visit and groups of high-spirited trainees to descend on the city pleasure spots. By day they paraded in ragged ranks to learn the rudiments of drill, by night they showed off their loose, pea-soup coloured, dull-buttoned uniforms and soft-brimmed hats to envious mates. These citizen soldiers did not take easily to army discipline, nor to the hectoring tone of the regular non-commissioned officers who instructed them, so 'it was a common sight at Blackboy Hill [outside Perth] the first few days, to see a sergeant or a corporal get a punch on the nose'. By late September they were ready to embark. However, the reported presence of German warships in Australian waters delayed the departure and it was not for another month that the fleet of transports assembled in King George Sound in the south-west corner of the continent. Twenty-six Australian troopships were joined there by ten New Zealand ships, and on 1 November their naval escorts led them into the Indian Ocean.[2]

Already the Australians had seen action closer to home. Just a few hours after the outbreak of war, there had been that moment of excitement at Fort Nepean, overlooking the entrance to Port Phillip Bay, when a German merchant vessel tried to put out to sea. After a shot across her bows, she turned and steamed back to Portsea. Then there had been the capture of Germany's Pacific colonies: acting on instructions from Britain, an Australian force had taken German New Guinea and the other smaller islands south of the equator. Meanwhile the Australian navy searched in vain for the German Pacific squadron, which in fact would be found some months later by British cruisers off the coast of South America. The first great triumph of the Australian fleet occurred instead in the Indian Ocean where the *Sydney*, escorting the troopships, intercepted and destroyed the German cruiser *Emden* on 9 November.[3]

These early successes bolstered Australian enthusiasm for the war and provided some relief from the disturbing reports of events in Europe. For as the Germans swept down through Belgium during the latter months of 1914, and into France, it became apparent that the Allies could expect no quick victory and that their very survival would require the fullest mobilization of resources. Yet the war had already brought serious disruption to the Australian economy. The closure of European markets, the shortage of shipping and the sheer uncertainty of what lay in the future paralysed much of the export industry. The stock exchanges closed to await future developments. The coal-mines of New South Wales and the base metal mines of Tasmania and Broken Hill laid off many men; pearling operations on the north-west coast were suspended; timber workers, trappers and a host of other bush-workers lost their livelihood. These adversities coincided with the return of drought conditions over most of the agricultural districts, so that the wheat harvest at the end of the year was only a quarter that of the previous season. To make matters worse for those thrown out of work, the unavailability of British loan funds severed public works: accordingly, the rate of unemployment increased from 5.9 per cent at the beginning of 1914 to 11.0 per cent by the end of the year. The unemployed made up a considerable part of the first AIF.[4]

The government acted promptly to restore business confidence. A premiers' conference met to provide credit guarantees to financial institutions and the federal treasurer increased the note issue, backing the augmented money supply with the reserves of bullion that heaped up now that gold shipments to London were suspended. A Royal Commission into Food Supplies and Trade and Industry put forward some anodyne recommendations for the economic controls that might be required if the conflict was prolonged 'for a year or more'. The Commonwealth expanded its factories to produce the uniforms, boots and limited range of armaments that were within its capacity. Beyond that, it was business as usual, except for the dramatic raid on the offices of the principal metal firms in November 1914. That most of Australia's lead, copper and zinc was sold by long-term contract to German combines (operating in some cases through British subsidiaries) was an open secret, and the importance of these materials for wartime uses was obvious. Armed with a Trading with the Enemy Act, Attorney-General Billy Hughes authorized the raid in order to seize documents from the leading Australian producers, including BHP, and in the following year he obtained further legislation to annul existing contracts. The chief beneficiary was an enterprising combination of London and Melbourne financiers, known by the address of their Australian office as the Collins House group, who already owned the further reaches of the Broken Hill lode. Their representative confronted this 'strange little gnome-like Welshman', and reached an agreement whereby the group acquired control of the Port Pirie smelter, co-operated in the establishment of an Australian metals exchange, and entered into extremely profitable contracts with the British government.[5]

Originally the Australians were to go to England for further training, but the camp sites on Salisbury Plain were deep in mud and Turkey's entry into the war persuaded the British command to land the colonial troops in Egypt. There they exercised and marched over the hot sand. They were joined in February by a second Australian contingent and organized in two divisions, one Australian and one made up of Australians and New Zealanders, all under the control of the English general Birdwood. A member of his staff coined the acronym Anzac for this improvized Australian and New Zealand army corps made up of the raw, unruly colonials. 'I think we have to admit that our force contains more bad hats than the others', the official war correspondent, C.E.W. Bean, confided to his diary after noting the high incidence of venereal disease and the frequency of drunken misbehaviour among Anzacs on leave in Cairo. In the most serious outbreak, the 'Battle of the Wozzer', an altercation in a brothel escalated into an orgy of violence, destruction and arson which the military police were powerless to control.[6] On the following day, however, the Anzacs began moving out to Alexandria where they filed aboard the transports that took them to the harbour of Lemnos, 60 miles from the Turkish positions on the Dardanelles. Their time had come. '[A]s we have been told that we will probably land under fire, we are full of joyous expectancy', an artilleryman wrote back to Australia; 'I am at present about to enter into the joy of my life.'[7]

The Australians were to land on the eastern side of the Gallipoli Peninsula as part of a larger operation involving some 75 000 men. Their task was to take the forts that guarded the Narrows Strait, the passageway between the Mediterranean and the Black Sea, so that the Allies might threaten the Turkish capital and relieve mounting pressure on beleaguered Russia. While the main force of French and British troops landed at Cape Helles on the toe of the peninsula, the Anzacs were to come ashore 13 miles further up, secure the hills on their left flank and proceed across land to the Narrows, a distance of some 4 miles. Their prospects were hardly favourable. This barren spit, a maze of gullies and ridges, was ideally suited to defence, and an earlier attempt to force the Narrows by naval assault had not merely failed but alerted the Turks to the likelihood of further attack, so they could hardly misunderstand the meaning of the flotilla that

Landing at Anzac Cove on the morning of 25 April 1915. The photograph was taken by C.E.W. Bean

gathered at Lemnos. Nevertheless, the Anzacs went in before dawn on the morning of 25 April.

As the first wave of men waded ashore in semi-darkness, Turkish fire came down from above. The Anzacs faced not the open country they had been told to expect but an abrupt and precipitous height—they had landed a mile too far to the north on a narrow beach that came to be known as Anzac Cove. Organizing themselves as best they could into their units, the men pushed up the slope. As they reached the crest, the Turks retreated across a small plateau and disappeared into the gullies beyond, for there was only a company of them. The Australians followed. Some scaled a second ridge, and the most adventurous pressed on until they could see their objective, the Narrows. But by this time the Turkish divisional commander, Mustafa Kemal, had come up with his reserves and the disorganized Anzacs were outflanked and killed or driven back. Furthermore, the British naval guns could not answer Turkish artillery which inflicted heavy damage with shrapnel fire throughout the day. When darkness fell, the invading force was clinging to a bare foothold on the second ridge, little more than half a mile inland. So precarious was their position that an alarmed General Birdwood sent a message to the Allied commander, Sir Ian Hamilton, and asked if the force should leave or stay. Hamilton established the fact that it would be impossible to re-embark before dawn and to do so presented the enemy with a defenceless target, so he advised Birdwood that 'there is nothing for it but to dig yourselves right in and stick it out'. That settled it and in the remaining hours of the night the clink of shovels could be heard everywhere. The Anzacs remained on Gallipoli because they could not be taken off.[8]

The Turkish counterattack came twenty-four hours later and failed to dislodge the invader. Both sides then consolidated their defences in a series of trenches, with special reinforcement of the strategic high points and passes. The Anzac position was dangerously exposed and separated from the enemy at some places by no more than the length of a cricket pitch; here the troops were occupied with sniping, mining and bombing each other's posts, but all attempts to win ground resulted in heavy casualties. The New Zealand and the 2nd Australian Brigade were taken down to Cape Helles early in May for an assault where the Australians lost a thousand men in an hour. The Turks launched a major offensive a week later and suffered so many casualties that it was necessary to declare a truce so that the heaped bodies could be buried. In August the Australians made fresh efforts to coincide with a new Allied thrust to the north. That enterprise ground to a halt while thousands fell in bloody dispute over the mount named Lone Pine, and at several neighbouring posts successive waves of men scrambled from the Australian trenches to be cut down before they progressed more than a few yards. Though skirmishing continued, the Allies had exhausted their powers of attack. Lord Kitchener came out from England in November to inspect the position and decided that the force must be withdrawn, the Australians first. Shortly before Christmas, the best-organized operation of the entire campaign, the withdrawal, was completed with only two casualties. The Australians were transported back to Egypt. On 400 acres of wasteland they left behind them 7500 graves.

These losses at Gallipoli were not heavy by the ghastly

Post office at Anzac Cove, 1915

standards of Europe. In this ill-fated sideshow, the French lost as many soldiers as the Australians, the British three times as many. Yet almost immediately Gallipoli became lodged in the popular consciousness as a special feat of Australian valour, an achievement beyond any other in the country's history and the embodiment of all that was worthy in the national character. Why was this? There can be little doubt that Gallipoli filled a deeply felt need for those Australians who had longed for the day when they could prove themselves in battle: thus Henry Lawson's prediction that 'the Star of the South shall rise—in the lurid clouds of war'. The fact that the Anzacs died in vain only enhanced their blood-sacrifice. In constructing the legend of Gallipoli, the Australians relied heavily on Englishmen, whose approval they specially desired: it was an English journalist who supplied Australian readers with their first account of the landing at Anzac Cove ('there has been no finer feat in this war'), an English poet and a Scottish novelist who provided the most fulsome tributes ('something as near to absolute beauty as I shall hope ever to see in this world', 'the finest body of young men ever brought together in modern times'), and who dwelt on the similarities of this campaign to that at nearby Troy three thousand years earlier ('they walked and looked like the kings in old poems'). Even the Australian war correspondent harked back to his English public school days when he likened the crack of rifle fire to the sound of the fives court or the cricket nets at Clifton College.[9]

An Australian officer described the condition of his men in the Gallipoli trenches more realistically as 'thin, haggard, as weak as kittens, and covered with suppurating sores'. Poor diet, lack of water, inadequate sanitary arrangements and clouds of flies that bred in the corpses all encouraged the spread of intestinal sickness, so that by July the Anzacs were losing a hundred men a day to disease. (Struck down by dysentery, even the ebullient Sir Ian Hamilton acknowledged a desperate longing to lie down and do nothing but rest; he comforted himself with the thought that this must be the reason why the Greeks were ten years in taking Troy.) Nor was it true, Bean admitted, that the Australians were uniformly resolute, determined to a man to fight on, and he scorned the popular accounts of wounded men insisting on returning to the front. In truth, he said, 'They dread it.' It was not unusual for the more hardy who pressed forward to meet weaker comrades making their way back to the beach. Again, there were some who exulted in 'the bloody gorgeousness of feeling your bayonet go into soft yielding flesh' and others who never forgot 'the awful look on a man's face after he has been bayoneted'.[10] If the Anzacs who sailed back to Egypt at the end of 1915 had come to share a common outlook, then it was perhaps a confidence in their own fighting capacity, a grudging admiration for the Turk, and a marked lack of respect for the British.

At home the events at Gallipoli augmented support for the war. The publication of the first list of casualties hardened attitudes and the evanescent enthusiasms of 1914 were replaced by a grimmer purpose. Certainly there were critics. The federal parliamentarian Frank Anstey had stated his misgivings at the outset and he repeated them just three days after the Gallipoli landing. Those Labor colleagues who shared his socialist, anti-militarist views were outnumbered in their own party, however, and subjected to increasing vilification. In July 1915, a pugnacious member of the opposition front bench, W.A. Watt, urged the House of Representatives to expel Labor's Frank Brennan for stating that he would be incapable of killing a Turk. (Brennan responded to Watt's accusations of cowardice by declaring his willingness to 'risk my pigeon liver on the hills of Gallipoli with that man' and inviting his accuser to meet him at the recruiting office; Watt failed to keep the appointment.)[11] Support for the war was overwhelming. The number of volunteers shot up from 6000 in April 1915 to 36 000 in July. Sporting competitions, which had generally continued during the first year of the war, were increasingly curtailed—though more commonly among the amateur associations of the middle class than the professional codes catering to working-class audiences. Publicans, who had weathered the temperance campaigns before the war, were brought under renewed pressure to shorten their hours of business: South Australia introduced six o'clock closing first, by referendum in 1915, and New

A display of war comforts prepared by the 'Busy Bees', the female workers of Swallow and Ariell, Port Melbourne

South Wales, Tasmania and Victoria followed by the end of 1916. Many organizations and societies followed the example of the King and forswore alcohol altogether until the end of hostilities.[12]

The failure of the Allied expedition to the eastern Mediterranean and the continuing deadlock on the Western Front made that prospect remote. Since Australia was now committed to a long and costly conflict, two questions had to be resolved: who would fight the war and who would pay for it? The answer to the first question seemed clear. Australian men of military age could be called up for domestic training but only volunteers served overseas.

The gender implications of such a division of labour were far-reaching. By drawing a third of a million men aside, putting them into uniform and subjecting them to protracted ordeal, the war reinforced an ethos of masculine superiority. The chivalry of the men and the defencelessness of women, central themes of recruiting material, were confirmed by females who shamed shirkers by distributing white feathers. 'The spirit which said "Return with or on your shield" is burning brightly in the land', rejoiced the Women's Compulsory Service League. It was expected that women would keep the home fires burning and provide comforts for their protectors. Fund raising, nursing, knitting socks and balaclavas, sending messages of support and various forms of self-sacrifice did in fact absorb the energies of many women, but in other respects the war loosened conventional restraints. It allowed females to replace males in the workforce, gave some women a respite from the duties of marriage and at least for the time being made single parents of those whose husbands volunteered. That military service became a convenient means of desertion was tacitly acknowledged when the government began remitting to wives the bulk of the pay of married men on overseas service.[13]

Freedom of choice is seldom absolute. The weight of expectation from family, friends or employer undoubtedly persuaded many a young man to enlist, while in some families it was accepted that this son would stay at home while that one served. Even so, the composition of the AIF was broadly representative of the general population. But would voluntary enlistment meet the country's needs? The army's demands for additional men were already threatening to exhaust the available supply, and it had become necessary to establish recruiting committees to persuade more men to come forward. In any case, was voluntarism an appropriate principle? Some said the perils of combat should be shared by all eligible to serve, that it was inequitable and possibly even deleterious to the vitality of the race to allow the flower of Australian manhood to sacrifice their lives while shirkers evaded their duty. Hence a Universal Service League was established in the latter months of 1915 to argue that this 'process of unnatural selection' be replaced by compulsion. And in September 1915 the government conducted a war census of manpower and wealth, in which all men were asked if they were willing to enlist and if not, why.[14]

As with men, so with money. The great bulk of war expenditure was financed by voluntary war loans, which took the Commonwealth public debt from £6 million in 1914 to £325 million by 1919, a sum not much less than the accumulated borrowings of the states. Some money was collected in 1914 and 1915 through the new land and income taxes, and probate duty, but these raised only a small part of the total spending. Thus in fiscal terms the chief burden fell immediately on those who chose to accept it, and since they were lending rather than giving, their actual sacrifice was small; the government rejected the suggestion of interest-free borrowing on the grounds that a loan 'without interest means confiscation'.[15] But of course the war had a greater impact on economic fortunes than this. By interrupting the inflow of population and capital, it cut at the taproot of economic growth; by halting family formation it paralysed the house-building and ancillary industries. The gross domestic product fell in real terms by 10 per cent during the war, and the decline in per capita consumption, excluding defence expenditure, has been estimated at 12 per cent.[16] Again, some fared better than others. Many manufacturers benefited from the stimulus given to production for military purposes and the new opportunities for import substitution. The infant steel industry—BHP's Newcastle steelworks began production in 1915—found orders beyond its wildest dreams. Special arrangements were made for primary producers. The government purchased the wheat crop from growers and stockpiled it; sugar was bought from the Colonial Sugar Refining Company, processed and then returned to the company for distribution, an arrangement from which it did not suffer; and meat and wool were sold by special arrangement to the British government.[17] Discontent was greatest among wage-earners since there was still widespread unemployment and earnings lagged far behind the rapid increase in prices. The 1915 census of wealth did little to forestall criticism of the wartime 'profiteer', or 'boodler', and patriotic speeches did not fill empty stomachs. The first wartime strikes began in 1915 and were denounced in turn by the propertied classes as acts of disloyalty.

As if to contain these growing rifts and affirm the threatened national identity, there was an upsurge of xenophobic repression. The first victims were German settlers, whose roots in this country were deep and who had previously been perhaps the most accepted of its non-British inhabitants. In the cities, they were well represented in industry, commerce and the professions, and their contribution to cultural life was considerable; in South Australia, Queensland

and Victoria, they had created stable and prosperous farming communities. Their loyalty was hardly in question. A spokesman at the outbreak of the war expressed the 'deepest gratitude to that Government under whose wise and benevolent rule we have always enjoyed protection, liberty, peace and prosperity', and on board the troopships that left Albany three months later there were groups of German-speakers wearing the King's uniform.[18] How quickly the government betrayed that trust! Within six months the 'enemy aliens' who were of military age, including men who had settled in Australia decades earlier and in many cases believed themselves to be Australian citizens, were rounded up and put in internment camps. Men and women with an established place in their local community were vilified, harassed and dismissed from their jobs. Anti-German leagues sprang up to dwell with lascivious relish on the murder, rape and pillage of Belgian civilians. Anonymous letter-writers signed themselves as 'Britisher' or 'Loyal Australian' to denounce a neighbour with a German surname. Once the poisonous process began, no one was safe from denunciation, not even John Monash (or Monasch as his enemies would have it) whose father had been naturalized in 1856, nor the New South Wales premier William Holman (Hoffman, the whisperers said) who resisted demands to dismiss all his German public servants. In 1917 it was declared an offence to anglicize one's name.[19]

The net spread wider to include all subjects of the enemy powers, and others besides. The Afghans of Broken Hill—they were known as Turks but in fact had come as camel-drivers from further east—were subjected to harassment until some could stand no more. One who had been stoned by local children took a friend and embarked on a two-man jihad: they opened fire on a trainload of picnickers travelling to the Manchester Order of Goodfellows' New Year celebrations, killing three and wounding several more. After these two tormented men were tracked down and shot, strenuous efforts were made to round up all the 'Turks' and Germans left in the town. Similarly, the Greek shops of another mining town, Kalgoorlie, were sacked in 1914 because it was thought that Greece was not a wholehearted supporter of the Allies. Tony Splivalo, the young Dalmatian, was accosted by a schoolfriend and asked, 'Hey Tony! Why aren't you at Rottnest? We don't want people like you running loose around here.' Shortly after, he was interned on Rottnest Island and later transferred to the Holdsworthy camp in New South Wales where 5000 men from the various nations of central Europe were thrown together to pursue their historic feuds. The definition of 'enemy subjects' was widened in 1916 to include those born in Australia of enemy parentage or grandparentage.[20]

By this time no German-born Australian was immune from the hatred of the Hun. Hans Heysen, the painter who had enjoyed vice-regal patronage before the war, was subjected to suspicion and insult. A.H.M. Heinicke, a fashionable teacher at the Adelaide Conservatorium of Music, was stopped by nine students who painted the Union Jack on his bald pate—he declined to seek their punishment and some time afterwards abandoned his grand Medindie home to live on a poultry farm outside Adelaide. Otto Krone, headmaster of Melbourne's Methodist Ladies' College, though defended by his predecessor, W.H. Fitchett (author of *Deeds that Won the Empire* and other patriotic homilies), was harassed to his grave in 1917. Even the place names that acknowledged the German contribution to Australia were obliterated. The South Australian township of Hahndorf became Yantaringa and Hamburg, Haig; New South Wales's Germantown became Holbrook and German Creek, Empire Vale; in Victoria Mount Bismarck was renamed Mt Kitchener.[21]

The bulk of the Australian forces, rested from their Turkish exertions and augmented by fresh reserves, sailed from Alexandria to Marseilles in the early European spring of 1916. They left behind the 25 000 men of the Light Horse, who would take the brunt of the Turkish attack on Suez and then push forward into the Holy Land to win glory in the last great cavalry battles of military history. The infantry divisions travelled from Marseilles, up the Rhone Valley, past Paris and on into Flanders, to the northern section of the Western Front, barely 20 miles from the English Channel. Here they exchanged their slouch hats for steel helmets and

Exchanging slouch hats for steel helmets; the Australians arrive at the Western Front in 1916

entered the line in early April. The war they would fight for the next thirty months was on a scale beyond anything at Gallipoli—the fortifications massive and elaborate, the soldiers counted in armies rather than brigades, their instruments of destruction more varied and lethal. The material and human resources of the most powerful industrial economies were mobilized to bear on a long snake of land just a few miles wide, twisting from Belgium down to Switzerland. Once Australia was drawn into the grip of this serpent, its demands would press the country to breaking-point.

Just seven weeks of fighting round the village of Pozières in the July and August of 1916 killed more Australians than had died in battle since white settlement. The Fifth Division suffered 5000 casualties in little more than a day, then the First Division took up the attack and lost men on a similar scale, and so on through the AIF. As a contingent came out of the line, an observer wrote, its survivors 'looked like men who had been in Hell ... drawn and haggard and so dazed they appeared to be walking in a dream and their eyes looked glassy and starey'. One survivor abandoned his attempt to relate what had happened and wrote simply, 'Mother, it's quite indescribable, it was just awful.' There were nineteen such attacks in all during this, the Allies' summer offensive, and they followed a pattern that would become all too familiar. First came a bombardment of the enemy position which might continue for days, usually turning the terrain into a morass and certainly destroying any element of surprise, but leaving the strong points intact; the infantrymen therefore suffered heavy casualties as they moved forward and even when they reached the German line, were pounded in turn by enemy artillery and counterattacked by reserves. At most they would win just a few hundred yards. And that was when the planning was accurate. One brigade of the Second Division was launched so far from the front that it had to begin its forward movement before the covering barrage commenced and consequently was mowed down by machine-gun fire while caught up in the uncut entanglements of barbed wire. 'You're not fighting Bashi-Bazouks now', the British commander-in-chief told the Australians in response to this blunder. Another operation, conducted at Bullecourt in the spring of the following year, took place only on the insistence of the British general staff over the objections of the Australian commanders and required the Fourth Brigade to advance across snow in mass formation towards uncut wire—of 3000 men, 2339 were killed, wounded or captured. Fromelles, Pozières and Mouquet Farm; then Bullecourt, Messines, Passchendaele, Menin Road, Polygon Wood and Broondseinde—these names entered the Australian consciousness in 1916 and 1917 as places of death.[22]

Such squandering of human life called for new quantities of soldiers. A second Anzac corps had been formed at the beginning of 1916 and by the end of the year there were 150 000 Australians serving on the Western Front. Such was the casualty rate, however, that more replacements were needed than were coming forward at the recruiting centres. The enlistment rate fell back after the early enthusiasms of Gallipoli to less than 10 000 per month by the end of 1915; it picked up again with a recruiting drive in the early months of 1916, then declined to a little more than 6000 in July, barely a third of requirements.[23] To press the army's needs, the Universal Service League stepped up its campaign for the

Wounded Australian soldiers return down the Menin Road, September 1917

compulsory participation in the war effort of all men eligible to serve. While the League's origins lay in the Round Table Movement and it drew funds and organizers from professional and business circles, support cut across party lines so that in Sydney it attracted the endorsement of the Labor premier, W.A. Holman, and both the Anglican and Roman Catholic archbishops. Nationally, its influence was considerable. Yet the hostility in the labour movement to conscription was marked and there was little chance that the federal Labor government would defy it while Andrew Fisher remained prime minister. Even when Fisher made way for Billy Hughes in October 1915, the opposition to conscription in Labor ranks was too powerful for him to confront directly. On his urging, the Universal Service League suspended its activities and held itself in readiness.[24]

Both in his raids on company offices and his free use of the power given to him by the War Precautions Act to make regulations 'for securing the public safety and defence of the Commonwealth', the new prime minister made it clear that the needs of the nation at war took precedence over its citizens' individual liberties. Sir Ronald Munro Ferguson, the governor-general, commented of this impatient, unscrupulous, irascible 'little man' with whom he would form such a close association that 'He probably considers himself a Socialist and in respect [of] his belief in Government interference with the individual he possibly has some claim to the name.'[25] Even this judgement needs to be qualified. Those planks in Labor's programme that Hughes judged inessential for war purposes or likely to divert the country from its purpose, he simply set aside. Thus within a week of taking office he abandoned a planned referendum to control prices—the very power for which he had fought so hard in 1911 and 1913—after five of the premiers agreed to give the Commonwealth that power for the duration of hostilities. (Only one of the premiers redeemed the pledge.) The effect of the war was to augment the nationalist orientation of Hughes's collectivist creed at the expense of any vestigial concern for the class from which he sprang. All sections of society, he insisted, had to stand together to defend the homeland and advance Australia's special interests; there could be no room for internal dissension. Hughes held the portfolios of both prime minister and attorney-general from the end of 1915. With the assistance of his solicitor-general, R.R. Garran, he used special wartime powers to conduct much of the business of government simply by decree, proclaiming whatever new regulations were necessary to carry his will. As he put it, 'the best way to govern Australia was to have Sir Robert Garran at his elbow, with a fountain pen and a blank sheet of paper, and the War Precautions Act'. Garran wrote afterwards that 'To all intents and purposes Magna Carta was suspended and he and I had full and unquestionable power over the liberties of every subject.' Such was the government's control of the press, to take just one of its powers, that when the First Division lost half its strength in the disastrous engagement of 19 July 1916, Australian readers were informed simply that 200 Germans had been captured. On one occasion Garran was asked by another official whether it would be an offence under the War Precautions Act to—. He got no further. Without waiting for the rest of the question, Garran replied 'Yes'.[26]

The prime minister left for England in January 1916 at the request of the British government. It was a triumphant return for the 'nervous, white-faced little wretch' who had sailed from London thirty-two years earlier: his fiery speeches were taken up by the British press, he was lionized in social circles and in the Cabinet meetings that he attended by invitation his forceful contributions struck a welcome contrast to the indolence of Asquith, the British prime minister.[27] The Australian insisted that a greater effort was necessary, that the defeat of the enemy demanded a sterner resolution and the complete mobilization of available manpower and productive capacity. Before he left London, Britain adopted military conscription.

Hughes returned to Australia at the end of July 1916, convinced that the same measure must be introduced in his country. But how? To simply declare it would be unconstitutional, according to confidential advice from the chief justice. To introduce it by act of parliament, as the opposition urged, would divide his party and bring down his government. He therefore decided to appeal over the head of the Labor Party directly to the people. This strategy would still endanger party unity but it might be possible by careful manoeuvre to carry the parliamentarians thus far and then secure the party's compliance by force of popular mandate. Throughout much of August the prime minister was therefore locked up in cabinet and caucus meetings, culminating in what he described to the governor-general as 'a week of civil war in the Caucus'. By the narrowest of majorities, his referendum proposal was accepted and announced to parliament. But at this point the plan failed. Hughes approached the Victorian Political Labor Council for support on 1 September and was rebuffed. Pausing only to instruct the censor to suppress this news, he hurried to Sydney to repeat his request—the prudent Victorians sent their own envoy. After a stormy session of the New South Wales Political Labor League adjourned at midnight, Hughes sought to resume the meeting in the small hours of the morning and sent for four supporters to make up a quorum. Unfortunately, they brought his opponents back with them. Here too the prime minister left empty-handed and on 15 September 1916 the New South Wales executive expelled him from the party.[28]

The unrest in labour ranks was caused chiefly by the hardship the war had brought to wage-earners. Prices rose rapidly while wages were pegged, so that real wages fell by 10 per cent between 1914 and 1915. Hughes's abandonment of the prices referendum was therefore received with great indignation as a sell-out to 'the Trusts and Combines, the food-monopolists, the sweaters of industry, and the whole vile crowd of Capitalistic Huns'; and even though he later used the War Precautions Act to control the prices of essential foodstuffs, the damage was done. The year 1916 saw a rash of industrial disputes among workers in key industries—the waterside workers, the metal miners, the coal-miners and the shearers. In all, 1.7 million days were lost.[29]

'We are living in the midst of a furious coal strike', the wife of the governor of Victoria wrote back to England, 'and it is having an appalling effect on all industry. We are always expecting some blazing up of rioting to take place. I am quite prepared for a flight to Varennes.'[30] Now Lady Stanley was no Marie Antoinette and the vice-regal summer residence at Mount Macedon was hardly Varennes, yet the view from Government House was not altogether fanciful. A change had come over the workforce. The new militancy was such that where union officials sought to keep men working, the officials were bypassed in favour of rank-and-file delegates. The pressure upon their livelihood and their inability to obtain redress through traditional means caused these workers to reappraise established methods. There was a marked impatience with the limitations of arbitration and political action generally, since the Labor politicians who held office federally and in a majority of the states offered no remedy for inflation or unemployment. Under war conditions it was more difficult to maintain the balance between their working-class base and the wider class coalition they sought to maintain, harder to realize the aspirations of wage-earners within an institutional structure of industrial regulation and social reform. In Tasmania, where Labor was voted out of office in 1916, it was stated that the industrial and political wings of the movement seemed to be moving 'in opposite directions', while in Queensland, where Labor had only just come to power, the Trades Hall told the temporizing state caucus that it was not worth 'a pinch of manure'.[31] The apparent volte-face of the labour movement over the issue of conscription—after all, Labor had presided over the introduction of compulsory domestic training just a few years earlier—is to be understood in this context. When a national congress of unions condemned conscription for overseas service in May 1916, it warned that it would be 'an instrument of working class subjugation'. Citing the 'suppression of speech and press, imprisonment of workmen, and unpunished brigandage on the part of the predatory interests', the unions predicted that conscription would mean 'control of the lives and working conditions of the wage earners by the bugle and the drawn sword of militarism'. As if to confirm their fears, the military raided the Melbourne Trades Hall to seize copies of this manifesto. Under such circumstances the membership of the IWW increased to perhaps 2000 by late 1916 and its newspaper *Direct Action* achieved a wide readership. Not all the readers shared the Wobblies' intransigent hostility, but where previously they had expected a Labor government to use the state as an instrument of benevolent paternalism, they had now become suspicious of its coercive authority.[32]

Hughes could still find considerable support for conscription in the labour movement and on the other side of the political spectrum opinion was almost unanimous. A torrent of conscription material produced by a National Referendum Council was distributed through schools, business houses, churches and the machinery of local government. So confident was the prime minister that the people would vote for conscription on 28 October 1916 that four weeks earlier the government began calling up for home service all unmarried men between the ages of twenty-one and thirty-five. Even so, little was left to chance. Hughes met the editors of the metropolitan dailies and agreed with them that 'it would be well to shut down on all resolutions of anti-conscriptionist meetings, as they were generally violent and mostly intended to do harm'. Anti-conscription literature was heavily censored and meetings broken up by men in uniform; few prominent opponents of conscription spoke without a detective taking shorthand notes, and one slip could bring a fine or imprisonment. With precise timing, the government raided the Sydney headquarters of the IWW in late September and charged twelve leading members with treason and arson; twelve more were arrested in Western Australia shortly before polling day. The commander of the AIF was pressed to appeal for a 'Yes' vote from the troops and in Western Australia an All British Association was permitted to inspect government files on naturalized persons in order to identify citizens now disenfranchised because of enemy origin. Finally, on the very eve of the poll, Hughes used the War Precautions Act to instruct returning officers to interrogate male voters of military age and establish if they had obeyed the Defence Act proclamation and entered camp. Here he overreached himself. Three members of the cabinet resigned in protest and when the prime minister sought to keep that fact from the voters, he discovered that one of them had used his power to pass the resignations through censorship an hour or two earlier. The instruction was withdrawn.[33]

Other excesses of the conscription campaign rebounded against their instigators. Protestant bigots who impugned the loyalty of Catholics revived the sectarian division. Hitherto Catholics had been fully involved in the war effort, and even Britain's brutal suppression of the rising in Dublin in the Easter of 1916 did not at first alienate Australians of Irish descent (Archbishop Carr of Melbourne condemned the rebellion as 'an outburst of madness', Kelly of Sydney agreed that it was 'anti-patriotic, irrational and wickedly irreligious'); but these reflections on their patriotism did. The prime minister's play on threats to Australian security were countered by allegations that conscripts were to be replaced at home by cheap alien labour—it was particularly unfortunate for the government that a boatload of Maltese immigrants arrived at the height of the campaign. Doubts cast on the virility of men who stayed out of uniform were answered by exaggerated reports of the increase of venereal disease among the AIF and its effect on the purity of Australian womanhood: 'The Red Plague'—the author of *The Curse of Cairo* employed a conventional euphemism—'has eaten into the very bones of the nation.' Conscriptionists who celebrated filial duty drew from their opponents an equally emotive appeal to mothers:

> They put a dagger into my grasp,
> It seemed but a pencil then

Even the feminist Women's Peace Army relied on a mixture of eugenics and conventional morality in its anti-conscription propaganda. Both sides were trapped within the same divisive antinomies.[34]

All across the country friends fell out and neighbours turned on each other. From Western Australia William Somerville wrote to George Pearce, his comrade in the labour movement for fifteen years and now minister for defence, accusing him of 'blowing the Labor Party to shreds'. He never spoke to Pearce again. In the northern Queensland town of Cloncurry a young man was proceeding down the main street on 28 October when a big fellow lunged out of a laneway and knocked another pedestrian flat with a stick.

> 'What did you hit him like that for?'
> 'What did I hit him like that for? Don't be a moron, sonny, he's a bloody conscriptionist.'

So the country went to the polls.[35]

'We have lost by a head! Ah! that head. How little yet how much.' Thus the prime minister wrote to England a few days after the referendum. It was indeed a close result. Asked to give the federal government the same power to require military service outside the Commonwealth as it already possessed within, 1 160 033 voters said 'No' and 1 087 557 'Yes'. Analysis of the voting suggests that women, the British-born and primary producers were more likely to favour conscription, wage-earners and Catholics to oppose it. Among the troops there was a narrow majority in favour, though informed observers believed that the men at the front were actually opposed to conscription and that the 'Yes' majority came from the Light Horse in Egypt and troops training in England.[36]

In the aftermath of the referendum, the Labor organizations in all but one of the states expelled the parliamentarians who had campaigned with the prime minister. Holman, the premier of New South Wales, was one of them; he joined with his former opponents of the Liberal Party to construct a National ministry. Crawford Vaughan, the premier of South Australia, was another who crossed the floor, but he was soon dumped by the triumphant conservatives. John Earle, who had been the first Labor premier of Tasmania, abandoned both the leadership and the party, as did John Scaddan who had been premier of Western Australia from 1911 to 1916. Labor lost these and others of its founders, but except in Western Australia the resolve was firm, as an interstate conference on 4 December 1916 confirmed; even the Western Australian branch of the party cleared out its conscriptionists in the early months of the new year. The federal caucus met on 14 November. Hughes, in the chair, came under immediate attack and, calling on his supporters to follow him, he walked out. Twenty-three answered his call.[37] From them he constructed an interim ministry while negotiating the terms of a merger with some of his erstwhile conservative opponents. Joseph Cook, the Liberal leader, and the indomitable John Forrest both coveted the premiership for themselves but as one Liberal intermediary discovered, 'The little devil won't listen to anything I have to say.' Hughes therefore kept the leadership in the Nationalist or 'Win-the-War' Party that was formed at the beginning of 1917, and when bribery failed to buy him a majority in the Senate, he called an election.[38] Neutralizing the conscription issue by assuring the voters that he accepted their verdict, he won control of both Houses.

8

A NATION DIVIDED

FROM THE BEGINNING of 1917 to the end of the war and some time beyond, the government of Australia was conducted in an atmosphere of almost perpetual crisis. Freed from the constraints of caucus, Hughes seemed a law unto himself—exploiting his special executive powers to the full, dramatizing issues, vilifying opponents and cutting across the normal lines of administration. The governor-general likened his prime minister to 'a jackdaw who pounces on everything and secretes it in his own nest', and even this warm admirer regretted Hughes's propensity to summon meetings of the Executive Council 'at any centre convenient to himself to pass highly controversial measures which had not the sanction of either Cabinet, or Parliament, or Governor-General'.[1]

But Hughes's apparent strength was deceptive, his supremacy precarious. The Labor Party, though severely reduced in parliamentary strength after the 1917 election with just 22 seats in the House of Representatives, had retained most working-class votes and 44 per cent of the poll. In the National Party that was formed after the election, Hughes's 12 ex-Labor members of the House of Representatives were therefore heavily outnumbered by 40 previous members of the Liberal Party. They needed the little firebrand, for none of them could match his energy and popular appeal, but the more conservative accepted him only with deep misgiving. They had criticized from the beginning the dubious expediency of his resort to referendum for the introduction of conscription; they were angered by his election undertaking not to proceed except by another referendum, and some made it clear that they did not accept such a restriction. As members of the 'Win-the-War' Party, they wanted strong government but strong government in narrow channels. Where Hughes used the wartime emergency to increase economic controls and extend public enterprises, they looked on state intervention as a temporary necessity at best and deprecated excessive public expenditure. Hughes could command their acceptance only as long as the emergency continued.

This uneasy relationship aggravated the prime minister's habits of secrecy and duplicity, and fostered his reliance on methods of political brinkmanship. To carry his party with him it was necessary to continue the polarization of political life—to magnify the dangers that beset the nation and the tasks before it—in order to maintain the fevered pitch of excitement on which he throve. Not that he found it hard to vilify his opponents: speaking at the Ipswich Town Hall in November 1917, Hughes denounced the 'insidious campaign to undermine the war spirit of the country'. In the United States, he noted, 'they had a very short way of dealing with men who carried on such treacherous work. There was the white wall—and no more.' Throughout this period the prime minister carried a revolver. For their part, Labor activists fixed their hatred and fury on the 'rats' who had betrayed the movement and, above all, on 'Judas' Hughes whose vindictive language and autocratic style made it easy to believe that he was bent on the destruction of their liberties and living standards.[2] This was not so. On conscription, as we have seen, he was exceeded in his own cabinet by those who wished to bypass the referendum procedure. On industrial questions, similarly, he showed by his settlement of strikes first at Broken Hill and then on the coalfields that he was prepared to concede workers' demands in order to keep essential war industries running. The bitter wartime divisions were caused not by the prime minister's inflexibility but by the unprincipled opportunism with which he exploited his powers.

To augment those powers, the government fixed on both real and imaginary threats. Following the Holman government's arrest of the twelve IWW members in Sydney and their sentencing, in December 1916, to terms ranging up to fifteen years' hard labour (reduced on appeal to ten), a series of further outrages was alleged—arson conspiracies, campaigns of sabotage in the naval dockyards, even a plot to attack Parliament House in Melbourne. Hughes rushed an Unlawful Associations Act onto the statute book at the end of 1916, whereby his government could declare such organizations illegal; an amendment in 1917 made membership an offence punishable by six months' imprisonment and attendance at meetings prima facie evidence of membership. By the end of the year over a hundred members had been imprisoned, ten deported and the IWW was effectively suppressed. Yet the various surveillance organizations that had sprung up during the war—the Counter-Espionage Bureau, later the Special Intelligence Bureau, Naval Intelligence and Military Intelligence, not to mention the special branches of the state police forces—found no lack of work. With information gleaned from detectives, informers and the inspection of mail, they compiled their reports on dissidents and sent them on to the prosecutors. Few critics, not even members of parliament, escaped prosecution for offences as vague as making a statement prejudicial to recruiting.[3]

By this time, when accusations of disloyalty could be directed against any obstruction of the war effort, an industrial dispute that began in the workshop of the New South Wales railways became a massive trial of strength between patriots and unions. The immediate issue was the introduction of a time-card system, which the craftsmen in the workshop suspected of being a prelude to speed-up methods; the underlying sentiment was identified by the *Worker* as a 'revolt against government tyranny'. The stoppage began early in August 1917 and spread quickly through the state's transport system to the coal-mines, the ports and other workplaces where unionists refused to handle 'black' cargo. In all 100 000 wage-earners went out and remained out for up to eleven weeks. There had been nothing like it since 1890—4 million days were lost and the direct cost to the workers in wages foregone was £2.5 million. For Hughes and the Nationalist acting-premier of New South Wales the explanation was simple: sinister forces of extremism were engaged in 'a great conspiracy' to 'prevent Australia rendering further assistance to Great Britain'. Speaking at the Sydney Domain to a crowd of 40 000, a member of the Labor Council answered them: 'Well, I'm here to say that we too are the win-the-war party, and the war we are waging is the great class war.' This particular battle the workers lost and the cost was heavy. Their resistance was broken in Sydney by the deregistration of their unions, the arrest of their leaders, the swearing in of special constables and the recruitment of volunteer labour from the university as well as farmers' organizations—the billeting of rural strike-breakers at the Sydney Cricket Ground caused impotent strikers to dub it the Scabs Collecting Ground. Hughes also enrolled strike-breakers to keep the wharves open and gazetted new regulations under the War Precautions Act that made it an offence to interfere with the movement of shipping, and withdrew preference from the Waterside Workers' Federation, the very union that had served as his stepping-stone to political fame. For their part, the employers 'put the boots in' with a vengeance, as the Labor Council secretary put it, by blacklisting activists, refusing to restore strikers to their old positions and maintaining 'loyalist' strike-breakers in key posts. By October those workers who could obtain their old jobs had accepted defeat.[4]

Clearly, war conditions called the conventional strategy of the labour movement into question. The government had, in effect, redefined the code of industrial relations and deprived the unions of their principal resource, the right to withdraw their labour collectively and peacefully. In so doing, it risked diverting popular discontent into more precipitous channels. The methods of the men in the great strike of 1917 can be instructively compared with the activities of the women. During this period a section of the women's anti-conscriptionist organizations led by Adela Pankhurst, a younger daughter of the English suffragette family, turned their attention to the hardship caused by wartime inflation:

> Men and women and boys and girls are in want of bread and butter ... It is sinful to waste, to fill storehouses with meat and wool, and warehouses with cloth and boots, while human stomachs are empty and human bodies want clothes.

Refused permission to speak to the prime minister about food prices, the Women's Peace Army organized demonstrations outside the Commonwealth parliament in Melbourne. Pankhurst and others were arrested but released on bail. She brought to the traditional methods of collective protest a new insight into the susceptibilities of the authorities: 'Whilst the people are suffering they do not seem to care a bit, you touch their pockets and you will immediately begin to get something.' A massive torchlight procession on the night of 19 September, when turned away from Spring Street, moved downhill to the business district where the demonstrators began smashing the windows of shops and offices. Damage exceeded £5000. Though emergency regulations and large contingents of special constables quelled the disorder, the fragility of civil order was clear.[5]

The other war remained as costly as ever. In central Europe the Russian army collapsed and with it the Tsarist regime; to the south both Italy and the Austro-Hungarian empire were tottering, while on the Western Front the combatants battered each other to a standstill. Australian casualties amounted to 38 000 in the European autumn of 1917 against less than 10 000 enlistments. Under heavy pressure from the right wing of his party, the prime minister announced on 12 November 1917 that there would be a new referendum at which the government would propose to make up the deficiency in recruiting by a ballot among the eligible single men. 'I tell you plainly,' he added, 'that the Government must have this power; it cannot govern the Country without it, and will not attempt to do so.'[6]

The stakes were higher in this the second conscription referendum, the antagonisms even more fierce. Hughes's strategy was to discredit his opponents as traitors: 'the forces arrayed behind the campaign against the Government's proposals could be divided into three sections—the Germans of Australia, the Sinn Fein and the IWW.'[7] Since Australians of German descent were by this time either interned or forbidden to hold meetings, and most were in any case denied a vote, it is difficult to see how they affected the outcome of the referendum. The reference to Sinn Fein was directed against the Catholics and especially Dr Mannix, now archbishop of Melbourne, who since his reference earlier in the year to the 'sordid trade war'—or had he said 'ordinary trade war'?—had become a lightning-rod for sectarian hatred. The Irishman, however, was as uninhibited a polemicist as the prime minister. What else did Sinn Fein mean, he asked, than national self-sufficiency? And was it not the case for all true Australian patriots as well, that 'Australia is first, and the Empire is second'? Why then should they condemn Irishmen who felt similarly? So wounding was his scornful invective that Hughes requested the Vatican, via the Foreign Office, to restrain its turbulent priest; and eventually Rome did send a reminder to Raheen that 'the office of a Pastor is to pacify souls, to allay discords and prevent their arising or becoming embittered'. This and other rebukes had little effect. Mannix continued to address enormous crowds and to rally Catholics to the anti-conscriptionist cause.[8]

The third element in the unholy trinity, the IWW, was a label Hughes pinned on even the most moderate of his former Labor colleagues. As during the previous campaign, their spokesmen were prosecuted and their publications censored, but there remained a Labor government in Queensland in which Premier T.J. Ryan, and his deputy, E.G. Theodore, were leading anti-conscriptionists. On 19 December Ryan made a speech in Brisbane from which passages were deleted by the Commonwealth censor. Ryan then arranged a debate in the state parliament and read the offending passages into *Hansard*, while Theodore, who took the opportunity to add two anti-conscription pamphlets that had been suppressed, ordered the Government Printer to publish 10 000 copies of this material as a special issue. Upon hearing of this, Hughes ordered that all copies be seized and while in Brisbane led a military contingent to the printing office. The Queensland government eventually retrieved its copies of *Hansard* and for good measure distributed 50 000 copies of a *Government Gazette Extraordinary* with an account of the events, but not before an egg-besplattered Hughes became involved in a mêlée on the Warwick railway station. Out of that incident he created the Commonwealth Police Force to uphold law and order 'in any State which refuses or does not enforce Commonwealth law'. 'The Warwick incident has done much good,' he told the governor-general. 'Everywhere I have had splendid meetings: there's going to be a *great* fight. Glory be to God for that.' He could not have been disappointed with his eve-of-poll meeting at the Melbourne Cricket Ground where 100 000 people and five bands gathered round three platforms. At the platform erected before the members' stand he obtained a hearing, but at the Richmond end the air was thick with eggs, stones and bottles, and the prime minister did not even attempt to speak.[9]

The referendum proposal failed by a wider margin than that of the previous year, the 'No' vote increasing by 21 000, the 'Yes' vote declining by 72 000. Only two states, Tasmania and Western Australia, voted 'Yes'. Among soldiers, who had been presented with a pro-conscription appeal entitled *All for Australia*, there was again a slender margin in favour; but, as before, observers agreed that the men at the front were opposed. Writing back from a military hospital, a former organizer of the Universal Service League explained that he had 'hardly met a man yet who will vote for conscription. They allege all sorts of queer reasons, but the commonest by far is that they would not ask their worst enemy to go to France—France is Hell.'[10]

So Hughes had again failed to obtain the power that he insisted was indispensable. What, then, of the undertaking that had been forced from him by his conservative colleagues, that he would not attempt to govern the country without it? He did not himself see any insuperable difficulty. After all, his personal secretary reasoned, 'Democracy by quitting the fight in Australia has done something immoral and must not in itself constitute a standard of measurement for what follows.' His colleagues took a sterner view. So after some days of equivocation, he submitted his resignation to the governor-general and a succession of taxi-cabs then deposited their hopeful passengers at Government House. Frank Tudor, the leader of the Labor Party, and bluff old Sir John Forrest each wanted to form an administration, but the governor-general countenanced neither: the ageing Forrest represented only one section of the Nationalist right wing; Labor was unthinkable. Consequently, before parliament could even consider the matter, Hughes and the old ministry were sworn in afresh.[11] There was then one final attempt to replenish the depleted ranks of the AIF. The governor-general convened a recruiting conference in Melbourne during April 1918, to which he invited federal and state ministers, Labor leaders and representatives of the employers and unions. It was a forlorn attempt. Voluntarism was dead, killed by the excesses of the past eighteen months. Labor was moving from criticism of the conduct of the war to criticism of the war itself, and would shortly embrace a policy of negotiated peace. Its representatives treated Hughes with open hostility and when he began to read a cable from Britain, they interjected, 'We've had enough of your forgeries.' In a last attempt to persuade them to join in an appeal for a whole-hearted effort to secure the necessary reinforcements under the voluntary system, the federal government undertook to review the operation of the War Precautions Act and release those imprisoned during the 1917 strike and referendum campaign. Labor bodies scornfully repudiated such an agreement. Civil liberties were not to be offered as 'bribes for lives', said the New South Wales Labor Council. Even the governor-general welcomed Hughes's departure shortly after, so that he could attend the Imperial War Cabinet, as a welcome respite from the constant jarring of nerves.[12]

'Never before have we faced so critical a moment in our history.' In thus opening his ill-fated conference, the governor-general could hardly exaggerate the gravity of the military situation. The Germans had launched a desperate last effort some weeks earlier in northern France and broken the Allies' line. In those grim spring days, the fate of the war hung in the balance. Once again the AIF was involved in heavy fighting. 'Fini retreat—beaucoup Australiens ici', one Digger boasted to a villager as his contingent prepared to relieve a shattered British division in the path of the German advance. And hold them they did. The Australian Corps—for in the previous November they had achieved their ambition to become a single unit—was stretched thin, so thin that three of its battalions were broken up to reinforce the remainder, but they met the challenge. By this time they constituted nearly a tenth of the British and dominion soldiers on the Western Front; their confidence in their ability was high and justifiably so since they were employed wherever and whenever an extra effort was needed. Under the leadership of Sir John

Monash, who was appointed in May to command the corps, the Australians stood firm and played a leading part in the Allied counter-offensive that began in July and then, with increasing momentum, pushed up to and beyond the Hindenburg line. Twenty-three per cent of the prisoners taken in that victorious thrust, 23 per cent of the guns that were captured and 21 per cent of the territory that was reconquered fell to the Australians. Monash, that painstaking engineer-turned-soldier, revealed rare qualities of leadership in his orchestration of a vast and complex military machine. 'A perfected battle plan', he wrote, 'is like nothing so much as a score for musical composition'.[13] The success of the Australians in this theatre was as much due to their extensive experience in the management of large-scale public enterprises as it was to the celebrated qualities of courage and individual initiative.

Elsewhere the hallmarks of the bush legend were more in evidence. During 1917 the men of the Light Horse had fought their way from Gaza to Jerusalem as part of the force commanded by the British general Allenby. In 1918, with a series of audacious advances, they pressed on through Palestine and into Syria. Chauvel, the senior Australian, led the Mounted Corps and his compatriots wore the emu-plumed hats that had their origin in the shearers' strike of 1891, when the men he commanded in the Darling Downs Mounted Infantry had, for variation, chased emus instead of strikers. These cavalrymen were a high-handed lot. They had put down revolts among Arab tribes, they were similarly employed in crushing an Egyptian nationalist uprising in 1919, and in between they committed the Surafend atrocity, an indiscriminate vengeance meted out to the inhabitants of a Palestinian village, killing or wounding thirty of them, after a nocturnal thief shot a New Zealander.[14]

The 'digger', as the Australian soldier had become known by the end of the war, was a citizen soldier. This fact determined many of his virtues and vices and, in turn, meant that the values formed in war would persist in peace. He did not respond easily to the discipline that was instilled in regular troops and extended to the conscript armies of other combatants in the First World War; he showed a marked reluctance to salute and an officer had to earn his respect. Protected from the normal provisions of martial law and immune from the death penalty except for mutiny, desertion or traitorous conduct, he was more fractious than the British Tommy. Of 182 men in the Fourth Army sentenced in December 1916 for being absent without leave, 130 were Australians; in March 1918, 9 Australians per 1000 were in field prisons as against 2 per 1000 among British units. He was involved in ugly scenes around the pubs. In this respect his behaviour in the company of his mates was rather like that of a football team on an end-of-season trip. And while they could not get used to women drinking in pubs or the failure of Englishmen to offer their seats to women on public transport, their own 'catting' resulted in a remarkably high incidence of venereal disease, one in seven receiving treatment.[15] On the other hand, the digger exhibited an exceptional degree of cameraderie. Among members of the AIF there were few barriers—as a batman told a new chaplain who asked about the religion of the other chaplain attached to the brigade, 'there ain't no religion out here, sir, we're all brothers'—and the war experience created bonds that long afterwards marked the men who went out from those who didn't. The statistics attest to their bravery. Of the 330 000 Australians who put on uniform during the war, and they constituted two-fifths of all men of military age, 59 000 died and 167 000 were wounded. The digger was bombastic and self-aggrandizing, given to contempt for the British officer and intolerant of others with whom war threw him into contact, yet not unresponsive to flattery. When the French premier Clemenceau visited the Australian troops, he received their cheers and remarked as he turned to leave, 'Des jolis enfants'.[16]

For what had they fought? Although the call to arms back in 1914 had been answered by an almost reflexive expression of imperial loyalty, four years of sacrifices and the arguments they engendered forced a clarification of objectives. Hughes, as the leader of the 'Win-the-War' Party and through his participation in the Imperial War Cabinet and Peace Conference, was able to define the Australian war aims almost single-handedly. In London and Paris for fully twelve months during 1918 and 1919, he ignored cables from his cabinet in Melbourne as easily as he failed to hear the objections of adversaries in the conference chamber. In his narrow obstinacy, his refusal to heed advice and his abrasive larrikinism, he was, as one of his party said, 'a perfect example of how not to behave'. But his authority was real. Asked by the American president Wilson to explain the basis on which a nation of 5 million could set itself against the 1200 million represented at the conference, Hughes is said to have replied that he represented 60 000 dead. No one other than the 'Little Digger' could have made such a claim—better than anyone else, he did represent the hopes and fears and prejudices of those who had fought the war. Through them, he told the Commonwealth parliament on his return from Paris, 'Australia became a nation ... We had earned that, or, rather, our soldiers had earned it for us.'[17]

Nationality here consisted in the assertion of distinct national interests. Throughout the peace negotiations, Hughes pressed what he regarded as Australia's legitimate demands and he pressed them with scant regard for their wider implications. Mistrusting the pious intention that the League of Nations should 'prevent future wars by establishing relations on the basis of justice and honour', he sought safeguards for Australian security, for trade restrictions and for the White Australia Policy—and to this end he blocked the Japanese proposal of a declaration of racial equality. Wanting his pound of flesh, he sought heavy war reparations and a fair share of those reparations as compensation for Australia's war bill of £300 million—in the event Australia received just £5 million.[18]

'The natural destiny of the Pacific Islands is that they should come under the control of Australia.' So wrote Sir James Burns, head of the trading firm of Burns Philp, in response to the governor-general's request for an opinion on the future of the region after the dispossession of the Germans. He and other Australians had long coveted the copra plantations and trade opportunities of Germany's Pacific possessions, which stretched northwards from New Guinea to the Marshall and Caroline Islands. Economic advantage and strategic anxiety alike convinced Australia that the islands should be theirs (except for Samoa, which was marked for

New Zealand). Some they had occupied already. Those north of the equator had been occupied by Japan, acting, they were told, on a purely temporary basis on behalf of Britain. The governor-general knew what they did not—the Colonial Office had in fact instructed him to prepare his ministers for the unpalatable news: Japan meant to hold these northern islands for itself. Fisher was the first to hear of this; Hughes probably did not learn it until his visit to London in 1916 and he kept the secret since it would have been dynamite in the hands of anti-conscriptionists. And while he protested two years later in meetings of the Imperial War Cabinet against the threat of Japanese expansion, his hosts gave him no comfort so he had to accept the *fait accompli*.[19]

There was a greater threat by this time, in any case. Following the entry of the United States into the war, President Wilson's Fourteen Points became the basis of the armistice. The spirit of these points was against annexations and in favour of international co-operation; the fifth of them stated that in all claims to colonies, 'the interests of the populations concerned must have equal weight with the equitable claims of the government whose title is to be considered'. If Wilson's view prevailed, there could be no guarantee that Australia's claim to any of the Pacific islands would prevail, nor could there be the absolute sovereignty and exclusive possession that Australia sought since such colonies would be held in trust on behalf of the League of Nations. Hughes's objection was that if Australia held German New Guinea as a mandatory territory only, the principle of the 'open door' would prevent her from excluding the Japanese. Small wonder, then, that the Australian fulminated against the American's glacial determination. The problem remained intractable until the Australian and British advisers hit upon the device of the 'C class mandate', which allowed Australia to administer the islands under Australian law subject to the weak control of the League of Nations. That sufficed. The Commonwealth was able to establish an Expropriation Board designed for 'kicking out the Hun'; to extend the Navigation Act so as to exclude foreign shipping; and to make 'closed door' provisions to preserve a 'White Australia'.[20]

What of the other nationality, the one to which Hughes

referred on another occasion when he reminded the members of the Commonwealth parliament that 'We are a nation ... [but] we are a nation by the grace of God and the power of the British Empire'?[21] That larger loyalty had been shaken by the revelation that an Anglo-Japanese treaty took precedence over Australian interests; it was strained further by Britain's failure to consult the dominions over the terms of the armistice. Australia's insistence on an independent voice at the Peace Conference and separate membership of the League of Nations signalled an appreciation of the fact that it would henceforth have to represent its own interests and so, reciprocally, did the adoption of Hughes's suggestion that dominion governments should be able to communicate directly with the British prime minister instead of through the viceregal representative and the Colonial Office. Deploring the devaluation of his role that the change entailed, Sir Ronald Munro Ferguson became all the more impatient with colonial sensitivities. It was necessary to take a firm line on the Pacific islands issue, he advised London, since 'when Australia is brought to face facts she is more likely to prove herself a chip off the old block than she has done under spoon feeding'. As for Australian complaints that they were denied shipping, this was the only appropriate response to referenda results showing that 'Australia declines to take her full share in the war'; in any case, 'if the Colonial Office has done nothing worse than keep freight for something better than Australian wines it will not have much to answer for at the day of Judgement'. His attitude was shared by his compatriots. British officers complained of the colonials' lack of respect, British hostesses found them gauche and boorish, British magistrates declared them a menace to public order.[22]

The Australian soldier repaid such abuse with interest. His confidence in the British army suffered an initial shock at Gallipoli from which it never fully recovered. For the ordinary Tommy in the line he had some sympathy, but for the hidebound British officer he had a savage contempt, reinforced by leave-time encounters with the class rigidities of British society. What hurt most—and it was this that betrayed an enduring symbiotic relationship—was the lack of recognition of his contribution to the fighting. The tug of emotions particularly affected those middle-class Australians, mostly officers, who had volunteered in a haze of Anglophile emotion: the more keenly they felt the imperial loyalty, the more they needed to confirm their sacrifice in the approval of the imperial masters. Yet their efforts went largely unacknowledged. The lawyer Frederic Eggleston, who before the war had joined the Round Table group in Melbourne, found that English newspapers gave all the credit for stopping Hindenburg's 1918 offensive to crack regiments like the Guards and passed over the role of the Australians. The truth of the matter, he said, was that the British had dissolved into a rabble and their generals had been 'rooted out of haystacks by Australians'. Eggleston was moved to fury by the self-deprecating reports that the Australian secretary of the Round Table sent to London:

> I remember Australia as I left it and cannot control my indignation. I remember its graceful social ideals and happy amenities of life, its almost unique chivalry for the weak in the economic sphere, the native generosity and hospitality which contrasts so favourably with the grudging and suspicious atmosphere around here. ... Why should a people with such qualifications and with such a record be so traduced?

Under such circumstances, it was hardly surprising that the Australian Round Table groups contracted and their project of closer imperial unity languished.[23]

On one side of politics there was a noticeable hardening of anti-British feeling. The fighting platform adopted by the Labor Party in 1918 included abolition of appeal from the High Court, an end to imperial honours and removal of state governors. On the other side the influx of ex-Labor men injected an assertively Australian tone into the Nationalist Party, exemplified especially by the prime minister. Whereas before the First World War nationalism had been radical in its orientation, the split over conscription allowed a conservative alternative to emerge, one more popular in orientation and more strident in tone than Deakin's earlier notion of the 'independent Australian Briton'. New perils demanded more vigorous remedies. It was clear to Keith Murdoch, journalist and wirepuller for the prime minister, that 'the old feeling of subservience to England' could no longer serve as an adequate 'counterforce to Bolshevism, Sinn Feinism and all the present disuniting, anti-Australian sections'; consequently, 'the only banner under which the truly creative forces in Australia can be collected is the banner of Australianism'. Hughes practised such a nationalism in London and Paris during 1919 and again at the 1921 Imperial Conference. But the Nationalist Party was also the home of conservatives who clung to the old imperial relationship, including Hughes's cabinet colleagues who rebuked him during one of his pyrotechnic displays in London for 'hanging the British family linen on the line'. Nor was Hughes the separatist that his histrionics might suggest. He was an imperialist by sentiment and conviction, only a pragmatic one. While he believed that Australia's fortunes lay within the Empire, he realized also that the constituent parts of the Empire did not have identical interests. It was therefore necessary to promote a closer consultation between the dominions and the parent country, so that the imperial view would give weight to Australian needs, and at the same time to accept that there would be times when Australia would have to speak with an independent voice. Fearing isolation and resenting dependence, Australia occupied a yo-yo position, moving towards or away from a superior being while remaining always on a string.[24]

The diplomatic settlement, if settlement it was, formed but the prelude to a period of intense domestic turmoil. The armistice solved none of the divisions that had become apparent in the conscription referenda, indeed it widened them. Hughes and the Nationalists had to meet the pent-up frustrations of the unions, to fit the soldiers back into civilian life, to reconcile the Catholics and find new ways of reordering a society racked by unrest. Nor did they always wish to smooth over the divisions. Emergency was the midwife of the Nationalist Party, that fragile child of two unlikely parents, ex-Labor and anti-Labor. Now that post-war circumstances revealed their internal differences over questions of economic and social policy, they found it all too tempting to revert to the politics of emergency by dramatizing discontents, denouncing scapegoats and generally encouraging that atmosphere of alarm in which mutual recrimination took precedence over all else.

There was first the industrial front. The release of a backlog of demand at the end of the war brought a surge of business activity, increased employment and renewed inflation—conditions that were conducive to industrial militancy. The unions sought restoration of their old living standards as well as a reduction of the working week to forty-four hours, and in many industries they were successful. But the mining and maritime unions met employers who had the determination and resources to resist these demands, resulting in the most protracted and costly industrial disputes that Australia had yet known. During 1919, 6.3 million days were lost in strikes and lockouts. The mining dispute at Broken Hill alone cost £2.5 million in lost wages and lasted for eighteen months, from May 1919 to November 1920. The central issues here were wages, hours and working conditions in a notoriously unhealthy occupation, and eventually the companies did concede pay increases, shorter hours and a scheme of compensation for those who contracted mining diseases. The other leading combatants were the seamen, who presented a similarly wide-ranging log of claims to the shipowners in April 1919. While their strike lasted only three months, its impact was even more profound since it deprived many industries of their raw materials, and food rationing had to be introduced in a number of outlying regions. The seamen gained most of their demands over the protests of Mr Justice Higgins, who once again saw government intervention as undermining the principles of arbitration.[25] There followed similar victories for the marine stewards, cooks and engineers, each of whom brought shipping to a halt; indeed there was a rash of localized stoppages by workers occupying strategic positions and therefore able to disrupt whole industries without risking the hazards of a general stoppage. Hughes's habit of cutting across the Arbitration Court to punish or cajole such workers became endemic. He antagonized Higgins irrevocably with his Industrial Peace Act of 1920 which enabled the government to create special tribunals in order to settle disputes and, if necessary, to vary the awards of the Arbitration Court. The Act remained a dead-letter except on the coalfields, but Higgins, whose term as president was about to expire, announced that he would not seek reappointment because 'the public usefulness of the Court has been fatally injured'. Lest his position remained unclear, he reiterated that 'A tribunal of reason cannot do its work side by side with executive tribunals of panic.'[26]

Speaking to shareholders in June 1920 at the general meeting in London of the Zinc Corporation, which was a member of the Collins House group, the chairman and managing director recorded that the company's Broken Hill operations had been closed down for the previous thirteen months. During all that time, he said, the company had been living on its 'fat', and it was fortunate that the fat was so substantial since he could see no settlement in sight. 'No one can "settle" a dispute with an insane man who has got you by the throat.' For this was no 'mere dispute for increased wages', it was part of a 'social struggle which pervades the world', waged between the capitalist and the revolutionary socialist.[27] The doctrines of industrial militancy certainly made great strides in the immediate post-war period. Focusing on a scheme for One Big Union, 'one class-conscious economic organisation to take and hold the means of production', the militants won control of the New South Wales Labor Council and secured expressions of support from other states. Their project was defeated in the end by those in the labour movement who had most to lose from such reorganization, the Labor politicians, the leaders of the craft unions and the powerful AWU. But the militants crystallized a mood of impatience with the orderly methods of industrial negotiation, arbitration and parliamentary reform. Again and again, industrial disputes that flared up in different parts of the country ignited a combustible mixture of discontents.[28]

In Darwin the local organizer of the AWU led opposition to the Commonwealth administrator, whose transgressions ranged from inefficient management of the hotels (they had been nationalized in the top half of the Northern Territory during the war) to favouritism towards Vestey Brothers, the British owners of the local meatworks (the administrator's secretary became Vesteys' chief clerk). Resentment culminated in the 'Darwin Rebellion' on 17 December 1918, when several hundred men marched on the Administrator's Residence to force his resignation. A navel vessel had to be sent to

Darwin to take him away.[29] In Townsville a strike among meatworkers, inflamed by the importation of blackleg labour, led in 1919 to a siege of the police station and a volley of shots that injured nine men. While a trainload of police rushed north from Brisbane (causing an extension of the strike to the railways), the men seized firearms from the local shops and commenced firing practice at pigeons in the main street. With singular folly, the manager of the meatworks chose this moment to emerge from the local newspaper office—he was roughly handled. Townsville remained in a state of siege for weeks afterwards.[30] And on the other side of the country, in Fremantle, striking waterside workers figured in equally dramatic scenes. Premier Hal Colebatch chose a Sunday morning to lead police down the Swan River in order to clear pickets off the wharves, but news of his mission preceded him so that as his launch passed under the Fremantle bridge, it was bombarded with road metal and old iron. There followed a battle between police armed with bayonets and batons, and unionists who tore iron pikes off the church railings. One lumper died of his injuries. That same afternoon the people of Fremantle returned to the wharves and swept the offices of the strike-breaking organization into the harbour.[31]

The second element of instability, as volatile as the first and scarcely less powerful, were the men who returned from the war. There were 170 000 of them in the northern hemisphere when the war ended and their repatriation at a time when shipping was scarce took some time. Even so, all but 20 000 were back in Australia within a year and all but 13 000 found employment in a buoyant labour market.[32] It was as well that they did since the ex-servicemen were already vociferous in demanding financial assistance from the government and preferential treatment from employers. In some cases the question of preference brought them into disagreement with the unions, although this was by no means always the case—there was, after all, considerable overlap between the two groups. The larger difficulty was the reincorporation into a world that had changed in their absence of men whose lives had been so completely disrupted and who brought home experiences that only they could share. They had left to almost universal acclaim, they returned to find unanimity of purpose

The protest march through Brisbane, 1919

had gone. One soldier was playfully teasing a girl when she abruptly turned on him with a crushing retort: 'Well, I've never killed anyone.' Ex-servicemen sought reassurance in their own company: 'The army was not going to disband because the war had ended', declared a gathering of returned soldiers in Hobart. They lashed out at those who depreciated what they had done, be they stuffed-shirts, socialist agitators or mere do-gooders: a prominent temperance campaigner found his meetings disrupted by diggers who took offence at his denunciation of the demon drink.[33] In particular instances they went further. Enraged by a march of radicals in Brisbane who demonstrated against the continuance of the War Precautions Act and carried red flags in defiance of one of its regulations, hundreds of ex-servicemen took to the streets. Nineteen people were wounded by bayonet, bludgeon or bullet. The Brisbane *Daily Standard* deplored the violence of the ex-soldiers, who responded by smashing its windows. In

Sydney, where radicals burned the Union Jack at a May Day demonstration in 1921, a huge crowd turned out at the Sydney Domain to see the diggers sweep the Bolsheviks from the stump and burn a red flag. And a former MHR living near Ararat was tarred and feathered by returned soldiers who had taken exception to his anti-war verse.[34]

Third were the sectarian antagonisms that had been mobilized during the conscription campaign and still remained fierce. Mannix certainly remained as obdurate as ever. On St Patrick's Day, 1918, he led a procession, which included Sinn Fein banners and a float commemorating the martyrs of the Easter Rising, through the streets of Melbourne. Leading Protestants responded by forming a Citizens' Loyalist Committee and Herbert Brookes, Deakin's son-in-law and a leading industrialist, took the platform at its inaugural meeting in the town hall to denounce a Catholic plot to let Protestant boys die so that the sons of Rome could marry their sisters. 'We'll see Mannix in Hell first', he said and wrote later of the 'glorious experience to feel yourself not yourself, but an instrument in the hands of that Power that works for righteousness'. John Wren made one of his racecourses available to Mannix for his reply. Not to be outdone, the Loyalists filled the Exhibition Building to sing 'Rule Britannia', 'The British Grenadiers' and 'Scots Whae Hae Wi' Wallace Bled' in a celebration of martial ardour. Mannix could play that suit also. For the St Patrick's Day procession after the war, he paraded fourteen Catholic winners of the Victoria Cross as a guard of honour. With some relief, the government then saw him depart on a tour of Britain (though not, they ensured, of Ireland) and some Nationalists scouted the proposition that he should be denied re-entry into Australia. Even during his absence, Catholics mocked a new regulation requiring that the Union Jack be carried at the head of any public procession by employing an elderly English-born derelict to perform the task on St Patrick's Day in 1921. Yet, as one of Mannix's less belligerent co-religionists lamented, each time the archbishop spoke out Catholic workmen were sacked and Catholic children forced to run a gauntlet to and from school.[35]

Finally, the atmosphere of disorder was compounded by a pandemic of influenza that swept into Australia in January

The 'little digger', Hughes, in Sydney on his return to Australia, 1919

1919. It carried off 12 000 victims, a fifth as many as had died in battle over the previous four years. Normal activity came to a standstill. Schools, churches and meetings were closed, the Anzac Day marches planned for April 1919 had to be cancelled, and strict quarantine regulations delayed the landing of the troopships. The pandemic also revealed the hollowness of the national identity proclaimed during the war—each state closed its borders to isolate the disease and framed its own regulations, regardless of their impact on other states. More than this, the powerlessness of the authorities to control the infection except by compulsory isolation seemed to heighten the atmosphere of panic and mistrust. In New South Wales, for example, the refusal to allow Catholic chaplains to visit the dying victims in the quarantine centres provoked the usually conciliatory Archbishop Kelly to present himself at the gates of the North Head station. He was turned away.[36]

Against this background of disorder, it is possible to discern processes whereby order was restored. The Nationalists legitimated their repressive policies by the construction of a mass base of support along the 'aggressively Australian' lines that Keith Murdoch had recommended. Just as the prime minister had cultivated such a style at the Peace Conference, so, upon his return to Australia in August 1919, he seized every opportunity to identify himself and his party with the love of country and fidelity to its most sacred traditions. Wherever he went, reported the governor-general, he 'was received by enthusiastic crowds of "Diggers" ... and the welcome everywhere took the form of seizing his person, draping him in flags, bonnetting him with an Anzac's hat'. It was not a dignified procedure in the opinion of the fastidious Munro Ferguson, but it was one calculated to win the support of the returned soldier.[37] Hughes used other means: gratuities, preference in public employment and, later, an ambitious scheme to settle soldiers on the land. Soon after his return to Australia, he devoted several days to reach agreement with the Returned Soldiers' and Sailors' Imperial League of Australia (RSSILA). This organization, formed in 1916 and reaching a membership of 150 000 by the end of 1919, was ostensibly above politics, but its first president became a Nationalist senator and the close connection with that party contributed to a precipitous loss of membership over the next few years until by 1924 it represented only 9 per cent of all ex-servicemen. The crucial fact remained, however, that it did represent them. Since the RSSILA enjoyed official recognition, it had sole access to ministers on questions of benefits or pensions and it alone could make representations on matters of national security. Furthermore, the RSSILA controlled the official commemoration of Anzac Day and, after conservative state governments declared it a public holiday, was able to shape it into an expression of national fulfilment.[38] As codified by the office bearers of the RSSILA and taught to generations of schoolchildren, the Anzac legend was also consistent with imperial loyalty. Australians had proved themselves a nation of warriors and they had done so in the noblest of causes, loyalty to the home country. To strengthen the larger allegiance, a string of notables was despatched from Britain, first Admiral Jellicoe, then the general who had commanded the Anzacs, and finally, the water having been tested, the Prince of Wales. They came in quick succession for the King's secretary judged that 'the psychological moment' for these visits was as soon as possible after the war, and the itineraries kept them clear of Labor strongholds. Their success was impressive and their purpose clear: when 400 000 lined the streets of Melbourne to welcome the Prince of Wales in May 1920, Herbert Brookes rejoiced that 'Mannix and his ulcerous brood are completely smothered'.[39]

The Nationalists stood on such a platform as defenders of a healthy Australianhood, one that had to be actively defended against its enemies and purged of its disloyal elements. It seemed for a time that the government had only one response to its opponents—deportation. In the same way that it attributed labour unrest to the influence of alien agitators, it singled out foreign-born members of the IWW for expulsion under the provisions of the War Precautions Act. Still using the Act despite the fact that hostilities were over, the government sent many of the internees back to their country of origin (their families also, unless the wife had been a British subject before marriage). There was even a new law that required British subjects entering the Commonwealth, on request, to subscribe an oath of allegiance. It was no accident that the most celebrated deportee was a Roman Catholic priest, Father Jerger, whose case was taken up by Mannix. 'Jerger must go', announced the prime minister and neither a High Court action nor a seamen's strike could save him.[40] Critics were tarred with the same brush. 'The honourable Senator is a good old Hun advocate', jeered a government member when a Labor representative took up the case of the deportees. Hughes himself went even further and expelled Hugh Mahon, a former ministerial colleague, from the House of Representatives for making seditious and disloyal utterances at an Irish protest meeting. 'As Australian citizens,' said the prime minister, 'we clearly cannot allow conspiracies against the Empire.'[41]

It is evident that wartime strains completed a realignment of Australian politics. The Labor Party, having lost power at the end of 1916, would not regain it for more than a decade—and then only briefly—and would fail to command a majority in both Houses of the Commonwealth parliament until 1944. The conservatives, whose pre-war programme had been aptly described by Deakin as 'a necklace of negatives', would hold the initiative for a full quarter-century. The change is associated with the new valency of Australian nationalism. Nationalism at the turn of the century and into the early years of the Commonwealth, was a force for change, an expression of self-realization for the nation that was to be. An activist in the early labour movement explained that 'I belong to the Labor Party because it is the only Australian party there is. All others are imperialists and imperialism is the real enemy of Australia.'[42] But once the Commonwealth was established and its institutions created, nationalism was increasingly identified with the *status quo*. The war strengthened such affirmative connotations and, if only in retrospect, mediated the tug of loyalties between Australia and the home country. The patriotism of the Nationalists was contained within their loyalty to the Empire, yet was no less powerful for that fact.

Calling a federal election shortly after his return to Australia, Hughes invoked all the loyalist themes, repeatedly questioned the patriotism of the Labor Party and called on the voters to 'Let our watchword be "Australia", and, as our splendid boys have fought for it and saved it, let us all live and work for it'.[43] He was rewarded with 38 seats in the House of Representatives, with a further 11 going to representatives of farming and pastoral organizations, against 26 to Labor; and the introduction of preferential voting gave him a clean sweep of the Senate. By this time there was only one state where Labor governed, Queensland, and Hughes's victory seemed to confirm his mastery of the federal sphere, just as in the following year a gift of £25 000 from his admirers showed where his popularity lay. The only problem was to give meaning and content to the forces he had unleashed. The margin of his success, the very breadth of his support, made it difficult for Hughes to span the non-Labor interests; accordingly, his efforts to direct the course of national development became less convincing until, finally, his very success in isolating and discrediting the Labor enemy made it possible for the conservatives to get rid of him.

The task of balancing the divergent interests of the propertied classes was especially difficult in the aftermath of the war. Could Australia revert to its close pre-war relationship with British investors and buyers? Would it be possible simply to scrap the economic controls and marketing schemes for primary products? What would happen now to the heavy industry that had made such rapid strides over the past few years? Hughes's own preference was for continued government intervention to promote a prosperous self-sufficiency, and he was prepared to supplement private enterprise with government undertakings such as the Commonwealth Shipping Line that was created during the war and the Commonwealth Oil Refinery that began in 1920. In accordance with his principle that the needs of the nation took precedence over sectional interests, he was publicly as critical of the selfish profiteer who held the community to ransom as he was of the militant unionist: 'I am against profiteering and against Bolshevism, and if I had my way I would shoot them both.'[44] Such a drastic remedy was unavailable but at the 1919 federal election he revived the referendum proposal abandoned in 1915, to give the Commonwealth power over trade and commerce; and for good measure he sought also the power to nationalize monopolies. Both proposals were narrowly rejected.

There was a strong element of theatre in all this but the business community took the prime minister sufficiently seriously to rebuke his 'socialist' backsliding. For them the end of the war was an opportunity to remove controls, reduce public expenditure and make the market the means of allocating resources. Or so they said during the buoyant trade conditions of the first eighteen months of peace. But by 1920 the backlog of demand was exhausted and the international economy went into recession. Australia experienced a sharp contraction of export sales, leading to a severe shortage of overseas funds, a problem exacerbated by the fact that the Australian currency was tied to sterling.[45] Now the businessmen sang a different tune in discordant voices. Manufacturers, who had made rapid strides during the war (the 1921 census revealed that they had displaced primary producers as the principal employers of the Australian workforce) sought increased levels of protection to rescue them from cheap imports. They obtained a new schedule in 1920 as well as a Tariff Board composed of businessmen with formal powers to investigate and recommend new duties, and an Australian Industries Preservation Act whose anti-dumping provisions were interpreted freely and used liberally. Even then, the small size of the domestic market and the high cost structure of Australian industry handicapped manufacturers. BHP suspended operations at the Newcastle steelworks in 1922.[46] Exporters, on the other hand, complained of the cost effects of increased protection in a period of falling world prices, and the mercantile and financial houses deprecated the whole panoply of economic nationalism. Yet while the merchants called for the dismantling of compulsory marketing pools in order that they could regain control of this profitable trade, the primary producers sought rather to maintain publicly financed boards, only under their control and on a voluntary basis. The boards were reorganized on these lines, only to lose out to the broking houses.[47]

The recession of 1920 therefore caused widely divergent demands, from the protectionism of the manufacturers to the grower syndicalism of the primary producers and the retrenchment strategy of the traders and financiers. On one proposition, however, businessmen were in agreement: labour costs were too high and ought to be reduced. The unions had sought wage increases to offset rises in the cost of living over the previous few years; the constant disputation was an expensive method of wage determination, so before the 1919 election Hughes appointed a royal commission chaired by A.B. Piddington to consider afresh the level of the basic wage. Piddington completed his investigation at the end of 1920 and found that to meet the needs of a family with three children a weekly wage would need to be £5 16s—at least 30s more than the current minimum for adult males. Piddington's solution to this embarrassing shortfall was to suggest a minimum wage of £4 and a benefit of 12s for every dependent child, but such familial welfarism was as unacceptable to industry as the wage increase. The strategy of the employers was to drive down wages and in a period of renewed unemployment (union returns in 1921 disclosed that 11.5 per cent of their members were out of work), employees were less able to resist them. The incidence of strikes as measured by working days lost declined from 6.3 million in 1919 to less than 1.0 million in 1921, and since arbitration served as a brake on the reduction of wage standards, there were renewed calls for its abolition. The judgement of the High Court in the Engineers' case, indicating that federal arbitration could apply to state enterprises, antagonized employers further.[48]

The working-class advance was at an end, an employer

Table 8.1: Working days lost in strikes, 1913–22[49]

1913	624 000
1914	1 090 000
1915	583 000
1916	1 679 000
1917	5 000 000
1918	581 000
1919	6 308 000
1920	1 872 000
1921	957 000
1922	859 000

counter-offensive was under way. Unions fought as best they could to defend wages and working conditions. In these unpropitious circumstances the project for One Big Union lost ground, while the formation of an Australian Communist Party at the end of 1920 came too late for it to do more than gather up some of the left-wing fragments. The Labor government in Queensland, where Theodore replaced Ryan on the latter's move into federal politics, was hamstrung in 1920 by the refusal of the London money market to provide loan funds, while that elected in New South Wales in the same year soon demonstrated its moderation. A federal conference of the Labor Party in October 1921 adopted a socialist objective in the watered-down version that the parliamentarians advocated.[50] These developments rendered superfluous Hughes's value as a leader drawn from the ranks of the workers and undermined his claim to represent their traditions against a rump that had fallen into the hands of extremists. His skills as a negotiator were no longer so relevant now that employers did not need to make concessions, while his temperamental erraticism had become an obstacle to the orderly administration they desired. The coalition that was the National Party began to break up.

The farmers were the first to leave. Their dissatisfaction with two-party politics was not new: they had complained for some time that it left no room for their needs and they had already established a separate political organization in some states. From a long-term perspective, it would seem that the growth of the cities (by 1921 just under two-thirds of the population was urban) sharpened the rural minority's feeling of neglect. Similarly, the growth of agriculture, with the family farm as its unit of production, meant that the rural capitalist and rural wage-earner, who had both been accommodated in the existing parties, were increasingly displaced by cockies, who could not. But this is to overlook the particular antagonism towards Hughes, whom farmers saw as favouring secondary industry at their expense. Wartime controls on the price of foodstuffs were bad enough, the prime minister's attempts to extend them was intolerable. Helped by the introduction of preferential voting (which was itself a recognition of Nationalist disunity), they returned eleven representatives to the House of Representatives who came together at the beginning of 1920 to form the Country Party. Under the slogan 'production first', they called for the reduction of expenditure and taxation, and abolition of duties on articles required for farm production. Their cohesion and effectiveness increased when Earle Page assumed their leadership in 1921. This Grafton doctor came to federal parliament in 1919 as a raw novice, punctuating his rambling speeches with a refrain of 'y'see, y'see', but he combined a keen political intelligence with the ability to ruffle the prime minister. He could see the opportunities presented by the Nationalists' lack of a parliamentary majority and he appreciated that his minority party could exert influence only by avoiding the enticements of office except under conditions that preserved its autonomy. He therefore resisted Hughes's offer of a coalition and insisted on expenditure cuts as a condition of parliamentary support.[51]

Among the conservatives, also, dissension increased. The treasurer, W.A. Watt, resigned from the government in 1920 after quarrelling with the prime minister; so did another minister in the following year, while the spent Sir Joseph Cook retired. S.M. Bruce, a young lawyer recently returned from the war to take charge of the family's importing business, suggested in parliament that the Commonwealth Shipping Line would be better sold to British operators. That Bruce was brought shortly afterwards into a reconstructed ministry as treasurer was testimony to Hughes's difficulties. The decision was almost certainly forced on him by the National Union, a group of Melbourne businessmen (among them, Herbert Brookes) who provided much of the finance of the Nationalist Party. Nor was this the only pressure group operating on the government. Its Sydney counterpart, the Consultative Council, was also becoming restless and in London, W.S. Robinson had brought together financial, pastoral and shipping companies into a committee of the British–Australasian Association to channel funds where they could do most good—£100 bought them a dossier from the prime minister's private secretary.[52]

During all this string pulling and muttering, the foibles of the prime minister were a recurrent complaint and such were the personal animosities he engendered that there is a danger of overlooking the larger issues. Those issues are demonstrated in the similar rifts and instabilities in state politics. In every state the war had brought a coalition of ex-Labor and anti-Labor forces together and in every state except Queensland they enjoyed office, yet by 1921 hardly one of these alliances remained intact. By then they had generally been replaced by a Nationalist Party pursuing policies similar to those of the federal party (though usually rejecting its centralism), a Country Party pushing rural interests, and, as likely as not, a Progressive or Economic or Liberal Party seeking retrenchment—the same range of positions as there were nationally. In Victoria and Western Australia this produced a succession of unstable non-Labor administrations; South Australia and Tasmania managed to maintain the tensions within a coalition while New South Wales fell to Labor. Indeed, Holman's fate offers an instructive contrast to that of Hughes. Like his associate from the pioneering days of the labour movement, he crossed the floor to form a Nationalist government after the conscription split, only to run up against the dissatisfaction of the Farmers' and Settlers' Association on the one hand, the Taxpayers' Association and Employers' Federation on the other. His failure to reduce expenditure or abolish public enterprises caused the financial backers of his own party to withdraw their funds, contributing significantly to his election defeat in 1920. Well might Sir Owen Cox, the convenor of this business group, boast that 'the power of commerce and finance [could] speak quietly but in no uncertain language'.[53] It is evident, therefore, that Hughes's difficulties were not unique.

Hemmed in by his opponents, the prime minister risked all on an election. His impenitent rhetoric merely emphasized his party's disarray and only 26 of its members were returned. Labor won 29 seats, the Country Party 14 and there were several other conservatives opposed to Hughes. In desperation, he opened negotiations with the Country Party, but Page insisted that his resignation must be a condition of any arrangement between the two parties. The game was up. In January 1923 Hughes tendered his resignation.

9

AUSTRALIA UNLIMITED?

ONE OF THE SUSTAINING national myths is that of cornucopia. Nineteenth-century visionaries had imagined a time when Australia would outstrip Britain in wealth and importance, when its open spaces would support rolling acres of farms and factories to match those of the United States. Some estimated the future population at 100 million, 200 million, or more. This belief in the national destiny was particularly strong in the early part of the twentieth century, its relevance heightened by a new urgency. The writer E.J. Brady spoke to it in his book *Australia Unlimited* (1918), a massive survey of the land and its prospects. Everywhere he went he found 'Wonder, Beauty, Unequalled Resource'. The policy of development was said by Brady to be already a 'fixed national ideal', and he claimed that Australia had learned from the recent struggle for international supremacy that it must realize its destiny and become 'the richest and most powerful, as she is now the freest and most prosperous, nation of the world'. This would require capital and labour, but Brady felt sure they were available; most of all it called for qualities of courage and determination to push forward the frontier of settlement. Previously, Australians had seen their land through Old World eyes, so that its vast interior appeared to them as wasteland. Now 'thoughtful people have come to doubt the existence of one actual desert within the wide borders of the Commonwealth'. Instead of a 'Dead Heart', there existed in reality a 'Red Heart, destined one day to pulsate with life'.[1] That vision and its corollary, rural settlement, were a dominant feature of the 1920s.

Doubters were unwelcome. Before the war the professor of geology and mineralogy at the University of Melbourne had made a sober appraisal of the inland region. He had warned in his book on *The Dead Heart of Australia* that even with expensive bore-sinking and irrigation, the region's potential for use was limited. In the buoyant atmosphere that now prevailed, the geographer Griffith Taylor aroused a storm of controversy when he confirmed that prognosis. Nor did he placate the boosters with his insistence that much of the country was 'barren desert'. His statements triggered a lengthy public debate and calls for his dismissal from the chair he occupied at the University of Sydney—and Taylor took particular pleasure when one prominent detractor, the representative of a rural electorate in the Commonwealth parliament, became lost on a backroad and was found some days later suffering from dehydration.

Not all critics were rebutted so easily. The senate of the University of Western Australia proscribed Taylor's textbook. Unabashed, he replied that 'A desert by any other name is just as dry; and if certain sensitive West Australians were to call the arid centre of their state a "Garden of Eden" it would not make it easier to develop it.' Among those who took up the cudgels was Bill Somerville, a member of the university senate and an assiduous correspondent. He sent Taylor a photo of a pastoral property near Yalgoo, in the north-west, and drew attention to its prosperity: 'You will see that the homestead is three hundred miles into what you have the colossal impudence to call desert.' Taylor replied in kind. While respecting the older man's sincerity, 'even if his possession of that virtue lets his ideals cloud his brains', he expressed the hope that someday the lay public would 'learn what is meant by a desert in scientific literature'. The dispute spilled over into discussion of immigration policy, for Taylor held that if intensive settlement was to be attempted, it would require 'a small influx of Chinese labourers'. 'You ask what is high in the White Australia ideal', Somerville rejoined. 'To so work that my children will be protected from competition that will degrade them to the level of a Chinese coolie ... is a very high ideal indeed.'[2] That Australia must be preserved for the white man, with all the discriminatory controls that implied, was another axiom of development policies.

The belief in Australia Unlimited was general. The Million Club, an association of New South Wales businessmen formed before the war with the object of bringing out that number of immigrants, now renamed itself the Millions Club. Since population meant security, to question the wisdom of these objectives was to cast doubt on one's patriotism. Sir James Mitchell, a Nationalist premier of Western Australia whose bucolic enthusiasms were so pronounced as to earn him the nickname 'Moo-cow Mitchell', dismissed sceptics out of hand: 'We must not listen to the croakers.' All politicians, Labor, Nationalist and Country Party, were agreed that the land must and therefore could be made to support greatly increased numbers. As expounded by Bruce, the new prime minister, at an Imperial Economic Conference in 1923, 'Australia's aim above everything else is to populate her country and advance from her position of a very small people occupying a very vast territory.'[3]

The logic of this strategy dictated the course of national policy during the 1920s. First, it was necessary to augment the country's human resources. Australia's enumerated population at the end of the war was 5 082 000 (with perhaps a further 80 000 Aboriginals). The process of demographic transition that had been apparent at the turn of the century was now more marked: the rate of marital fertility in the 1920s was half that of the 1880s. Consequently, the Commonwealth and state governments embarked on schemes of assisted immigration that brought more than 200 000 settlers to Australia. Almost as many came unassisted and altogether more than 300 000 persons settled permanently in Australia, helping to increase the population to 6 414 000 by the end of the decade.[4] Second, massive spending programmes had to be undertaken in order to fill the empty spaces. Official estimates of the cost of putting a family on the land ranged from £1000 or less in moments of excessive optimism, to £2000 or more when realism prevailed; then there were the expenses of railway and road construction, irrigation, schools and other services for the new communities. For these and other purposes the Commonwealth and states borrowed more than £300 million in the post-war decade.[5] Third, farmers needed to find customers for their produce and the limited size of the domestic market meant that much of it—two-thirds of the wheat and dried fruit, a third of the sugar and butter, a quarter of the meat, all but a fraction of the wool—had to be sold abroad. Here again the government helped in the search for customers and underwrote marketing schemes. Fourth, the government gave an express undertaking that the process of development would not be at the expense of established living standards. 'I believe in the Standard of Living', insisted the prime minister. 'Except the "White Australia" policy, the Standard of Living is the last article of political faith which I would sacrifice.'[6]

The first three elements of policy were conceived as complementary aspects of Australia's place in the post-war world. Bruce brought them together in the formula 'Men, money and markets' that he enunciated to the Imperial Economic Conference of 1923. A sparsely populated region of recent settlement, Australia lacked the means to develop its natural resources unaided: therefore he sought immigrants and capital. A trading nation, it needed to increase and diversify its output: therefore he searched for markets. Preoccupied with its territorial and racial integrity, Australia still felt the need for a special relationship with the homeland: therefore the prime minister looked to Britain. This remained the basis of his government's vision of the national destiny. As Bruce never tired of repeating, 'He was more than ever convinced that men, money and markets accurately defined the essential requirements of Australia.'[7]

Despite the prime minister's assurance that he would safeguard living standards, his last element of policy was more tenuous. It was widely held that all Australians stood to benefit from a prosperous rural export sector. At a post-war conference of Commonwealth and state ministers, Hughes had endorsed the statement of a Labor premier:

> Mr Theodore said, and he was quite right, that if you put one man on the land, he provides a job for one man, and probably for two men in the city. If we attend to the first part, the other will adjust itself.[8]

Table 9.1: Gross domestic product, 1920/21 and 1928/29 (£m at 1910/11 prices)[9]

Industry	1920/21	1928/29
Pastoral	43.7	45.4
Agriculture	44.4	25.8
Dairying, forestry, fisheries	14.6	21.6
Mining	10.7	10.3
Manufacturing	44.3	68.7
Construction	23.9	34.2
Distribution	76.7	75.3
Finance	6.7	8.8
Railways, other public undertakings and government services	35.7	42.3
Other services	43.6	46.4
Rents	31.2	39.8
Other	3.6	3.8
Total	379.1	421.4

But would it? The workforce and production statistics recorded in tables 9.1 and 9.2 show considerable growth in the non-rural sectors of the economy. However, closer inspection reveals that the increase in the national product lagged behind the increase in the workforce—the average worker produced £189 per annum in 1920/21, £179 in 1928/29. Furthermore, much of the diversification represented a shift towards less productive forms of economic activity—notably manufacturing, where annual output per worker was only £137 at the end of the decade. Maintenance of urban living standards therefore depended on tariffs to protect jobs, arbitration to protect wage standards, and more public borrowing to provide urban amenities; and these devices imposed a growing burden on the more efficient export industries. This was the cost of the greater self-sufficiency that the Commonwealth pursued for the sake of 'balanced' economic development and national security, as well as sheer electoral expediency. The cost was accepted for the time being, but already the protective impulse of domestic expectations tugged against the imperial thrust of men, money and markets.

Table 9.2: Workforce by industry, 1920/21 and 1928/29 ('000s)[10]

Industry	1920/21	1928/29
Rural	487.8	530.8
Mining	54.0	43.5
Manufacturing	434.4	503.2
Gas, electricity, water	20.4	28.3
Construction	159.4	246.0
Transport	169.6	188.8
Commerce	272.7	347.4
Community and business services	110.5	140.2
Finance and property	36.6	50.9
Other	262.4	271.5
Totals:		
Workforce	2007.8	2350.6
Population (excluding Aboriginals)	5412.3	6336.8

For Bruce's plans rested on an imperial division of labour. Australia, the dominion, was to develop its resources in harmony with the needs of the Empire. It would send Britain foodstuffs and raw materials, and in turn would receive people, capital and manufactures. 'Men, money and markets' expressed this relationship from the colonial standpoint, but the metropolitan perspective was rather different. Post-war Britain was indeed interested in fostering economic links with the dominions because its staple industries—textiles, coal, iron and steel, engineering—faced increasingly stiff competition in foreign markets. If some of the people who were no longer needed in the domestic workplace could be used to populate the empty spaces of Canada, Australia, New Zealand and South Africa, they would at once contribute to the wealth of the Empire and augment the export market for British manufacturers. In 1924 the Queensland Rhodes Scholar P.R. Stephensen (one of a remarkable generation of radicals to graduate from that university which included the writer Jack Lindsay and the communist member of parliament, Fred Paterson) visited the great Empire Exhibition at Wembley. He found to his disgust that the Empire was nothing but 'a vast trading concern'.[11]

At the Imperial Conference of 1921 the British government offered to establish a fund to further its aims. Under the

Empire Settlement Act of 1922, Britain allocated money to assist immigration and to finance rural settlement on the understanding that the dominions would share costs. New South Wales, Victoria and Western Australia submitted schemes, of which the proposal to establish 6000 family farms in the south-west of Western Australia was the most ambitious. It soon became apparent, however, that the settlement schemes were undercapitalized and in 1925 Britain and Australia negotiated the '£34 million Agreement', which was expected to bring 450 000 migrants in a decade. The £34 million was available to the states in cheap loan funds (at a rate of £75 for every assisted immigrant and £1000 for every family farm) and the states were invited to submit proposals to a Commonwealth Development and Migration Commission. But already the momentum was slowing. While Western Australia put forward plans for 3500 additional farms, it actually established 200. Altogether the Development and Migration Commission authorized the expenditure of only £8 million.[12]

As with men and money, so with markets. The expectation that Australia would sell Britain growing quantities of food and raw materials ran up against the harsh realities of international trade. The long era of free trade was past and the world's producer countries competed fiercely for the remaining markets. A high-wage economy like Australia's was at a comparative disadvantage in the production of labour-intensive commodities. Other countries could place butter, meat, fruit and sugar on the British market more cheaply than Australia. What then of the special imperial relationship? Australia acknowledged it with a preferential tariff that gave British goods a 5 per cent advantage over other imports, increased to 12 per cent in the schedule of 1921. Britain also had introduced duties on a limited range of goods at the end of the war, with a small imperial preference for sugar, dried fruits and wine. A Conservative government proposed to extend imperial preferences in 1923 but was voted out of office. The result was a persistent trade imbalance. In the five years from 1923–24 to 1927–28, Australia bought 43.4 per cent of its imports from Britain and sold Britain 38.7 per cent of its exports. Wheat and wool made up more than two-thirds of all Australian exports. Reliant still, and to a dangerous degree in an increasingly autarkic age, on just two commodities, Australia turned to other customers. While Britain remained the largest buyer, sales to the rest of Western Europe were of a nearly equal magnitude by 1929, and Japan was increasingly important. Furthermore, the United States was a fast-growing rival supplier of manufactures.[13]

These developments might have been expected to loosen the imperial ties. In 1922, when Australia found itself expected to support Britain in a threat of war against Turkey, there was certainly keen resentment of Lloyd George's lack of consultation. And when the Imperial Conference of 1923 acknowledged the right of dominions to appoint their own diplomatic staff and negotiate their own treaties, it was only giving belated recognition to the reality that their interests and concerns did not always coincide with Britain's. Largely to ensure that Australia was better informed, Bruce arranged for an Australian liaison officer to be attached to the Foreign Office in London. The post was filled by Richard Casey, elder son of the businessman R.G. Casey, who had the education, the confidence and the private means to fit into this exclusive milieu. From Melbourne Grammar School and the University of Melbourne he had proceeded to Cambridge. After enlisting in 1914, he had been attached to the staff of General Birdwood—an appointment that gossips alleged had been made possible by his father's present of a Rolls Royce to the British officer. Richard Casey senior had resided in London during the war and was a generous donor to conservative causes. He died in 1919, leaving an estate of £112 000. Now, from his rooms in the British cabinet secretariat, the younger Casey sent the prime minister a flow of information about what was happening in the corridors of power. Casey shared with Bruce a shrewd appreciation of the more ceremonial aspects of the British presence in Australia. 'He is a good type and should go down well', he reported of a governor destined for South Australia, 'not too intellectual, nor too sporting, nor too anything else ... His military record, of course, is first class.'[14]

These were merely the trappings. Australia was still bound closely to Britain by preference and necessity. It was the Canadians and South Africans who found the remnants of British sovereignty most irksome. It was for them, rather than for the Australians, that Balfour came forward in 1926 with his ingenious definition of Britain and the dominions as 'autonomous communities within the British Empire, equal in status, in no way subordinate one to another in any aspect of their domestic or external affairs, though united by a common allegiance to the Crown'. Australia would have preferred to leave well alone. 'What could the Dominions do as independent nations that they cannot do now?' Hughes asked at the 1921 Imperial Conference.[15] To put the relationship into words could only demean it. Therefore Australian representatives resisted the new mood and would not even ratify the Statute of Westminster, which gave statutory effect to Balfour's formula, until 1942. Their defence planning, similarly, continued to rest on the military capacity of the Empire. The British were building a naval base at Singapore and this, despite delays and shortcomings, was the fulcrum of Australian strategy.[16]

Members of the Australian ruling class looked askance at what they perceived as a growing 'Americanization' of Australia. The advent of the phonograph and cinematic soundtrack were especially significant—it came as a shock to realize that the idols of the silver screen spoke with American accents. Hence the guardians of Empire loyalty deprecated American films and American music, a Nationalist senator even alleging that the movie distributors were intent on undermining Australia's relationship with the home country by portraying the British as 'cowards, criminals, fools and drunken degenerates'. One of the great scandals in 1928 was a police raid on a flat in Melbourne where five young women were found in compromising circumstances with the black entertainers of Sonny Clay's vaudeville company. 'These boys are Americans and they are interesting', one of the women told the court, but that did not protect her guests from deportation, after which the cabinet announced that no further entry permits for foreign performers would be granted to Negroes.[17] Businessmen appreciated the industrial might of the United States, whose investment in Australia reached £60 million by 1929; they tried to adapt some of its business methods and the Commonwealth accepted the need to appoint a trade commissioner. But there was no attempt to develop this post into a diplomatic mission. To do that would have betrayed the cosy family circle that was the Empire.[18]

Popular attitudes were more antagonistic. Jock Neilson, now in his fifties but still working as a casual labourer, found that the assisted British immigrants 'completely swamped the labour market as far as rural work was concerned'. One urban newcomer, a garment worker from Manchester, encountered among his new workmates a similar hostility to things British: 'if you were a Pom and wanted to remain English in your outlook, they didn't care for you'. Yet acceptance was possible, 'The quickest way to become accepted was to become "Australianised" as quickly as possible, which I did and fitted in.' Before the winter was out, he was standing with his mates at a Saturday afternoon game of Australian rules football.[19] His perception was accurate—conformity was the key. There was, of course, the long-standing fear that immigrants threatened employment and wage standards, and the emphasis during the 1920s on rural settlement was meant to allay such misgivings. But fear of the newcomer went deeper than this. It attached itself to speech or dress or habit, but most of all to race. An influx of Southern Europeans into Australia in 1924, after immigration restrictions had been imposed by the United States, aroused an immediate upsurge of xenophobia. Within a year the Commonwealth government imposed new restrictions on 'foreign' immigrants. Yet even here there were acceptable and unacceptable foreigners. In Queensland, where several thousand Italians settled in the sugar-growing districts, a royal commission drew a distinction between the 'knife-wielding, inferior, racially-minded Sicilian' and the 'blonde, intelligent, hard-working, assimilatable Alpine'.[20]

The intention was that the newcomers would go onto the land. Under the Empire settlement schemes, the states created farms, either by alienating Crown land or by subdividing estates bought from private owners; they lent the settler money for housing, fencing, equipment and stock; they watched over his endeavours, and they expected him to begin repaying his debts once the land was brought into production. Yeoman ideology and actuarial calculations alike dictated that the farm would be small. The tract of land was meant to satisfy the needs of a family from which it would also draw its labour. In any case, even at reduced interest rates, a settler could hardly be expected to service a debt of more than £1500, and the cost of land was at a premium during the buoyant post-war years when the schemes were worked out. At best a wheat farmer could expect to begin on a few hundred acres, while a dairy farmer would have no more than a hundred and a fruit grower in an irrigated area even less. Except that the allowance for initial working capital was a little more generous, similar conditions applied to the soldier settlement schemes that were undertaken by all the states during this same period.

The arrangements were absurdly optimistic. Too little attention was paid to the suitability of the land or the aptitude of the settler. Even in 1921, when prices were high, the new farms were small and hopelessly undercapitalized—the working capital needed to operate a successful dairy farm was at least £2000. And by cruel irony, those who battled against the odds to clear their blocks achieved full productivity just as the price for their products fell away. Wheat, meat, fruit and butter commanded higher prices for the first three post-war seasons, then stabilized at a lower level until 1925 when increased world capacity brought further decline.[21] Settlers were therefore burdened with repayment commitments that were beyond their reduced means. The future leader of the Country Party, John McEwen, who took up a block in the Murray River irrigation area at this time, commented succinctly: 'The whole thing was, of course, ludicrous.'[22]

Nowhere was the contrast between promise and reality more cruel than in the jarrah forests of the south-west corner of Western Australia. Immigrants were lured from Britain by pictures of attractive homesteads set on prosperous dairy farms. After landing in Fremantle, they were taken by rail and then by dirt road to be dumped before primitive shacks built on bare earth in virgin country miles from the nearest shop or doctor. 'My wife thought she had come to the last of her days', one settler recalled. 'Have we to live in this?' she asked him. They did. For the next two years this man and nineteen others endeavoured to fell massive trees and grub out undergrowth with axe and pick. Theirs was a group settlement scheme in which a score of families worked collectively at below basic wage rates to establish their farms, then balloted for twenty-five-acre blocks which were encumbered with mortgages for the cost of the land and the sum of the accumulated advances. Milk cheques went straight to the bank. Barely more than half the original 'groupies' lasted the first three years.[23]

By such means a new generation of small farmers came into being in the established areas and new land was brought into cultivation. The statistics attest to an expanded and more diverse primary sector. The wheat acreage increased by a third, while sugar, dairy and fruit-farming made undeniable advances. But at what cost? The expenses of assisted settlement increased far beyond original expectations. The Commonwealth and state governments spent £40 million during the 1920s buying land for settlers and lent them a further £30 million. It proved impossible to maintain the initial schedule of repayments since some of the debts had to be written off when settlers abandoned their holdings and further advances were necessary to save many of those who remained. The Western Australian group settlement scheme incurred losses of £3 million by 1930 and the average cost of a farm increased from £1000 to £3404. Of 37 000 soldier settlers across Australia, fewer than 27 000 remained and the accumulated losses amounted to £23.5 million.[24] Not even these sums sufficed to secure the viability of many of the new farms and few of the newcomers were able to take advantage of the more advanced technologies and techniques that export producers in other countries adopted: where North American wheat farmers used a tractor, most Australian growers still relied on horses; where New Zealand dairy producers had milking machines, the great majority of Australian farmers milked by hand. Consequently, the new rural industries depended on government assistance. They were established at interest rates subsidized by the Commonwealth and states; their transport costs were subsidized by the state railways; to find overseas markets, the Commonwealth provided marketing support schemes and bounties; some of the least efficient, such as dairying, were even protected in the domestic market. This was the price that had to be paid for pushing settlement into the marginal areas and for diversifying from an efficient, large-scale staple industry into less efficient, smaller-scale forms of production. It was fortunate indeed that the pastoral industry managed to expand its output from 550 million pounds of wool in 1920 to 930 million pounds by 1930, and doubly fortunate that wool commanded high prices for most of the decade.[25]

The final cost is more difficult to calculate. It fell on the men, women and children who lived in makeshift accommodation of hessian and galvanized iron, subsisted on rabbit and parrot stew, worked from sun-up to sun-down seven days a week, and still found their financial position worsening from year to year. Settlers' handbooks were full of helpful suggestions about how the necessities of domestic life might be fashioned from salvaged materials. Two poles and wheat bags improvised a bed; packing cases could be turned into cupboards, tables and chairs; kerosene tins could be reworked into buckets, meat safes, even cutlery. But the manuals did not explain how to cure the barcoo rot that was brought on in the group settlement blocks by malnutrition, nor did they warn settlers of the emotional tensions induced by the wracking strains of this culture of rural poverty. One soldier settler's wife in Victoria testified to the Settlement Board that she had taken on all kinds of manual work, including rabbiting, in order to get a few shillings with which to buy clothes for her children.

> One day I helped to hold fencing wire to thread through posts—unfortunately I got a 'kink' in the wire which touched a vital spot, and before I knew where I was my husband threw me down by catching hold of my throat with his muddy hands.

Above all, there was the demeaning dependence on a remote and all-powerful officialdom. 'How can you sit there in the office & live on the swett of a returned soldier ... and see [him] flogg his life away, then send in a demand for rent which is impossible to pay?' asked a desperate settler in Tasmania.[26] From resentments such as these the Country Party drew much of its strength.

People drifted from the bush to the towns. It was not a dramatic movement—rural dwellers remained slightly more than a third of the total population and even increased slight-

The Newcastle steelworks, a nodal point of industrial growth in the 1920s

ly in absolute numbers—but amounted in all to perhaps a quarter of a million people over the decade. Some of them were settlers, recently arrived from Britain in many cases, who abandoned the unequal struggle. Some were the children of established farmers, attracted by the bright lights, the freedom and enhanced opportunities of the metropolis. Others were rural labourers, displaced by the spread of the family farm and the increased mechanization of large-scale farming. The city offered all of them wage levels higher than obtained in the country (where many occupations were still not covered by an arbitration award); furthermore the city possessed amenities such as sewerage, public transport, gas and electricity that made such a difference to living standards. There was also a subtle but unmistakable shift in the rural identity. As country districts yielded their restless sons and daughters, they lost much of their vitality. As the churches surrendered their role as leaders of opinion and foci of social activity to the RSSILA and Country Women's Association, so the bush towns took on a more restricted, insular feel.

Factories established in the inner suburbs employed more

advanced technologies to produce a wider range of commodities. Metal manufactures, textiles, electrical goods and the production or assembly of consumer durables all made rapid strides in these years. Unlike so many of the older trades which relied on the processing of primary products, the import replacement potential of the newer ones allowed room for faster growth. The basis for industrial diversification had been laid during the war when the Newcastle steelworks began production and many imports were unavailable. Now there was a spurt of growth. Whereas Australia produced just 14 000 tons of steel in 1913, output in 1929 was 400 000 tons. The number of motor vehicles on Australian roads increased from 9000 at the beginning of the 1920s to 571 000 at the close; while the engines and chassis were imported, body-building, assembly and the manufacture of components was carried out in this country. By 1930, when most houses were connected to mains, electricity consumption had quadrupled and there was a strong demand for domestic appliances such as irons, radiators and vacuum cleaners. Around such growth sectors the urban workforce expanded, and this in turn generated growth in the construction, retailing and service sectors.[27]

The new industries produced for a domestic market. Their cost structure and limited scale of operations prevented them from finding overseas customers; within this country, even with tariff assistance, they could not command their field of operations. The two steel producers, BHP and Hoskins, satisfied only two-fifths of Australian steel requirements and many of the more important steel products were still wholly imported. Australia failed to generate its own automobile manufacturing industry and imports of electrical goods and textiles still far exceeded local production. The habitual response was to call for greater protection. The Tariff Board established in 1921 to investigate such requests was kept busy throughout the decade. 'Although we are a fact-finding and non-partisan body', one of its members confessed,

> our facts are sought only with the object of improving the protectionist system our country has adopted . . . We are a non-partisan body because we have been selected by a Government of a country 95 per cent of whose representatives are protectionists. We are four protectionists—God helping us and, you will add, God helping our country.

By the end of the 1920s, more than 250 items carried a duty of more than 40 per cent.[28] The tariff was a crude instrument. It could safeguard the existing local market, or at least a portion of it, but it could not create the enlarged market that would have been required to support more advanced industries. Nor did the tariff distinguish between indigenous producers and foreign ones that established operations in this country only to avoid the tariff and take advantage of the preference given to domestic tenders for government contracts. In the petroleum, electrical and automobile industries, for example, British and American capital dominated the field and tailored local activities to a strictly limited volume and range of products.

Beyond these shores, the Australian entrepreneur was seldom found. In Malaya and elsewhere in South East Asia, small and medium-size mining companies run from Sydney or Melbourne were simply brushed aside by large-scale British and American ones, leaving Australian entrepreneurs the more restricted opportunities for mining, planting and trading in the islands of the south-west Pacific.[29] The only businessmen to play a significant role on the world stage were those of the Collins House group, the conglomerate of companies based in Melbourne but with mining and smelting interests stretching further afield: several of its directors lived in London and were active in the metal and money markets there. Their activities can be contrasted with the strategy of the BHP Company with its vertically integrated operations in metalliferous and coal-mining, shipping and smelting for the domestic market. Since the major industries with more advanced technologies were suited to large-scale operations—by the end of the decade half the secondary industrial workforce was concentrated in less than 5 per cent of all factories—oligopolistic practices flourished.[30] Yet even with these advantages and with tariff protection, it was hard to compete with the mass production methods of overseas manufacturers; 'For us to introduce the "Owen machine" would be like putting a racehorse in a dray to do dray work', explained the leading manufacturer of glass bottles when asked why he could not achieve the same economies of scale as overseas suppliers.[31]

The expansion of secondary industry, then, was uneven. Throughout the 1920s the level of unemployment ranged between 6 and 11 per cent, with a permanent residue of unskilled and casual labour. The urban economy was still susceptible to seasonal fluctuation and still tied to the fortunes of the primary industries; its growth came to an end in 1927, just as export prices fell away.

Given these limitations, how was the Australian producer to compete? In the full flush of expansionary enthusiasm for Australia Unlimited, some looked to a wholesale reorganization of business methods in order to increase productivity. Henry Holden's motor body works in Adelaide were commonly cited as an example of what could be achieved. Branching out of a coach-making business at the end of the war, the Holden Motor Body Builders Limited was formed in 1920 with a capital of £130 000 and within six years increased annual output from 3000 to 36 000 bodies. This was done by breaking down the work process into a series of simplified operations that could be performed by the workers as bodies passed along an assembly line. Since the momentum of the line dictated the tempo of work, any hold-up in the flow could immediately be traced to inefficient labour or bad organization and rectified. Furthermore, the assembly line rendered redundant the skills of craftsmen who had previously carried through a piece of work from inception to completion and who exercised a corresponding degree of control over the production process. The new technique was modelled on American practice and Holden's was advised by the General Motors Corporation which took the company over in 1931.[32] Partially understood and frequently denounced—according to the New South Wales Labour Council, 'it reduced men to nothing more than a tool-using animal, a mere cog in the wheel of production'—the American gospel of scientific management was preached widely. If only by reading Henry Ford's *My Life and Work*, which sold 47 000 copies in Australia during the 1920s, many Australians were familiar with its principles. In the pursuit of efficiency, it sought greater managerial control with a new emphasis on cost accounting (this period saw the introduction of calculating machines and card record-keeping systems) and a constant search for improved methods of production. Wherever the scientific manager saw slow or wasteful work practices, there he was to simplify the task into a series of routine operations, introduce a pacemaker and establish new targets with incentive or piece-work methods of payment. 'It is all in sections and one man simply does a part', the secretary of the Coachbuilders' Union reported of Holden's and, 'the men complain about monotony and fatigue and speedups'.[33]

A number of Australian businessmen embraced this creed. Herbert Gepp, who returned from selling zinc concentrates in the United States during the war to become general manager of the Collins House group's Electrolytic Zinc Company and later the head of the Development and Migration Commission, was one of an 'advance guard of progressive business thinking'. Essington Lewis, who inspected American steel plants in 1920 before his appointment as general manager of BHP the following year, was another and his company, like a number of other large concerns, recruited American managers. There were others with first-hand experience of how it was done in the world's most dynamic economy. Harold Clapp had worked in the United States from 1900 to 1920 before becoming head of the Victorian railways, and he sent more than a score of his supervisors to gain American experience (every one of them, the union warned, a 'potential scientific manager'). So too did the American W.A. Webb, commissioner of the South Australian railways from 1922. In each case the results fell some way short of the orderly precision that was specified in the scientific management texts. Webb, described in state parliament as 'a migrated American Mussolini', was in constant conflict with his ministers, his workers and the public until his resignation in 1930. Clapp fought on but with almost as many frustrations. Gepp, for all his love of elaborate organizational charts, was incapable of delegation, while the remorseless Lewis, forever tabulating production schedules for every aspect of BHP's activities, fought a lifelong battle against human imperfection.[34]

The power of such men was immense, their prestige undeniable. They moved easily from shop floor to boardroom to ministerial suite, never doubting that they were in the van of progress. Lewis and Gepp symbolized the transfer of effective control away from the shareholders and directors of the enterprise to its executives. While they possessed a scientific training far beyond that of an R.G. Casey, they commanded higher authority than mere technical proficiency would bring. They made themselves into the hub around which all else revolved. Their elaboration of management as a science and an ethos therefore served to legitimate their role, and it was here, rather than in any transformation of the process of production, that the real significance of the doctrine lay. The businessmen of the 1920s took up the gospel of efficiency that the progressive administrators had preached before the war—indeed it was their chastening wartime experience of rancour and division that caused such progressive publicists as Meredith Atkinson and Elton Mayo to turn from the state to industry as a more propitious site for transformation. Yet here again results fell short of expectations. As implemented by most Australian employers, in fact, scientific management usually meant piece-work and speed-ups in conditions more akin to an old-fashioned sweatshop than any citadel of science. At the Pelaco shirt factory in Richmond, 'Home on the Hill' for nearly a thousand women and cited frequently as a showpiece of the new industrialism, the effects of noise, speed and monotony were such that 80 per cent of the workforce left within a year.[35] And since arbitration awards hedged the freedom of the employers to impose piece-work, the opportunities for increasing productivity by such means were limited.

Scientific management and industrial efficiency therefore operated more as aspiration than achieved practice, betokening an idealized world from which all sources of discontent, all friction and all impediments to maximum efficiency had been eliminated. In this spirit, the new methods of organizing and rewarding labour were put forward as a means of avoiding industrial conflict since, in the words of the general manager of BHP, they would give employees 'the opportunity of earning up to the full extent of their working capacity'. Some of the more progressive employers went further in their conviction that a whole new spirit of co-operation was necessary. 'We live in dangerous times', wrote W.S. Robinson to another of the Collins House directors.

> The industrial question is not going to be settled by the old system of bare fists and a fight to the finish ... [U]nless the employing class will at this late hour make an honest effort to convince the

employee that it is not wholly selfish and soulless, the industrial revolution will be far more dreadful in its results than any political or social disturbance which has preceded it.

Speaking to this fear, some intellectuals even extended the principles of scientific management from the enterprise to the nation at large. They looked for a reorganization of economic and social life along lines of national efficiency that would supersede class antagonisms, and they sowed these ideas in the labour movement at conferences and classes of the Workers' Educational Association.[36]

Corresponding notions of services and fellowship were spread among business and professional men at the weekly luncheons of the Rotary clubs that were brought to Australia from North America. The first clubs were established in Melbourne (where Clapp, Sir John Monash of the State Electricity Commission, Herbert Brookes of Australian Paper Mills, D. York Syme of the Melbourne Steamship Company, and Sir Robert Gibson, ironfounder and later head of the Commonwealth Bank, were among the foundation members) and Sydney (where Sir Denison Miller, the governor of the Commonwealth Bank, Sir Henry Braddon of Dalgety's pastoral company, C.C. Jones, the proprietor of the David Jones emporium, and Hugh Denison of *Sun* newspapers were included) in 1923. Within seven years there were thirty of these clubs. Yet neither the industrial nor the civic initiative achieved its desired results. A national conference of employers and unions convened by the prime minister in 1922 saw the former recommending piece-work payment and wholesale changes to the arbitration system, the latter declaring that socialization of industry with workers' control was the only solution for the impending capitalist collapse. And the first national conference of Rotary issued a testy call to 'clean up the industrial position'. The battle-lines remained as before.[37]

Both the economic changes of the 1920s and the ethos that they expressed shaped modes of consumption. In ever-growing cities (Sydney passed the million mark in

1922, Melbourne in 1928), residential development extended several miles beyond the pre-war outskirts. The essential homogeneity of these new suburbs was remarkable: street after street of standardized bungalows set on quarter-acre blocks, their minor differentiation by size and ornamentation quite overshadowed by a general conformity to a cult of home and garden. Inside, the ornate furniture, hangings and bric-à-brac of an earlier generation were replaced by a new functionalism of built-in fittings, clear surfaces and labour-saving technology. Where the sideboards had been littered with knick-knacks, ornaments and *objets d'art*, there was now a simple side-table on which the telephone sat in splendid isolation. Upon the dining-tables the first of the processed 'convenience' foods replaced items that had previously called for laborious preparation. The wardrobes held a wider variety of clothes made up of lighter fabrics that were easier to clean. With more time for leisure and with their own transport, the inhabitants now had access to a greater number of commercial entertainments—including those powerful stimuli to further consumption, radio and cinema. Under their influence, Australian 'flappers' learned to bob their hair and dance the Charleston, young men to imitate their film idols.[38]

The meaning of these images of the 1920s is revealed in the processes that disseminated them. This period saw the establishment of the advertising industry to promote consumer goods aimed at a mass market. Its founders were among the most ardent apostles of American business methods and they brought to their profession a portentous confidence in the ability of advertising to 'mould public opinion, advance human progress and influence all habits of civilized life'.[39] There was also the rapid development of the cinema from inauspicious origins, purveying cheap thrills and jerky melodrama to a restricted working-class audience in makeshift halls, to the full-length 'movie' shown in the grandeur of the picture palace. Luxury theatres were built in the 1920s with exotic décor, sumptuous fittings, massive lighting displays and retinues of uniformed attendants—the State Theatre that was opened in Sydney in 1929 cost £1 million. Simultaneously, the industry came to be dominated by American productions offering the patron a vicarious glamour and excitement; hundreds of films were imported annually and it was calculated in 1927 that one Australian in three saw a show each week. The advent of radio was hardly less dramatic. At the end of 1923 broadcasting began, by 1929 there were 300 000 licensed listeners. Here was another potent medium for advertising and communication. The impact of the mass media on leisure activity should not be exaggerated—more money was spent on gambling than on cinema, and more people found companionship in the pub than gathered round the radio in the front room. Even so, the beginnings of a new popular culture that was narrowly controlled, privatized and treated the audience as a pliant consumer were already evident.[40]

Among the early radio programmes were sessions for the housewife, advising her how to organize her tasks and offering information on motherhood and household management. This was just one means for the propagation of augmented expectations of women's domestic responsibilities. Through advertising, education and exhortation a sphere of activities that was once assumed to be part of a women's innate make-up was carefully defined and elaborated. Child-rearing could no longer be left to instinct or popular lore; it was now a subject for expert instruction. A shift from breastfeeding to bottle-feeding reinforced the new emphasis on hygiene and the establishment of regular routines. Cooking, shopping and other aspects of housework were put on the same footing. The prospectus of the Emily McPherson College of Domestic Economy, opened in Melbourne in 1927, caught the new approach: 'It is intended to develop and lay special stress on the higher scientific aspect of household economics.'[41]

An explanation for these changes is deceptively obvious. The movement of women into the paid workforce and the growing separation of home and work meant that by the twentieth century many women entered marriage without basic cooking and housekeeping skills. Moreover, the increasing reluctance of girls to become domestic servants—as a proportion of all female employees, they had fallen from 31 per cent in 1901 to 21 per cent by 1921—forced many middle-class women to fend for themselves or make do with the limited assistance of a 'daily'. So many of the innovations in domestic technology—hot water on tap, the gas stove, the built-in fittings that did not need such frequent cleaning and dusting, the carpets and vacuum cleaner, even the refrigerator and washing-machine for the particularly well-to-do—were sold as labour-saving devices that would liberate the woman of moderate means from drudgery.[42] But there is more to the ideology of domestic management than this. The detailed prescription of daily routines to a meticulous standard of hygiene and efficiency shackled the housewife as completely as did the new science of infant care with its rigid schedules of sleeping, waking and feeding. Through their elaboration of the domestic sphere, these practices actually intensified gender divisions and increased the privatization of the family.

How widespread were they? It is evident, first of all, that a large minority of women did not conform to the domestic ideal. One-third of them went out to work, most as factory workers or domestics but a growing minority in offices and shops; however, it remained the convention to give up work upon marrying and even in working-class communities it was taken as a sign of desperation if a girl stayed on at a factory after marriage. Some eschewed maternity. From an annual rate of 28.6 births per 1000 on the eve of the war, there was a drop to 24.9 births per 1000 immediately after and then a continuous decline until by 1934 the rate reached an unprecedented low of 16.4—though of course it was the very restriction of family size that allowed a mother to lavish time and affection on her children.[43] Some contested the inequalities of employment and education, but they were largely ineffectual. The more serious impediment was economic. The prospect of owning a home or acquiring the objects of the new lifestyle remained far beyond the means of many families. Even in the cheaper working-class suburbs, a bungalow cost £800, requiring a deposit of at least £80 and repayments of well over a pound a week. Car prices at the beginning of the decade were higher than they became when the volume of production increased, but even in the late 1920s, when a car could be bought for £200, running costs remained high. Hire purchase or 'time payment' brought some of the consumer durables within reach of the ordinary wage-earner. Nevertheless, £8 per week may be taken as the dividing line between those able to enjoy the full comforts of the new lifestyle and those who lived in more straitened and precarious conditions in the inner suburbs.[44]

Wage rates held up well during the 1920s and there is general agreement among economic historians that real wages were higher in 1929 than in 1920. A skilled tradesman could earn in excess of £6 per week, a semi-skilled hand around £5. For an unskilled labourer, however, the male basic wage was a little more than £4, while the rate for a female was barely half. There was no guarantee of steady employment and especially in the winter months, when casual jobs cut out, the struggle to make ends meet became acute. Families in such circumstances could afford only cheap rental accommodation, with the children sharing a bed and wearing hand-me-downs. Dad got to work on a push-bike, mum was tied to the immediate neighbourhood. Not for her the contraceptive devices that were increasingly available to the middle-class woman, and her unwanted pregnancies could be terminated only by a backyard abortion. They washed once a week in a tub, saw a doctor only in dire emergencies, and their diet was a far cry from the balanced nutrition of the domestic science manuals:

> the staple diet [of such families] consists of white bread with either dripping, jam or treacle, and tea. There is a certain amount of meat scraps and potatoes, perhaps rice occasionally, but the staple diet the children get, two or three times a day, is such as I have described.[45]

Glamour might touch their lives if they could find a few shillings for the local cinema, but that was a momentary escape from hardship and uncertainty. At least one unemployed wharfie and his wife took their pleasure in cookery books, reading about and imagining the dishes they would never eat.[46]

10

HOLDING THE CENTRE

'AT THIS MOMENT we are doing nothing spectacular. We do not believe that Australia wants the spectacular now. Australia, we believe, wants now, above all things, a period free from political turmoil ...' With these words the new prime minister marked off his administration from that of his predecessor. The strident vituperation and restless urgency of Hughes gave way to the Olympian calm and studied detachment of Stanley Melbourne Bruce. By background and training he was destined to rule. Born into Melbourne's mercantile élite, connected by marriage to a Western District pastoral dynasty, educated at Melbourne Grammar and Cambridge, where he had rowed for the crew, decorated for valour in the war, Bruce accepted leadership as both a right and responsibility. With his strongly held code of right and wrong, he found it difficult to adapt to the shifting morality of some of his colleagues and supporters. His fundamentally hierarchical instincts were ill-suited to courting public opinion. A propensity to argue from first principles limited his patience with the democratic process. He liked to present himself, not as a politician, but, as he had put it in 1918 when he first offered himself to the electors, 'as a plain businessman and a plain soldier'.[1]

Everything about him, from his spats and Rolls Royce to his aloof demeanour and Wodehousian turn of phrase, sug-

A gentleman among the players: Bruce displays himself with his Rolls Royce, 1923

gested that this was a gentleman among players. He travelled with a £50 note pinned in his pocket, 'for emergencies', and it was the duty of his valet to transfer this item from suit to suit. The house he built at Frankston, with non-union labour, cost £20 000. Yet on two occasions Bruce had to lay his affairs on one side in order to retrieve the fortunes of the family business. He worked harder than his unhurried manner implied, arriving early at his office and taking papers home in the evening to his dictating machine. He had, too, a formidable ambition and an obstinacy that was no less real for being concealed behind the carapace of good manners. Even the spats that were his trademark were only adopted because the press made fun of him for wearing them to the football to protect an old ankle injury from the chill of a Melbourne winter.[2]

Nor was he the compliant crony of the plutocrats that his opponents alleged. He could be as sternly critical of business cupidity as he was of trade unionists who held the community to ransom; and when the oil companies threatened to pass a tax increase on to the consumer, he condemned them for 'endeavouring to take over the government of this country'.[3] Believing himself to be entrusted with the national welfare, it was his aim to reconcile the legitimate interests of all. The party that he led, the Nationalists, was a loose federation of state bodies whose impressive enrolments (the New South Wales branch of the National Federation claimed 30 000 members in 500 branches, and there were just as many in the Victorian section of the Women's National League) were secured by nominal membership fees. Their role in policy formulation was weak. The Nationalists drew support from business organizations like the employers' federations and the chambers of manufactures, but these too found it hard to speak with one voice. The paymasters of conservative politics remained the more exclusive committees of big businessmen (in Victoria alone the National Union subscribed £50 000 for the 1925 election) and accordingly they enjoyed privileged access to cabinet members. 'As we pay the piper we think we have a right to call the tune,' was how the president of the National Union responded to complaints that his cabal was riding roughshod over the mass organization, the National Federation. But even that group could not always get its way with the prime minister. In 1926 the same president of the National Union was refused an advance copy of the government's Crimes Bill and Bruce cut short a stormy interview, insisting 'I am committed to the party and the country'. He relied more on the advice of party organizers than parliamentary colleagues and did not disguise his disdain for the placemen and hacks that local committees often favoured with safe electorates.[4]

As treasurer and leader of the other party in government, Earle Page enjoyed special access to the prime minister. 'He used to come down from his office to mine practically every morning', Bruce recalled. The Country Party man was wont to exaggerate his influence. True, he was full of ideas ('nearly always half-baked', commented Bruce) and his group of fourteen members of the House of Representatives initially commanded five places in a cabinet of eleven. But under Page it became clear that the Country Party was something less than the independent force it had seemed when it burst into federal politics at the end of the war. The populist fervour could not be sustained indefinitely, the grass-roots resentment of urbanism lost much of its impetus once the Country Party gathered support from the established and more hard-headed organizations of farmers and graziers. Faced with a choice between preserving its independence in order to drive the best bargain, or aligning itself with the Nationalists on the conservative wing of politics, the Country Party chose the second alternative and bound itself into the coalition with a reciprocal agreement not to contest each other's seats. Radicals who chafed at these compromises—'How could they serve Collins Street and the man on the land also?' asked Percy Stewart, Mallee farmer and minister for works and railways until 1924—were cast aside. Instead of fighting the protectionist incubus that benefited the urban manufacturer, the farmers' representatives chose to 'break into the vicious circle themselves' with bounties, freight concessions and price support mechanisms.[5]

The catch-cry of the Bruce–Page administration was efficiency. The government was committed to building up the population and industries of the country while preserving its living standards, and this demanded an increase in productive effort:

> The future of Australia depended on efficiency. If to producers, primary and secondary, could be brought a realisation of the vital necessity for efficiency in management, control, finance and marketing; if to the workers could be brought an understanding that the standard of living and wages and of comfort they enjoyed depended on efficiency, Australia would be far towards the solution of its great problems, and could look forward with confidence to the great destiny that lay before it.[6]

As part of the drive to eliminate waste and slack, the government embarked on what the treasurer described as a 'spring cleaning', selling off various public enterprises such as naval dockyards and woollen mills. 'We were guided not by ideological motives, but by strict business principles', he insisted. This was a somewhat disingenuous description of his government's actions with respect to the Commonwealth Shipping Line. Its reorganization in 1923 had impaired its capacity to compete with the British shippers, and its sale in 1928 (to a British purchaser who subsequently defaulted) allowed the British combine to increase freight rates immediately.[7]

The Commonwealth also reorganized other undertakings to insulate them from public pressure and make them more amenable to business practice. In 1924 the Commonwealth Bank was removed from ministerial control and placed under a board drawn almost entirely from private enterprise. In 1926 H.W. Gepp was recruited from the Collins House group at a salary of £5000 a year to chair the Development and Migration Commission. In Gepp's words, since 'the hard and fast divisions between Government and Industry seemed no longer possible', it was all the more important to remove these important activities from the political arena. The composition of the Tariff Board, with its direct representation of both urban and rural capital, illustrated the same pattern.[8] The National–Country Party administration was hardly a parsimonious administration. It accepted that the state had a major role to play in the processes of economic development. It borrowed heavily to finance immigration, rural development, road building and other schemes. It made a major initiative of the Commonwealth Institute of Science and Industry, reorganized and greatly expanded in 1926 as the Council for Scientific and Industrial Research.[9] But the thrust of its activities was consistently in favour of producers.

The prime minister had a habit, when under pressure to extend his government's welfare activities or expand the sphere of its operations to meet popular needs, of launching a lengthy investigation:

> If somebody wanted something, you appointed a Royal Commission, or somebody to investigate it. Then you received a report, and most likely you made a speech about it (£20,000,000 housing schemes, national insurance schemes, etc., dropped off his lips with the easy fluency of a lover paying compliments to his sweetheart). Eventually something else would turn up, and, in any case, the people who started the bother would soon be tired of it.

Thus a Royal Commission on National Insurance spent more than two years gathering evidence and preparing recommendations, but nothing more was heard of its proposals. A Labor critic remarked that Bruce delighted the masses with his unlimited promises and the wealthy by the fact that he never fulfilled them.[10]

Yet in other respects Bruce's government was as watchful and as interventionist as that of his predecessor. It upheld racial exclusiveness with unswerving consistency and was not averse to accusing the labour movement of betraying White Australia by flirting with the principle of international solidarity—no 'dirty, greasy foreigner', said Bruce, should interfere with Australia's industrial system. It kept Aboriginals on federal territory in a state of abject dependence. It maintained a close watch over the morality and loyalty of its subjects: a regulation proclaimed in 1921 under the Customs Act, prohibiting the importation of literature 'wherein a seditious intention is expressed', enabled the government to prohibit more than 200 publications by 1929.[11] Indeed, the constant iteration of the need to defend a 'clean, wholesome nation' betrayed an official mentality that verged on the pathological—the alien agitator took on the same significance in the government's deliberations as did modernism in the pronouncements of the guardians of culture. Bruce's warning of the dangers of foreign sedition was echoed by the artist Julian Ashton who spoke of the leading European painters as 'scum rising to the surface of the melting pot as a result of the turmoil caused by the war'. In politics, as in art and literature, Australia became a haven from the modern frenzy. The universities, the galleries and libraries remained quite removed from the stirrings of awareness among a new generation of intellectuals—indeed, they saw themselves as defending the existing order against the scribblers. This clinging to the old and resistance to the new offers a valuable insight into the limited impact of the doctrine of scientific efficiency. Just as an admirer of Streeton's pastoral canvases claimed that they enshrined 'the way in which life should be lived in Australia with the maximum of flocks and the minimum of factories', so there was widespread resistance to the standardized, levelling world of the machine. But conservatism went beyond mere romanticized nostalgia. A threatening world had to be kept at bay, a beleaguered civilization had to be guarded from its enemies by constant vigilance.[12]

Despite his proclaimed intention to still the turmoil, Bruce's refusal to accept dissent beyond narrow limits of legitimacy perpetuated the bitterness and division of public life. The recent unrest had created a deep-seated unease. The spectre of Bolshevism called forth a variety of right-wing movements dedicated to the defence of the established order and events such as the Melbourne police strike in 1923 confirmed their fears and augmented their ranks. Beside the public activities of bodies such as the Empire and Loyalty League, the King and Empire Alliance and Constitutional Associations, there were the covert operations of private security organizations. Leading lawyers and businessmen, respectable clerks and insurance agents, sworn to secrecy and mostly sharing military memories, slipped from suburban homes after dark to rehearse plans for the defence of families and homes against the anticipated uprising of the unwashed.[13] Bruce played on such fevered imaginations during the federal elections of 1925 and 1928. The first was called over a maritime strike when the government announced its intention to deport two leaders of the Seamen's Union. One Nationalist poster presented the Labor Party as a donkey ridden by a Bolshevik, another showed revolutionaries shooting down citizens before a burning church. From the beginning, Bruce campaigned on the Red Scare: 'At the period of our greatest prosperity and most glowing opportunity there are wreckers who would plunge us into the chaos and misery of class war.'[14] He was rewarded with 51 seats for the coalition in the House of Representatives to Labor's 23. It was little consolation for Labor when the High Court ruled that the deportation of the union leaders was invalid, because J.G. Latham, the new attorney-general, drafted fresh legislation to bring the promoters of 'dissension, unhappiness and discontent' to heel. Latham explained that 'some of these men are not open to intellectual conviction; they require criminal conviction' and his bill outlawed revolutionary and seditious associations.[15] The 1928 election was dominated similarly by the government's draconian proposals to deal with a strike among waterside workers. This time, however, the strategy was less successful and the government majority in the House of Representatives was reduced to nine.

At the state level the labour movement recovered more quickly from its wartime reverses, so much so that when the Bruce–Page government swept the polls in 1925, five of the six states had Labor administrations. Admittedly, this was a rather unusual conjuncture. Queensland remained a Labor stronghold throughout the decade and the same dominance was achieved in Western Australia after 1924; but success was episodic in New South Wales, South Australia and Tasmania, while Victoria had only two Labor administrations during the 1920s and both lacked parliamentary majorities. Even so, the contrast with federal politics is striking.[16]

The Labor Party represented essentially the same aspirations as had informed its founders before the war. It shared the enthusiasm for national development but gave it a distinctly labourist and masculine twist, encouraging those projects that were most likely to create jobs and regulating the labour market to protect the wages of working men. Since work came first, the states were of primary importance. They eclipsed the Commonwealth in mobilizing funds for capital works: in the 1927–28 financial year, when public capital formation reached a peak, the states and local authorities raised £142 million, the Commonwealth £21 million.[17] The projects to which these funds were chiefly applied, road and railway construction, and the provision of electricity, gas, water and sewerage, gave employment to thousands of labourers and provided orders for many local businesses, while leaving the more profitable fields of enterprise to private capital. Indeed, it was the employment-creating effects of rural settlement that allayed the labour movement's deep-seated misgivings over immigration. Thus with the rapid development of the Western Australian wheat belt, teams of men were put to work building railways and roads, loading the wheat at sidings and wharves, and operating the State Shipping Service and state railways. The same logic was consistent with the encouragement of private as well as public enterprise. It was a Labor government in New South Wales, after all, that had urged BHP to base its steelworks in Newcastle and had itself established a state dockyard there. The principle of public ownership was less important than the immediate question of wages and conditions, with unions expecting Labor governments to act as model employers and to ensure that state tribunals improved industrial awards. Since the Nationalists prevented the Commonwealth Arbitration Court from granting the forty-four-hour week, Labor did so in New South Wales in 1921 (and again in 1926, after a period of non-Labor government), Western Australia in 1924 and Queensland in 1925.

The masculine preference for work and a living wage quite overshadowed public provision in the fields of housing, health or social security. A Labor government in New South Wales did introduce child endowment in 1927, but this was meant to compensate for the fact that the basic wage no longer met family needs and the measure was received with limited enthusiasm by the unions, which would have preferred a wage increase.[18] Such public assistance was thought appropriate for mothers and children since they were deemed unable to support themselves. Male unionists still sought to restrict female employment and still urged Labor governments to fix the earnings of the female wage-earner on the basis of her needs alone, though some were not averse to adopting the principle of equal pay as a way of making female labour less attractive to employers. Women wage-earners received little organizational assistance from male colleagues, and their efforts to improve their lot horrified the middle class. 'It is impossible for me to associate in my mind the usual refined girl waiters with the militant creatures of Saturday night's exhibition,' wrote an indignant Western Australian who encountered pickets from the hotel and catering industry.[19] To have recognized claims for equal pay would have been to grant female autonomy; to encourage their struggle would unsex them.

Able-bodied men, on the other hand, were expected to support themselves and it was the business of the state to make sure that they did. Hence the general aversion to providing for unemployment except by the creation of public works. It is true that Labor in Queensland introduced a scheme of unemployment insurance in 1923, but this was tailored to the seasonal work patterns of the sugar industry in that state and meant primarily to tide the cane-cutters through the slack months. The submissions of unionists and even representatives of the unemployed themselves to the Commonwealth Royal Commission on National Insurance revealed general agreement that 'doles' were demeaning to the male breadwinner. Men wanted 'employment rather than charity', said the leader of the coal-miners, and the secretary of the Melbourne Trades Hall Council explained that there was 'nothing more objectionable to the average man than to offer him something for nothing'; another witness revealed the gender bias of his position more clearly still with the protest that 'a man loses his manliness under such a system'.[20]

These values were sufficiently widespread to determine the framework of state politics, for the ALP mobilized enough support with its policies of developmental labourism to place its opponents on the defensive. True, the non-Labor forces were sometimes able to respond with projects of their own: one example was 'Moo-cow' Mitchell's promotion of an Arcady in the West, another was the Lawson ministry's ambitious attempt to use the Victorian brown coal deposits to make the Latrobe Valley a second Ruhr. True, they could choose instead to deride union demands and offer themselves to the electors as prudent administrators. But either strategy was likely to cause dissension in the non-Labor ranks. The solid federal alliance of the National and Country parties could not be repeated since the compromise of 'protection all round' was not available at the state level. Thus urban interests condemned profligacy while rural interests resented parsimony, and big business sought to reduce the role of government while small business often depended on it. Moreover, the low calibre of men attracted to state politics imposed a serious handicap on the conservative side. Bavin, the leader in New South Wales, grumbled of his colleagues that 'if I could get rid of about 60 per cent of them, and fill their places with decent men instead of political hacks and deadbeats, I'd gladly do it'.[21]

Ironically, Labor's success was most marked in those states where the opportunities for rural development were greatest because in these regions secondary industries were tailored to primary ones and it was possible to construct a broad alliance of wage-earners and the self-employed, a popular front of those who rolled up their sleeves. Yet the very circumstances that favoured Labor in the tasks of state-directed development also imposed limits on its freedom of action. Since the all-important public works and public enterprise were funded by overseas borrowing, it was necessary to satisfy the lenders that their funds were secure. This ruled out radical measures such as the attempt of the Queensland government to raise pastoral rents and break up the big estates. When first introduced, the City of London warned that the reforms would 'tend to destroy the confidence of British investors'. At first Theodore stood firm in denunciation of the 'bondage of despotism of the money lenders of London', and went instead to New York in order to borrow there. But the cost of American funds was high and by 1921 Theodore's government was forced to retrench. By 1924 he had dropped his plans, made peace with London and obtained new loans.[22]

The same circumstances fostered a pragmatic moderation among Labor politicians. Fulminate as they might against Mr Fat and the Money Power, they made no attempt to upset existing property relations. At best, they promoted mildly redistributive measures. Even that limited objective was ruled out by the Labor premier of Tasmania, Joseph Lyons, whose insistence on balanced budgets and the amiable reasonableness of what he called 'purely national non-party work' won the co-operation of the conservatives. Insofar as the Labor premiers made an effort to implement the socialist objective that the ALP had adopted at its 1921 federal conference, they interpreted it to mean little more than the extension of state enterprise. Steady employment and good wages, the overriding objectives of the labour movement, were thus pursued within the existing class structure. Even then, electoral considerations demanded that the aspirations of the unionist be balanced against the interests of the customer of public utilities as well as the taxpayer. Thus Labor premiers in South Australia and Tasmania resisted the introduction of the forty-four-hour week. In Queensland, where Labor's grip was strongest, the increasing tension between cabinet and aggrieved unionists culminated in a lock-out of public employees and the expulsion from the party of several industrial unions. Bill McCormack, who had succeeded Theodore as premier on the latter's transfer to federal politics, spelt out the logic of his position to caucus:

> If any member here believes that a principle of unionism ... is at stake, and that because he is a member of a union he is compelled to give allegiance to some outside body, and not to this parliament, then he ought not to be in this parliament.[23]

Even in New South Wales, where the radicalism of the 'Trades Hall reds' meant that the unions controlled the political wing during the early 1920s, the redoubtable Jack Lang, a former estate agent, was able to establish his supremacy as parliamentary leader before the end of the decade.[24]

Then there was the continuing problem of the upper house. These second chambers of the state legislatures, either nominated or elected on a restricted franchise, were implacably conservative and they offered the shrewd Labor premier a convenient lightning rod for popular discontent. The Queenslander Theodore abolished his in 1922, Lang went through the same motions in 1925 when he stacked the Legislative Council of New South Wales and Joseph Lyons challenged the pretensions of the Tasmanian upper house in 1924. Elsewhere the Legislative Council remained intact and unbowed, an obstacle to legislative reform.[25]

The dominance of the political machines might seem an unusual feature of a party based on the trade unions. After all, union delegates made up a majority at most party conferences and union members usually determined the outcome of the pre-selection ballots that were then the usual procedure for choosing parliamentary candidates. It is true that spoils of politics, at both state and local levels, presented considerable opportunities for wirepulling. The liquor trade was a generous subscriber to party funds and a number of myths had gathered already round the notorious John Wren, who had survived the closure of his betting shop to own racecourses and newspapers. One of his critics spoke in parliament of the sinister parasitic interests that corrupted the Labor Party and various parliamentarians were certainly beholden to him. There is strong evidence that he paid several thousand pounds to induce a federal parliamentarian to vacate a seat for his business partner Theodore in 1927—not that Theodore was short of money after his government had paid £40 000 for the Mungana Mining Company, in which he held an undisclosed interest. Yet Wren's branch stacking and bribes could scarcely match the voting power in Labor assemblies of the affiliated trade unions.[26] The unions themselves were susceptible to dubious practices, as was notorious in New South Wales where ballot boxes with sliding panels were only one nicety of the political scene. But the labour movement in New South Wales was exceptional not so much in its sharp practice as in its instability—the ruling 'inner groups' were so greedy in the exercise of power that they were repeatedly overthrown by a coalition of the aggrieved elements that they excluded.[27]

Elsewhere, control of the party rested more securely in the hands of politicians and party administrators with close links to the biggest union of all, the AWU. The AWU had expanded from its original pastoral base, to take in the Queensland sugar industry, the metalliferous mines of Queensland and Western Australia, the smelting and ironworks of New South Wales and South Australia, construction workers on public works, and even small farmers. By 1928 it had 160 000 members, or one in six of all trade unionists. Its special strength in Queensland, South Australia and Western Australia meant that it dominated the party office in those states; its extensive rural coverage gave it control of pre-selection for most country electorates, and its provision for parliamentarians to hold the union ticket gave it a powerful voice in the state caucuses. In the Labor governments of those states especially, the AWU held the key portfolios of public works and industrial relations. Exploiting them to the utmost, silencing critics and eschewing militancy, its officers practised a ruthless but effective patronage. Projects and jobs went to those who co-operated. While the techniques of jobbery and kickbacks may not have been new, the rich spoils of developmental labourism gave them a fresh potency and, when wedded to the discipline of the party machine, such practices brought undeniable results.[28] The union's general secretary would claim in 1929 that half the Labor members of the House of Representatives held AWU tickets. At the other end of the industrial spectrum were the left-wing unions, notably industrial unions like the coal-miners, the railwaymen and the waterside workers, and the militants of the Sydney and Melbourne trades halls; but they had been set back by the collapse of the One Big Union and only began to reorganize with the creation of an All-Australian Council of Trade Unions in 1927.[29] For the time being the initiative rested with the pragmatists.

While the states still bore the greater responsibility for public provision of goods and services, the balance was shifting. Increasingly, their needs outstripped their means. The income of the Commonwealth grew to approximate parity with the states but the formula for reimbursing them for lost customs dues remained fixed at a flat annual payment of 25s per head of population, a shrinking proportion of federal revenue. Already the Commonwealth had begun to allocate grants for specific purposes, notably road construction and rural settlement, but these commonly required a matching state contribution and consequently reduced the states' freedom of manoeuvre still further. So too Commonwealth instrumentalities like the Development and Migration Commission were used to scrutinize the states' internal affairs. It was possible for the states to finance activity by borrowing, but this option also was closed when the Bruce–Page administration became concerned that excessive loan-raising was placing too great a pressure on interest rates in London. In 1927 the Commonwealth abolished per capita grants and thus forced on the states a Financial Agreement whereby a Loan Council was established to co-ordinate and control future borrowing; in return, the Commonwealth took over the accumulated states debts, some £672 million in all. The Financial Agreement was ratified by referendum in the following year.[30]

The judicial interpretation of the Constitution also turned in favour of the Commonwealth, as, following the retirement from the High Court of Sir Samuel Griffith, who had been a consistent upholder of states' rights, the doctrine of implied immunities fell into disfavour. In the Engineers' case of 1920 the court decided that workers in state instrumentalities fell within the jurisdiction of Commonwealth arbitration. Whereas in that year only 100 000 unionists worked under Commonwealth awards against 670 000 under state awards, by 1924 the Commonwealth covered 550 000 and the states just 225 000. But there were limits to the concentration of power. Many workers in the Labor states could obtain better conditions from local bodies than those handed down by the judges that Bruce appointed to the federal tribunal; some unions were in the habit of transferring claims to the court that best suited them. Bruce attempted in 1926 to close off these loopholes when he brought on a referendum to obtain all-encompassing industrial powers. Both conservative localists and Labor leaders denounced this as a case of 'the Federal spider spinning a web to catch the unfortunate State flies' and the voters rejected the proposal.[31] The same suspicion of a remote and unresponsive central authority was apparent in criticism of the move of the Commonwealth parliament from its temporary home in Melbourne to an unfinished national capital in the high country of southern New South Wales. 'It is just a windswept, cold, miserable place in poor country that would not keep a bandicoot', said one Nationalist senator and few who have stumbled off the overnight train at Yass to complete the journey to Canberra would challenge his verdict.[32]

Anti-federal sentiment was greatest in the isolated and underpopulated states, or the 'small states' as they were usually known, even though one of them, Western Australia, covered a larger area than any. Tasmania and Western Australia had received special Commonwealth grants since before the war, South Australia joined them before the end of the 1920s.[33] The rationale of such assistance was compensation for the high cost structure that benefited the secondary industries of the large states at the expense of the primary industries of the small. Beyond that, a more general principle can be discerned, one that operated also in Sydney's response to the growing 'new state' agitation in outlying regions of New South Wales. The expectation that Australian governments would protect citizens' living standards had taken on a spatial as well as a social meaning—it applied to rural as well as urban dwellers, small states as well as large. By protection all round, as well as regulation and allocation of resources, the Commonwealth and state governments were committed to a shifting, incomplete but nevertheless significant principle of equity.

The concern for balance was at the same time a search for authority. Bruce and Page, Theodore and Lang, despite differences in their notions of the good society, shared a remarkably similar understanding of the need for a guiding hand in the task of national fulfilment. The inchoate energies of the people had to be given purpose and direction, their egoistical impulses harnessed to constructive endeavours. It was a highly pragmatic mentality, expanding the obligations of citizenship not out of any desire to glorify the state but simply because compulsion came readily to hand. Nowhere was this more evident than in the passage, after the briefest of debates in 1924, of legislation requiring all eligible persons to vote in federal elections. With similar indifference to ultimate purpose or the niceties of how it was achieved, the same leaders all sought to impose order on social unrest. The major political parties, despite the fact that they represented different classes, shared a belief that the antagonistic effects of class must be brought into institutional adjustment. Tariffs and bounties, wage awards and subsidies, public works and benefits, made up a rough and ready calculus for this purpose.

It was not a generous endeavour. When the national vision shrank to material dimensions, the readiness to experiment in other fields was lost, so that the reforming zeal that had been applied to education, for example, succumbed to parsimony and institutional inertia. With Labor confining its energies to a limited range of well-worn expedients and the cautious optimism of the Deakinite liberals giving way to conservative mistrust, the capacity for innovation that had been apparent in the early Commonwealth was well-nigh exhausted. The political leaders aimed not so much to resolve conflict as to confine it, not to liberate energies but to fashion institutions that would operate as far as possible beyond the reach of the democratic process. The use of judicial bodies, statutory boards and commissions, as well as the elaboration of bureaucratic procedures, expressed a mistrust of allowing political representatives to control matters which so closely touched the interests of the electors.

Recourse to such devices did not, however, silence the popular clamour. The Arbitration Court, for example, was subjected to mounting attacks from both left and right, with militant unionists denouncing it as a tool of the capitalists and disgruntled employers condemning it for surrendering to workers' demands. Similar attacks were directed at the Tariff Board for giving in to importunate demands. The board itself expressed unease at the way employers and unions colluded to obtain increased support of local industry, then hurried off to the Arbitration Court to arrange a wage increase. A judge of the court protested against the expectation that he and his colleagues 'should be merely recording agencies of their demands and determinations, and complacent judicial tools for the legalising of their pronouncements and edicts'.[34] And even when such importunities were rejected, the applicants merely turned to parliament to override the tribunal's decision. A host of lobbyists specialized in buttonholing parliamentarians for this purpose. 'They swarmed about King's Hall,' one member testified, 'sneaked down corridors and even attempted to invade the rooms of ministers.' Invasion was hardly necessary—the minister for trade and customs was himself the former president of the New South Wales Chamber of Manufactures.[35]

The growth of 'protection all round' therefore brought a swelling chorus of criticism. Many of the complaints came from aggrieved claimants who were merely pressing their own concerns against those of rivals, but over that cacophony two voices sounded sharp and clear. They require consideration not because they prevailed at once against established decision-making procedures but because their critical diagnoses gathered relevance and cogency as Australia's economic difficulties mounted until, eventually, they could no longer be resisted.

The first and most beguiling voice was that of the expert. The species came late to prominence in this country, yet by the 1920s there were unmistakable signs that it had done so. Advances in the technologies employed in secondary industry were one indication, the growing importance attached to the Council for Scientific and Industrial Research was another. While the sociologists could not sustain their ambitious claims, other social sciences stamped their influence on public policy, so that anthropologists, for example, shaped Australia's administration of New Guinea from the 1920s onwards. Above all, there was the establishment of the economics profession. The handful of professors who had taught, year in and year out, the same syllogisms from the same textbooks, gave way to a younger generation more closely attuned to overseas developments in the discipline, confident of the validity of its analytical tools, and anxious to apply them to Australian conditions. A school of commerce was established at the University of Melbourne in 1924 and immediately enrolled 300 students. In the following year the evangelists created the Economic Society of Australia and New Zealand, with its own journal, the *Economic Record*, which proclaimed the need 'to promote economic knowledge, and more especially to encourage a respect for such knowledge, and for the scientific approach to economic problems which bulk so largely in the political life of Australia'.[36]

In 1929 the first research chair was taken up by Lyndhurst Falkiner Giblin, perhaps the most influential and certainly the most attractive of the new school. Born of a leading Tasmanian family, he had studied at Cambridge (where he represented England at rugby), prospected in the Klondike, and worked as a teamster, lumberjack and sailor before returning to his home state. He was a Labor parliamentarian until he volunteered for service in the First World War where he was twice decorated for valour, and then ran an orchard before taking up the post of government statistician. With his dubbined boots and a tie cut from a strip of red cloth, he was an unlikely figure to wield influence among politicians and businessmen, but the power and originality of his mind, coupled with a striking turn of phrase, made him an important adviser for the next twenty years.[37]

Giblin and his colleagues quickly established links with business and government. They were increasingly to be found on statutory bodies and committees of inquiry into areas of public policy, where they found much to criticize. Loan policy, public works, wage determination and the labour market, overseas trade, the tariff, rural settlement, all were weighed in the balance of their specialist knowledge and found wanting. The findings of the geographer Griffith Taylor were taken up, and married with the principle of diminishing returns to pour scorn on population policies: the popular slogan 'A million farms for a million farmers' was judged 'so mad that it should have laughed its author out of public life into an asylum'. The reliance on the export earnings of primary industry was analysed and likened to a vision 'of Australia as one enormous sheep bestriding a bottomless pit, with statesman, lawyer, miner, landlord, farmer and factory hand all hanging on desperately to the locks of its abundant fleece'. Wages were found to be excessive, and the method of their fixation vitiated by extraneous considerations. The tariff was compared 'to a powerful drug, with excellent tonic properties on the body politic, but with reactions which make it dangerous in the hands of the unskilled and uninformed'. Here again the stern dictates of economic reality elbowed aside the expectation that the state could make material development conform to national ideals.[38]

It is plain that the economists did not shrink from polemics. Much of their impact derived from an ability to translate their ideas into graphic metaphors that were impossible to ignore—hence Edward Shann at the University of Western Australia described Australia's predicament in 1929 as that of a stranded steamer:

> We have fixed so many costs and standards. The markets we serve have fallen and left our fixtures as high and dry as a steamer on the Nor'West coast tied to a jetty when the tide is out, though with the difference that the tide may not return.[39]

Seized with the zeal of the neophyte, convinced of the urgency of their insights, they loosed their jeremiads in all directions. The recurring theme, however, was that their advice was based on a higher wisdom that could not be gainsaid and that the popular expectation that government could abrogate the operation of the market was simply misconceived. Economic science transcended the passions of politics, the 'crude nationalism, political warfare and class consciousness' that impeded the adoption of realistic policies, and it was therefore imperative that politicians defer to the economists' superior and disinterested advice. 'Just at the present moment it so happens that the economist is (or should be) king in this as in every other country', claimed the head of the University of Melbourne's school of commerce.[40] Seen thus, Australia's problems became a problem of authority, centring around the inability of the state to resist the excessive demands that were made of it. Other intellectuals took up the theme, most notably the historian Keith Hancock in his influential survey, *Australia*, written between 1927 and 1929. In the absence of a class or a tradition that could withstand the assault of numbers, he explained, 'Australian democracy has come to look upon the State as a vast public utility, whose duty it is to provide the greatest happiness for the greatest number'. Too many eggs had been put into the political basket, with the result that when some went bad, the odour penetrated every corner of national life.[41]

While Hancock paid a layman's respect to the authority of the economists, he drew also on the experiences of some men of affairs who were out of sympathy with their times. There was his South Australian friend and companion of the Round Table, Charles Hawker, a blue-blooded and high-minded pastoralist who made light of the terrible injuries he had suffered in the war, and there was the prim lawyer, fellow-Round Tabler and Victorian minister, Frederic Eggleston. Both were on the conservative side of politics but were as much concerned with its deficiencies as with the failings of the Labor Party. They regarded public life as an avenue of service in which duty and honour should be uppermost, and they felt a corresponding distaste for the pursuit of advantage that so many of their colleagues exhibited. It was a common complaint among tories of the Round Table persuasion that the times were out of joint and that too many members of the propertied class were chasing business or financial success at the expense of public service. The organizer of the National Federation liked to recall how when first he approached Bruce to stand for parliament, the prime minister had declined. The organizer put it to him that as he had served his country in war, so he should protect it against the Bolshevik menace that threatened civilization in the aftermath of the war.[42] There were too few prepared to make Bruce's sacrifice. Denouncing the inefficiency, expediency and jobbery that seemed inseparable from the politics of the 1920s, these Cassandras made up a second voice of dissent. Examples abound, none more scathing than the impressions that the irascible Wilfred Kent Hughes set down when he first sat in the Victorian parliament:

> ye gods!—taken in the mass our side are a lot of boneheads, and the other side a lot of uncouth, semi-educated, ill-mannered, narrow-minded boors, except for a half dozen or so. At least they work as a team, well-organized, which is more than we can say for ourselves. No handle but that of the parish pump is ever grasped—barring that of the beer pump. It is all what is best for ME, MYSELF, the great god Ego.[43]

Kent Hughes entered politics in 1927 as part of the reform group of Young Nationalists that included the rising young lawyer Robert Menzies, but within two years he was sitting in the cabinet room himself. This was a common fate of initiatives launched during the decade for 'clean' politics. Just as the radicals in the labour movement made little headway against the entrenched moderation of the Labor leaders, not least because parliamentary life extinguished the fire of many a militant, so too the Nationalist Party absorbed and tamed its critics. Since those who held the purse strings enjoyed the rich spoils of political patronage, they carried the day against the purists.

By themselves the dissidents were unable to deflect Australian policy. The Commonwealth and the states continued to borrow in order to promote development and they maintained, even extended, their support for organized producers. But the limits of economic expansion became apparent in the second half of the decade. As output stagnated while debts mounted, the government was forced to pay heed to critics. As early as 1926, London financiers drew attention to some disturbing features of Australian borrowing. The accumulated foreign public debt had risen by then from £419 million in 1920 to £562 million, interest charges from £7 million per annum to £26 million. It was stated that 'In the whole British Empire there is no more voracious borrower than the Australian Commonwealth. Loan follows loan with disconcerting frequency.' Too many borrowing proposals were deemed unsatisfactory. Details were lacking; it was not always clear that funds were to be applied to productive projects, and even then there was insufficient provision for sinking funds; too often the borrowed money appeared to be used to pay off maturing loans or even to meet interest payments. 'It is, in fact, high time to ask the question—Is Australian finance sound?'[44]

Faced with such charges on his visit to Britain for the 1926 Imperial Conference, the prime minister had little alternative but to satisfy the creditors that all was well. R.G. Casey, with his city contacts, put the matter in a nutshell: 'It was no use saying that the London money market was "wrong" in being critical. It was our place to satisfy the lender that we were right.'[45] Bruce therefore invited a delegation of British businessmen to come out to Australia and set their minds at rest. Unfortunately, he waited so long to secure the businessmen he wanted that they did not arrive until September 1928, when conditions had deteriorated to such an extent that a favourable report was hardly to be expected. Even so, the severity of the British strictures was extremely disturbing. The Economic Mission declared that Australia 'has been mortgaging the future too deeply'. Loan funds had been misapplied chiefly because of the 'pressure of sectional interests', and the result was 'a dead weight burden of debt'. Moreover, the cost structure of Australian industry was too high and must be reduced:

> We have been strongly disposed to the view that the combined operation of the tariff and of the Arbitration Acts has raised costs to a level which has laid an excessive and possibly even a dangerous load upon the unsheltered primary industries.[46]

Bruce had by this time reached similar conclusions. The downturn in rural prices from 1927 convinced him that the export industries had to be freed from the burden they had been made to carry. 'We have gone tariff mad in Australia', he told a member of the Tariff Board.[47] Most of all, he was sure that labour costs were too high. The commitment to maintain living standards, on which had been predicated the policy of 'Men, money and markets', was effectively revoked with this judgement and the government prepared to force down wages.

The decks were cleared. In 1926 the government had amended the Crimes Act to make it an offence to obstruct maritime transport, and it had appointed three new judges, with increased powers, to the Arbitration Court. Now the Arbitration Act itself was overhauled. Provision was made for secret ballots. Penal powers were strengthened: it became an offence to prevent a person from working in accordance with an award and unions were held responsible for the actions of their members. Furthermore, the tattered convention that wages were determined on the basis of workers' needs was now explicitly qualified: in making awards the court was required to consider their economic effects. All these measures were meant to establish an effective Commonwealth jurisdiction within the constitutional limitations left by the failure of the 1926 referendum—in Charles Hawker's apt phrase, they amounted to 'stuffing the corpse in the best way'.[48] Bruce was more than ever convinced that the country's difficulties were caused by agitators, and he was determined that they should be prevented from working their mischief. He did not understand that his draconian measures cut clean across any possibility of voluntary co-operation from the unions, and their immediate result was the collapse of a national Industrial Peace Conference.

More than this, the attempt to reduce wage costs and stamp out resistance triggered a series of bitter and protracted industrial disputes whose significance is evident in the number of days lost in strikes during 1929 and 1930. The figures in table 10.1 understate the extent of the turmoil since they do not include those workers who were locked out by employers and denied re-employment when the dispute ended. The major conflicts were brought on by employers seeking to drive down wages, increase hours and impose more exploitative conditions, while the unions were fighting to defend their position. Not all employers could afford to embark on such a costly exercise; not all unions were able to stand firm. In three major industries, maritime transport, coal and timber—industries that still had not succumbed to the rhythms of the factory but that had a far-reaching significance as suppliers of basic materials and services for the national economy—determined owners met obdurate workers. The results were bloody.

The strike of the waterside workers was triggered by an award of the Commonwealth Arbitration Court in August 1928. The award removed a number of concessions that the men had won over the past few years and, above all, instituted a second pick-up. Previously, the men offered for work in the morning and if they were not engaged, could look elsewhere for a labouring job. The pick-up was a degrading procedure at the best of times: men stood like beasts in a stockyard to be scrutinized by the stevedoring foreman and those he did not like went away with nothing to show for their attendance. Now, for the convenience of the shipown-

Table 10.1: Working days lost in strikes, 1923–31[49]

1923	1 146 000
1924	919 000
1925	1 129 000
1926	1 131 000
1927	1 174 000
1928	777 000
1929	4 461 000
1930	1 511 000
1931	246 000

er, they were expected to reassemble at a second pick-up later in the day. The union officers were reluctant at first to declare a strike against the award but the men stopped of their own accord in several ports. If nothing else, one union official remarked, the government had ensured that the rank and file controlled their own affairs—they struck! Yet under the provision of the Arbitration Act, the union was held responsible and fined. By September the stoppage was general. This had been anticipated all along by Judge Beeby, the author of the award, who revealed his intentions in an astonishing confidence to the federal attorney-general:

> I realised when I made the award that there would be a strike but felt that the power of an arrogant faction which controlled the union must be broken. The conflict should result in a reorganized union from which the unreasonably militant element and other undesirables will be excluded.

To expedite the process, the government rushed new legislation through a parliament that was about to dissolve for the forthcoming election. The Transport Workers Act, better-known by the wharfies as the 'Dog Collar Act' since, as one put it, they were to be 'licensed like dogs', allowed the government to revoke the work permits of those who did not toe the line.[50]

This weapon quelled resistance in most ports. Since the stevedoring companies were able to replace members of the Waterside Workers' Federation with volunteers drawn from the ranks of the unemployed, the fear of losing a livelihood caused all but a few branches of the union to capitulate. It was evident that the mounting numbers of unemployed men willing in their desperation to cross picket lines severely weakened the unionists' ability to defend their living standards. This would be a recurrent phenomenon over the next few years, as the popular lament of the 1890s was heard again:

> The Lord above, send down a dove,
> With wings as sharp as razors,
> To slit the throats of bloody scabs
> Who cut down poor men's wages.

As Beeby had anticipated, some employers took advantage of the new provisions to blacklist unionists and deny work to them altogether. Here the struggle continued with particular intensity. There were pitched battles between the strikers and the police who each day escorted strike-breakers from the railway station to the wharves. Stragglers were ambushed and beaten up. Italian and Greek migrants, since they were readily identified, were a target for particular violence. In Adelaide the unionists were defeated by the superior strength of an armed Citizens' Defence Brigade. In Melbourne the daily experience of being left standing while newcomers were taken on proved too much for the unionists, who tried to rush the ships. The police opened fire and one unionist died of his injuries. Eventually, resistance was broken and as work fell away in the winter of 1929, there were only a few hundred members of the Melbourne branch of the Waterside Workers' Federation scrambling for work that was left by the scabs.[51]

Hard on the heels of that judicial assault on working conditions came another, this time in a timber-workers' award handed down by Judge Lukin in January 1929. Citing the industry's difficulties, he raised the working week to forty-eight hours, widened the provisions for piece-work, increased opportunities for replacing adult with juvenile labour, and reduced wages. It was, said the ACTU, the most iniquitous award ever proposed by any arbitration court in Australia.[52] The timber-workers refused to accept it and were duly fined; the court issued them with ballot papers and they either burned them, along with an effigy of the judge, or marked them in favour of the strike. Indeed, the progress of the dispute showed that the government had called into being the very spectre it had raised so frequently over the past few years. The Communist Party had declined in internal acrimony during the 1920s to the point of extinction, with just 250 members by 1928 and negligible industrial influence. Now, through its industrial Militant Minority Movement, it was able to offer workers the aggressive leadership that was so conspicuously lacking in the unions.[53] Whereas the union officials sought to confine the dispute to timber-workers only and keep them working a forty-four-hour week, the militants stepped up picketing and raised support from other unions. The changing atmosphere was evident in the large crowd that marched through Sydney singing that familiar dirge, 'The Red Flag', the Wobblies' 'Solidarity Forever', and a new anthem, 'The Internationale'.[54] In keeping with its sentiments, the strikers defied court orders and fines, and maintained a guerilla war against police and armed delivery drivers. Their last resistance was not broken in New South Wales until October.[55]

By then the northern coalfield of New South Wales had become another battleground. That the industry there was in urgent need of reform was not in contention: the growth of other fuels had resulted in overcapacity in the industry (the average miner worked only 168 days in 1928) and the restrictive practices of the Coal Vend allowed too many small and unproductive pits to continue production.[56] The owners proposed to resolve the problem by cutting wages, and pursued their objective at conferences held during 1928. Unable to obtain the union's agreement, they did not bother about judicial sanction but simply informed 10 000 miners in February that they would be dismissed unless they conceded wage reductions, accepted the owners' control over hiring and firing, and abandoned their right to hold pit-top meetings. Again the picket lines were established, again the police broke them to escort strike-breakers into the pits (and before the end of the year there was another union fatality, this time at Rothbury where the police opened fire on pickets).[57]

The intention to starve the miners into submission could hardly have been clearer—one owner told his workers that he would force them to eat grass before letting them into his pit. Here, surely, the Commonwealth was obliged to apply the same sanctions against employers in breach of an award that it had wielded against striking unions. A prosecution was in fact initiated early in 1929 against John Brown, one of the principal owners, but the government meekly withdrew it when the owners made this a precondition of further negotiations. The favoured treatment of John Brown flew in the face of the government's rhetoric of even-handed enforcement of industrial law. Here was a man of enormous wealth, one who could spend £2000 or more on racing horses and whose estate was declared at £640 000 when he died in the following year and a generous portion of it was bequeathed to the chief justice of the High Court, which caused many to query that court's impartiality in the recent litigation over the validity of awards.[58] Yet Brown denied work to men on the grounds that he could not afford to mine coal unless they gave up a shilling a day: Latham, the attorney-general, publicly defended Brown's action, explaining it is no offence to refuse to give employment if it is impossible to carry on at a profit, but he advised the prime minister in a secret memorandum that the withdrawal of prosecution was 'politically unwise' since it would encourage resistance to the law by strikers.[59] The publication of that memorandum in a newspaper on 14 August 1929 set in motion the events that brought Bruce down.

The government survived an immediate motion of no confidence in the House of Representatives, albeit with a reduced majority, and the sulphurous Billy Hughes was expelled from the Nationalist party-room along with another dissident. But the government's credibility in the field of arbitration was exhausted. In fact Bruce had decided already to abandon it. He had tried to impose order and he had met with no co-operation from the states and resistance from capital and labour. The leading employers saw little merit in an institution that they claimed had pampered workers to the extent that they had become 'bowelless tyrants and unmitigated ruffians towards the industries from which they draw their sustenance'. They wanted a free hand. 'Away with it, and let us get back to the clear, open, economic ring.'[60] The unions were even more antagonistic. Far from preventing strikes, it seemed to Bruce, the imposition of penal sanctions merely embittered them and caused Bruce's name to be execrated. He therefore decided that the Commonwealth

A Labor comment on Bruce's arbitration proposals, 1929

would hand responsibility for arbitration back to the states, retaining jurisdiction only over federal public servants and the maritime industry. The decision was confided to the officers of the Nationalist Party in May 1929 along with a request that they should 'steady the great commercial community' since it would be impolitic if the news 'were greeted with paens of joy and triumph by the employers' federations'. Hawker, as head of the South Australian branch, relayed the message to local employers and was able to reassure Canberra that they had not 'lost their heads with such exultation as might endanger the situation'.[61]

Not all the employers, indeed, received the news with the unmixed enthusiasm that Bruce, the Flinders Lane merchant, anticipated. Small employers, especially, lacked the strength to engage in a costly free-for-all and many manufacturers suspected that the abandonment of arbitration presaged a further retreat from the protectionist compact on which they depended. They lobbied vigorously against the Bill to dismantle the court. The Labor members were certainly critical of recent decisions of the court but they wanted reform of the Arbitration Act, not repeal. There were also some government members of the House of Representatives with scores to settle. Apart from Hughes and his fellow-malcontent, there was Percy Stewart and another Country Party radical, and a blind Melbourne lawyer affronted by the expediency of Bruce's volte-face. Then there was a wealthy New South Wales investor with links to the film industry, which was up in arms against the goverment's proposed entertainment tax. This member was thought to be the least resolute of the rebels so Hughes and Stewart kept him playing billiards away from the ministrations of his party whip. Lastly, there was the speaker and former attorney-general, Sir Littleton Groom, who had been dumped from the ministry back in 1925 because he had bungled the deportation of the two seamen's leaders. Groom's refusal to vote for the government meant that its Bill was defeated.[62]

Bruce called an election on the issue. But whereas he had campaigned twice before as the champion of law and order, this time it was the Labor Party that presented itself as the defender of living standards and industrial peace. Only fourteen Nationalists were returned to the House of Representatives and Labor, led by James Scullin, won a majority of seventeen members. Victory was all the sweeter because E.J. Holloway, who as secretary of the Melbourne Trades Council had been fined for his part in the timber-workers' strike, defeated the prime minister in his own electorate. He and the other Labor members gathered in Canberra to take office for the first time in twelve years, confident that they could restore prosperity and remedy injustice. A Nationalist senator would claim when the new parliament assembled that he had heard them singing 'The Red Flag' in the Hotel Kurrajong.

11

A CHANCE OF 'A BAD SMASH'

IN THE AUTUMN of 1927, as seasonal employment in the back-blocks fell away, Jock Neilson sought work in the metropolis. 'The depression was beginning to be felt and there were thousands of unemployed in Melbourne at that time.' With the onset of spring he tried again but 'The depression was now setting in properly and I found it impossible to get anything to do.' He did not work again in 1927. The year following brought no better fortune: '1928 was the start to the depression which has raged more or less ever since.' And in 1929, he recorded, 'the depression started in earnest'.[1] Neilson's recollection was not mistaken. Casual workers found work elusive during the 1920s, be they unskilled manual labourers or carpenters, and many of the features associated with acute distress were apparent long before the decade was out—thus the 'humpy', the rough accommodation improvised by the homeless, made its appearance in Newcastle by 1925. If a doctor's wife could remember a moment in 1929 when 'all of a sudden everything seemed to stop', a bush-worker's experience was quite different: 'Depression! I never knew nothing else! The 1920s was just as bad.'[2]

The term itself contributed to the confusion. Businessmen habitually used the metaphor 'depression' or 'trade depression' to refer to movements in the trade cycle—a stockbroker's newsletter reported early in 1929 that 'From "set fair" a few months back the barometer has swung round to "stormy"', and an economist explained that these 'highs' and 'lows' of business activity were like the 'highs' and 'lows' of barometric pressure. There was a belief not merely in the inevitability of such fluctuations but even in their salutary character. Thus the president of the Commercial Travellers Association told his members in May 1931 that

> The community cannot escape from the natural aftermath of extravagance and after an era of fictitious prosperity there must be a day of reckoning, just as the crest of a wave will leave a trough behind it before the water reaches its level again.[3]

But by then more than a quarter of all wage-earners were unemployed. The domestic product had fallen by 10 per cent, exports by more than a third and Australia was struggling to meet its financial commitments. More than this, the economic malaise paralysed government and threatened the social fabric. Even in its capitalized form, the term 'Depression' is hopelessly inadequate to convey the magnitude of the disaster. Yet it was with this beguilingly simple term that the emergency was understood and its remedy debated.

How did the Depression manifest itself? It was the custom of Australian governments to finance their activities by running up overdrafts with London financial institutions which were then cleared by floating a loan issue. The shortage of British loan funds in 1927 had resulted in the visit of the British Economic Mission and delivery of its unpalatable report. At the beginning of 1929 the situation became more difficult. The loan issued in January was seriously undersubscribed and 84 per cent had to be taken up by the underwriters, who refused to issue the next quarter's loan. Apart from forcing cuts in public works, the shortfall in London funds caused a growing difficulty in meeting external commitments since the debt now had to be funded from export income alone. Australia's overseas reserves dwindled to a dangerous level. Immediately after the federal election, the new prime minister, James Scullin, met with the chairman of the Commonwealth Bank who placed before him the figures regarding Australia's financial position in London. 'I was staggered,'[4] he said. By the end of the year the government had called the gold bullion of the private banks into the coffers of the Commonwealth Bank so that it could be used to meet overseas commitments, but even so its reserves continued to disappear. The Australian currency was losing value against sterling.

Worse was to follow. The prices of wheat and wool, Australia's principal export commodities, were declining even before the Wall Street Crash in October 1929: following that blow to the international economy, they tumbled steeply. From a high point in the late 1920s they would decrease by more than 50 per cent by the middle of 1931, a level that spelt ruin for many farmers. In 1930 Australia therefore lost £40 million from an export income of £139 million in the previous year. The loss was magnified by the inability to borrow and increased again as the shortfall worked its way through the domestic economy—for the spending power of consumers dropped still further as the banks restricted advances and as business activity fell away. Even though governments reduced their expenditure (and it has been estimated that the cuts in public works forced by the contraction in loan funds threw 200 000 out of work), deficits remained and further borrowing only increased public indebtedness. From £32 million at the end of 1929 the short-term overseas debt rose by August 1930 to £38 million, on top of long-term debts of £566.7 million overseas and £521.2 million internally. By then almost half the export earnings were needed simply to meet interest payments to external creditors.[5]

Table 11.1 estimates the fall in aggregates adjusted for price changes and it is an imperfect measure since it covers two years of rapid movement in prices, but it does indicate the impact of the immediate shock to the Australian economy. In current prices the fall was much greater, from £720 million in 1929/30 to £599 million in 1930/31. Table 11.2 also suffers from limitations but shows the dramatic collapse of construction and manufacturing. The level of unemployment rose inexorably, from 9.3 per cent of the working population at the beginning of 1929 to 14.6 per cent a year later and to 25.8 per cent at the start of 1931. It would reach a peak of 30 per cent in the middle of 1932.[6]

Australia's internal weaknesses and mismanagement may

Table 11.1: Gross domestic product, 1929/30 and 1930/31 (£m at 1910/11 prices)[7]

Industry	1929/30	1930/31
Pastoral	50.9	51.5
Agriculture	26.7	27.7
Dairying, forestry, fisheries	24.0	24.3
Mining	10.0	6.8
Manufacturing	67.6	54.1
Construction	24.4	23.9
Distribution	79.8	63.1
Finance	9.2	8.5
Railways, other public undertakings and government services	42.0	41.2
Other services	47.0	44.3
Rents	38.8	40.7
Other	2.2	2.0
Total	422.6	388.1

well have contributed to its misfortune; the severity of the crisis in Australia was certainly pronounced.[8] But as a trading economy that was still developing its resources, and accustomed to borrowing freely in order to do so, Australia could hardly have avoided the effects of the most severe capitalist crisis to afflict the world economy during the twentieth century. The stark challenge confronting the Labor government that assumed office at the end of 1929 was one of survival. Yet since the remedies on offer required allocating the burden of sacrifice among the principal economic classes, the very search for a solution raised well-nigh intractable disagreements.

The new administration sought at first to insulate Australia from world conditions and to apply traditional Labor remedies with almost autarkic assumptions. Assisted immigration was scrapped. Taxes on non-essential items were increased. Compulsory military training was suspended as an austerity measure. The pressing need to reduce the trade deficit gave a new cogency to the protectionist arguments of

Table 11.2: Workforce by industry, 1929/30 and 1930/31 ('000s)[9]

Industry	1929/30	1930/31
Rural	535.5	539.5
Mining	42.1	47.2
Manufacturing	470.0	384.0
Gas, electricity, water	27.2	22.6
Construction	194.0	139.8
Transport	179.5	167.0
Commerce	348.1	330.1
Community and business services	145.3	144.8
Finance and property	52.4	51.2
Other	272.6	273.8
Totals		
Workforce	2264.7	2100.0
Population (excluding Aboriginals)	6414.4	6476.0

manufacturers and trade unionists, and a Canberra journalist was hardly exaggerating when he stated that wherever two or three people were gathered together in a quiet place, it was an easy wager that one of them was a Labor member and the others high tariff advocates. More than a thousand applications for protection were received in the first two months of the new administration and during one day the prime minister was said to have received a deputation every five minutes.[10] Tariff schedules were raised repeatedly during 1929 and 1930. The same encouragement was given to primary producers when the prime minister launched a campaign to 'Grow More Wheat' and announced plans for a compulsory pool that would guarantee the growers 4s a bushel.

But the limits of the government's power were soon demonstrated. During the election campaign, which had been fought on the Bruce–Page government's record of industrial relations, several Labor spokesmen had promised that they would resolve the grievances of the unions. The coal-mines would be opened, strike-breakers would be cleared from the wharves. Not only did the Commonwealth fail to open the mines, it was powerless to prevent the New South Wales government from opening the Rothbury mine with volunteer labour, and the shooting of a picket was followed by police violence and intimidation against the mining communities. But the prime minister refused even to visit the coalfield. Nothing in James Scullin's long apprenticeship as a Labor politician who eschewed all extremes equipped him to deal with this or the other challenges he would encounter. The miners, who were eventually forced to accept the owner's terms, execrated him. While his government suspended the issue of licences under the Transport Workers Act and issued new regulations to restore preference to members of the Waterside Workers' Federation, the non-Labor Senate disallowed those regulations. After eight months of procrastination the government declared them again, and the Senate rejected them yet again. The cruel farce continued throughout 1931, with the government repeatedly promulgating regulations that remained in force until the Senate quashed them, and unionists accordingly receiving preference on some days, non-unionists on others.[11]

The government's lack of a majority in the Senate would prove a crippling restriction. The most radical member of the cabinet, Frank Anstey, argued from the beginning for an immediate double dissolution to capitalize on the new administration's popularity. He was disregarded by colleagues who were loath to fight another election and lulled by the initial docility of the opposition. It was only as the government exhausted the goodwill of its supporters that the Senate hoisted its true colours and commenced depradations against the legislative programme: a new Transport Workers Bill was blocked in May 1930, the Wheat Marketing Bill perished in July and a Central Reserve Bank Bill was hijacked in the same month. By then another limitation on the government's sovereignty had become painfully apparent. Control over the money supply rested with the Commonwealth Bank and this, as a result of Bruce's legislation of 1924, was run by a board of business and financial leaders appointed by the government but not responsible to it. The chairman of the Commonwealth Bank Board was Sir Robert Gibson, a cautious old Scot who regarded the Labor government with ill-concealed distaste and yet exercised an almost mesmeric influence over the prime minister. Gibson brooked no interference in his management of the country's finances. Behind him stood the private banks, even more hostile to what they described as political interference in financial affairs. They arranged for the Senate to prevent the passage of the Central Reserve Bank Bill, which proposed merely to separate the trading bank activities of the Commonwealth Bank from its limited reserve bank activities, but alarmed the business community because of its provision for the appointment of a new reserve bank board.[12]

The episode increased Labor's mistrust of the banks. Among trade unionists and radicals there was an entrenched hostility to the Money Power, dating back to the bank failures of the 1890s and heightened by subsequent occasions, such as when Labor reforms brushed up against the interests of finance capital in Queensland during the early 1920s. This portrayal of the banks as parasites battening onto the wealth of the toilers was strongly nationalist, with anti-Semitic tones. It expressed a populist belief in the essential congruity of interests among Australian producers—manufacturers, farmers and wage-earners—and it generated a conspiratorial explanation of the economic crisis. In the past the Labor publicists had explained working-class hardship as a consequence of the anarchy of capitalist production, and attributed unemployment to the workers' lack of purchasing power in a profit-making system. Now they spoke of 'the manufactured depression that high finance has engineered'. They located the crisis not in the operation of capitalism as a world system, but in the financial mismanagement of a knot of greedy men. All too often, during the fateful moments of the Depression, denunciation of the Money Power diverted attention from the need for reconstruction to a search for scapegoats.[13]

The belief in conspiracy gained plausibility from the circumstances surrounding the immediate financial crisis of 1930. At the beginning of the year, following the Loan Council's inability to borrow in London and the dangerous depletion of overseas funds, the Comonwealth asked the British government for temporary deferment of payment due on the war debt. The chancellor of the exchequer referred the request to the Bank of England, which in turn asked some disturbing questions about the debtor. To answer those questions the bank sent to Australia one of its senior officers, Sir Otto Niemeyer. His mission was given the cosmetic appearance of an invitation from the prime minister but was arranged beforehand between the Bank of England and the Commonwealth Bank; Scullin admitted subsequently that Niemeyer was nothing more than a glorified bank inspector surveying a property for the assessment of an overdraft. In London the former prime minister Bruce was in close touch with the governors of the Bank of England and he wrote to his former colleague Latham, now leader of the opposition, that they

> are not prepared to give their assistance unless they are satisfied that Australia is taking steps which will, within a reasonable time, lead to her getting on a sounder financial and economic basis. Niemeyer's job will be to convey this without appearing to dictate to Australia.[14]

Niemeyer disembarked at Fremantle on 14 July and, apart from a short visit to New Zealand, where his mission was similar, he remained in Australia until the middle of November. His travels took him to all the states, where he alternated his inspection of the books and private discussions among business leaders with sightseeing and tepid enjoyment of the hospitality his hosts thrust upon him. Politicians, businessmen and journalists hung upon his most casual utterances since, as the wife of the governor of New South Wales put it, 'If *he* speaks at all cheerfully, it means a lot, of course, as he has the Bank of England behind him.'[15] But wherever he went he preserved an affable public reticence, restricting himself to polite pleasantries. In the privacy of his diary he recorded the amusement afforded by members of the Melbourne Stock Exchange who were 'obsessed with the exploded doctrine of the enormous potentialities of Australia', and he allowed the mask to slip when the speaker of the House of Representatives inquired politely if his visit was proceeding satisfactorily. 'That depends on whether you do as you're told', he is said to have replied.[16]

By August Niemeyer was ready to deliver his judgement to Commonwealth and state representatives who were assembled for a premiers' conference in Melbourne. He had made it clear already that the question of relief from London could not be discussed without considering the larger question of government financial policy, and Sir Robert Gibson reinforced that message when he warned Scullin on the eve of the conference that there would be no further advances from the Commonwealth Bank unless 'a clear and definite financial scheme' was adopted. The purpose of the Melbourne meeting was therefore to accept the Englishman's findings. Flanked by Gibson, Niemeyer now presented them. 'By a series of accidents, chiefly the liberality of lenders and accidental high prices for Australian exports', the country had been able to enjoy a standard of living beyond its means. Furthermore, it had used protection and arbitration to stray from its proper imperial relationship as a producer of raw materials for British manufacturers and a customer for their products. The consequences were apparent. 'Australia is off Budget equilibrium, off exchange equilibrium and forced by considerable unfunded and maturing debts, both internally and externally.' The one alleviating factor, the economist Giblin remarked sardonically, was the goodwill of the Bank of England, to which Niemeyer's presence testified. But there could be no recovery without reducing the costs of the primary producer and no relief unless the country's finances were put in order. 'Australia must reassure the world as to the direction in which she is going.'[17]

The chastened representatives of the state and Commonwealth governments adopted forthwith a series of undertakings which became known as the Melbourne Agreement. They resolved that they would henceforth balance their budgets, borrow no more overseas funds until the short-term external debt was dealt with, confine their internal borrowing to income-producing schemes and publish monthly accounts to demonstrate their good faith. None of those present, Labor or Nationalist, resisted this draconian programme and none challenged its basis in Niemeyer's diagnosis. No doubt some of the treasurers appreciated the sheer impossibility of a reduction in public expenditure of this magnitude and sections of the labour movement were certainly outraged: the New South Wales branch of the Labor Party expressed its amazement at Labor politicians accepting the dictates of 'loan mongers and capitalists', and a special federal conference of the unions and ALP called on the Scullin government to resist cuts and mobilize credit.[18] But Scullin, along with the Labor premiers of South Australia and Victoria, regarded his hands as tied. The problem therefore became a political exercise of making the pious declaration of the Melbourne Agreement palatable to the party membership. Niemeyer observed scornfully how the federal acting treasurer sought to preserve appearances: 'Lyons, in a typical aside, asked me if I would mind not coming to the meeting the next day because he was afraid of criticism from his own people that he was acting under dictation.' Niemeyer was only too happy to oblige him.[19]

Niemeyer was preceded to England by a harassed prime minister whose attendance at an Imperial Conference offered him some months' respite from his critics. He did have one small success there when he insisted that an ungracious King George appoint an Australian governor-general, Sir Isaac Isaacs. Since the retiring viceroy, Lord Stonehaven, symbolized the link with the home country (he confided to Latham that all Australia's problems had been caused by 'imported agitators and agents from Maynooth and Moscow'), this show of independence increased conservative hostility towards the Labor administration.[20] But the prime minister left behind him what Anstey described as 'only the stuffed effigy of a government',[21] a party that was in open disarray. Theodore, who as treasurer had exercised an undeniable authority, was forced to resign the post in July when a Queensland royal commission found that he was guilty of fraud and dishonesty in the sale to the state government of the Mungana gold-mine while he was a minister of the Crown. In Theodore's absence the disquiet over economic policy could no longer be contained. The radical members of the caucus denounced the Melbourne Agreement and called on the ministers to overcome the opposition of the bankers and resist any cuts; they were encouraged by Labor's victory in the New South Wales election of October 1930, which Lang fought on that basis. Labor's right wing, including Joseph Lyons, the acting treasurer, and James Fenton, the acting prime minister, took the Melbourne Agreement as binding the government to policies of deflation and reduction of expenditure; they took heart from the prime minister's reappointment of Gibson to the Commonwealth Bank Board. 'The only way', insisted Lyons, 'is the traditional way of keeping faith with the lenders.' These were the extreme positions. Scullin and the majority remained prisoners of the Melbourne Agreement while shrinking from the cuts in wages and pensions that its implementation would necessitate. Thus when caucus carried Anstey's proposal to defer the conversion of a £27 million loan, Lyons cabled Scullin that he would not implement such 'absolute repudiation' and Scullin agreed that compulsory conversion would be 'dishonest and disastrous'. The matter was allowed to await the prime minister's return and Lyons carried through a successful voluntary conversion.[22]

Control of policy was fast slipping wholly out of the government's grasp. In rejecting a request that the Commonwealth Bank release credit to cover the budget deficit and finance public works, Sir Robert Gibson reminded the cabinet that 'political exigencies must not govern those charged with the responsibility of maintaining a sound financial and monetary system'. While insisting that they had not 'the slightest desire to dictate the policy of governments', the trading banks joined him in calling for a drastic curtailment of government expenditure. Early in 1931 the Bank of New South Wales began buying sterling at a price higher than the other banks and by the end of January £100 of British currency was worth £130 of Australian. The other banks were forced to follow in an effective devaluation that took place independently of the Commonwealth Bank, much less the Commonwealth government. In the same month the Arbitration Court reduced all award wages by 10 per cent, this on top of earlier reductions caused by the fall in prices and despite the submission of the government. One union official present in the court shouted 'To hell with these judges' and joined his members in giving three cheers for the social revolution.[23]

Devaluation and wage cuts constituted the first steps in a programme of recovery that the economists had been urging for the past six months. While by no means unanimous in their policy recommendations, they shared a common approach to Australia's predicament. The national income—and their popularization of this concept was crucial—had suffered a once-and-for-all loss of £50 million in exports and investment. That loss had fallen directly on the farmers and the unemployed, and if the imbalance was allowed to remain, it would work through the domestic economy and might eventually reach as much as £150 million (almost a third of the national income). If, however, the loss was spread evenly across the community it would be contained, and if the burden of taxes, wage costs and high interest rates was lifted from the primary producer, equilibrium would be restored and recovery could occur. 'Equality of sacrifice', which became a general catch-cry, was therefore advanced by the economists with a specifically economic, as well as a more general ethical, significance.

The economists did not pretend that they were presenting government with an immediate solution to the nation's difficulties. Just as the Depression was the consequence of past follies, so the adjustment would be all the more painful for previous self-indulgence, and part of the remedy was a reduction of the size and scope of the public sector. Even so, the more creative among them, like Giblin and Dyason, were extremely critical of the deflationary orthodoxies of Niemeyer and Gibson which, they thought, would needlessly intensify the severity of the Depression and might even endanger social order. Giblin told Lyons that he feared 'a bad smash with a chance of revolution and chaos'. The economists urged a judicious mixture of deflation and inflation. The inflation had been achieved by the market-led depreciation of the Australian currency, in which process one of their number, Edward Shann, played an important part as economic adviser to the Bank of New South Wales. A start on deflation had been made with the Arbitration Court's wage reduction, where again the testimony of the economists was an important influence after the ground had been prepared by Giblin in his popular *Letters to John Smith* ('so long as you believe that higher wages can always come like manna from heaven merely by asking loudly enough for them, we shall never get any further'). The urgent need now was to complete the process of adjustment by reducing government spending and interest rates. This was the advice the economists gave in their joint statement of January 1931.[24]

The devaluation, the wage cuts and the economists' manifesto haunted the prime minister when he returned to Australia in the new year. He was met also by sharp reminders that his government could procrastinate no longer. The farmers were harvesting the record crop they had planted in response to his appeal to 'Grow More Wheat', but the price of wheat had slumped to little more than half that he had proposed to guarantee a year earlier and the government had still not established its wheat pool. In the cities there had been ugly scenes when police broke up unemployed demonstrations and two days after his return there occurred the Adelaide 'Beef Riot', a police attack on a crowd protesting against the replacement of beef by mutton in the ration allowances of workless South Australians.[25] A premiers' conference was due in the following month where the federal and state governments would have to address the problem of public finance.

Bereft of ideas, Scullin turned to Theodore and reinstated him as treasurer. During his time on the backbenches and even while he was fighting to survive the Mungana allegations, Theodore had reappraised the government's financial policy in the light of his increasing scepticism towards economic orthodoxies. He had become convinced that a new and expansionary monetary policy was needed to restore prices to their pre-Depression level, stimulate production and reduce unemployment. These were the ideas that he prepared for the premiers' conference but the immediate cost of his reinclusion was heavy. Joseph Lyons, who as acting treasurer had established close relations with the business community, was appalled by the prospect of inflation. He set off from Canberra, intending to announce his resignation from the cabinet when he reached his home state of Tasmania. A colleague rushed to the railway station, imploring him: 'For God's sake, Joe, don't do it!' But Joe did and his colleague, Fenton, resigned with him.[26]

On 6 February the premiers gathered in Canberra to hear Theodore explain his plan. The crucial response came not from them but from Gibson who informed him that the co-operation of the Commonwealth Bank would depend on the reduction of all government expenditure, including expenditure on public salaries, pensions and benefits, and from the private banks who regretted that they could not join in any scheme that was 'not on sound banking or economic lines'. But before these replies were received, another hat was thrown into the ring. Jack Lang, premier of New South Wales and Theodore's principal opponent in the bitter factional conflicts of the Labor Party, told the assembled ministers that it was time to end the 'shilly-shallying'. He had a plan that went directly to the heart of Australia's difficulties, the crushing burden of debt. The Lang Plan consisted of a suspension of overseas interest payments, a reduction of the interest rate on all domestic public loans to 3 per cent, and an abandonment of the gold standard for one based on the wealth of Australia which he called 'the goods standard'.[27]

It is hardly necessary to add that the Commonwealth and other state representatives recoiled in horror from proposals tantamount to repudiation of debt obligations and debauchment of the currency. Lang did not expect otherwise for he possessed what Giblin described as 'a remarkable faculty for stating a perfectly reasonable proposition in such terms as to excite acute distrust'.[28] Since assuming office in the previous year, he had hardly protected New South Wales from the ravages of the economic crisis and he had himself reduced wages and social services along with the rest of the country. He maintained popularity by his capacity to coin a phrase of revolt and to manufacture dramatic confrontations. In the bond-holder he found the scapegoat for all the people's misfortunes and he used the Lang Plan as a rallying-cry in a federal by-election in East Sydney, which was won by his candidate, Eddie Ward. It was a fateful victory. Ward was denied membership of the federal Labor caucus on the grounds that he had not stood on federal Labor policy, and seven New South Wales members followed him from the caucus-room on 12 March. Hence in the space of six weeks there had been two defections from the Labor ranks—or as Scullin put it, 'five of my party went over the starboard side and five went over the port side'—and the government had lost its majority. Two weeks later a special conference of the ALP expelled the New South Wales branch. Undaunted, Lang announced that he would go it alone and to show that he was in earnest, New South Wales stopped paying interest to London.[29]

On Friday 6 March a rumour passed around the little wheat town of Ouyen in the Mallee district of Victoria: the communists had seized Sydney and even now their counterparts were advancing on Melbourne from Mildura. This was a remarkably indirect route but as Ouyen lay in its path, the farmers and local businessmen turned out with firearms, dug trenches, laid sandbags and kept watch all through the night. Similar precautions were taken elsewhere in the Mallee as well as the Wimmera and across into Gippsland. So widespread was the unease that Thomas Blamey, commissioner of the Victorian police, issued a statement to reassure citizens that they were safe and condemned those who had spread these 'childish and absurd rumours'.[30]

Country areas in other states experienced their own false alarms during the *grand peur* in the autumn of 1931 when the morning sun found the menfolk clutching a rifle in one hand, rubbing sleep from their eyes with the other. Yet the scale and precision of the Victorian 'stand to' enable us to unravel some of the links between historical context and human agency. Against a background of financial crisis and political deadlock, the visible presence of small armies of unemployed men could easily trigger hostility and panic—with the end of the fruit-picking season up on the Murray, the Victorian unemployed were expected to travel like a plague of locusts through the countryside on their way back to the metropolis. The sixth of March, furthermore, had been chosen as the date for a national demonstration of the unemployed. The evidence is overwhelming that some jittery member of the clandestine White Army in Melbourne, anticipating that this was the catalyst for a communist uprising, telephoned the country branches. Blamey, as one of the senior members of the White Army, would undoubtedly have known of this unauthorized communication and its unfortunate consequences—the turn-out of the rural zealots ripped the veil of secrecy from his organization—but he could hardly have acknowledged this fact. He and the other Melbourne leaders were better placed to appreciate that the eminently moderate state Labor government was more than capable of controlling the demonstrations. Moreover, the White Army's preparations for alerting the urban bourgeoisie with factory hooters and light planes were so thorough that they had no need to cry wolf.[31]

Such a clandestine organization was not new, and the earlier activities of the secret armies were noted in chapter 10, but since the purpose of those enthusiasts was to defend civilization against a left-wing *putsch*, their doomsday fantasies did not impinge directly onto public life. This was less clearly the case among new movements that sprang up in the Depression. In Melbourne some upper-class toughs calling themselves the Order of the Silent Knights began employing violence against working-class activists. In New South Wales an Old Guard was formed but soon gave way to a more interventionist New Guard, whose declared aims included 'the suppression of any disloyal and immoral elements in Governmental, industrial and social circles'. As recorded by its founder, the solicitor and army reservist Lieutenant-Colonel Eric Campbell DSO, the origins of the New Guard read like a Bulldog Drummond adventure story. Campbell and his friends, clean-cut, well-bred, resourceful and plucky to the last man, gathered with tankards of beer at their elbows in the Sydney Imperial Services Club ten days after Lang disclosed his plan. There they pledged themselves to redeem their country, safeguard its loyalty to the British Empire and build 'a new Australia, honourable, self-respecting, thrifty and secure'; within nine months 50 000 rallied to the flag. The reality was more disturbing: here was a tightly disciplined group of armed men (the number of pistol licences taken out in New South Wales doubled during these years) who were not averse to beating up left-wing speakers or assaulting Jock Garden in his home. Campbell himself would slide towards whole-hearted fascism after his visit to Germany and Italy in 1933.[32]

The scale of these activities far exceeded any threat from the left at which they were ostensibly directed. The Communist Party, which was only beginning to recover from divisions and defections, claimed 1500 members by 1931; its Unemployed Workers Movement, formed in the previous year, claimed 30 000. But the Communists were isolated from the broader labour movement by their sectarian excesses, and the local branches of the Unemployed Workers Movement found it hard to maintain continuity of membership or effectiveness in the face of police harassment. If the ACTU issued fiery denunciations of capitalism, its members were incapable of resisting wage cuts, let alone taking the offensive, and the Labor Party was completely ineffective in all states but New South Wales.[33]

The conservative mobilization of 1931 was directed not so much against any real threat to established authority as against the shortcomings of the existing system of government. Sometimes the criticism was directed at the federal Constitution which, it was alleged, meant that the Commonwealth raised the funds and the states spent them with little heed to prudent financial management. At the same time, the hardship of the primary-producing states of South Australia, Tasmania and Western Australia increased their discontent with the Scullin government's regimen of high tariffs, while even within the industrially advanced state of New South Wales the rural regions demanded to be released from the shackles of Sydney. Ten thousand gathered on the banks of the Murrumbidgee in February 1931 to launch the Riverina Movement and soon linked up with the separatists of New England and the inland plains.

Sometimes the fault was seen to lie in the very principle of representative government: 'The inaptness of the machinery of political democracy to the economic management of our society is now ... universally conceded', wrote a leading constitutional lawyer in June 1931.[34] The underlying basis of this attitude, namely that the assault of numbers threatened good government, had of course been apparent in the times of plenty, but with the Depression the danger intensified. The bankers appealed to precisely such a fear when they justified their control of monetary policy on the grounds that if the control of credit and currency were placed within the grasp of elected rulers, they would have 'no choice but to transmit the pressure of constituents for easy money'.[35] As the political consensus fell apart under the impact of the crisis, it could no longer be assumed that democracy was the best possible political shell for capitalism.

The dominant reaction was against the mode of party politics. By its weakness and irresolution the Labor Party had shown its unfitness to govern, but the Nationalists were hardly free of blame for the present predicament. Now, when sound government was required to save national honour and salvage national solvency, the interest groups that lay behind the political parties were squabbling for sectional advantage. A spokesman for the Riverina Movement judged that 'The party machines in Australia appear to have harvested nothing for many years but weeds.' What was needed was some catalyst for the frustrations and fears of the middle class.[36]

One such movement began in Sydney among a group of Rotarians who met in January 1931 to establish an All For Australia League. Their objects, embracing national unity, balanced budgets and a return to the Anglo-Saxon virtues of thrift and self-reliance, have been described aptly as more a set of moral injunctions than a blueprint for economic action. Above all, they wanted to 'clean up politics' and one of the league executive thought that proven business ability should be a prerequisite of parliamentary membership. The league was launched at a public meeting on 12 February. Looking at the audience, a reporter was 'swept back on a wave of memory to the camps and enlistment depots of 1914 and 1915. Clerks, bank managers, labourers, small shopkeepers, accountants, barristers, a mixed audience but all inspired by a wave of patriotic ardour.' By June 1931 the All For Australia League claimed 130 000 members.[37] The origins and attitudes of the Citizens League of South Australia were similar, its founder a failed bus proprietor with a fierce hostility against the old gang—'We hate all the parties'. It had 30 000 members and a full-time staff of thirteen organizers by the end of 1930. But events in South Australia reveal an important shift in the mobilization of the middle class. In March 1931 the master of the Anglican university college, Archibald Grenfell Price, was approached by his friend Charles Hawker and asked to take the lead in forming an Emergency Committee. Hawker and his fellow Liberals (the name taken by the Nationalists in that state) were concerned that the Citizens League might crystallize into a separate political party and split the conservative vote. Price did convene such a committee, which embraced the Liberals, the Country Party, the Citizens League and leading businessmen, and he found that 'the non-party people were children in the hands of ... the skilled political men'.[38] In Melbourne the All For Australia League was controlled, from its first meeting in the Collins Street chambers of a leading stockbroking firm, J.B. Were, by the paymasters of the National Union.[39]

Thus when Joseph Lyons began to appear on public platforms of these organizations and offer himself as an alternative to the party machines, it behoved John Latham to guard his back. Lyons had already been assisted by a group of Melbourne businessmen, including his old friend Staniforth Ricketson, the head of J.B. Were, who provided him with a secretary, and Keith Murdoch, now a newspaper baron, who promoted him assiduously. Why Lyons? The homely Tasmanian appealed to them as a man of safe opinions (Charles Hawker perceived that 'he has the really conservative habit of mind which twenty-five years' democratic training has quite failed to alter'[40]) with an electoral appeal that the glacial Latham could never match. He possessed the further advantages of a Labor background and a Roman Catholic faith, and these qualities enabled his promoters to claim that they were breaking the mould of conservative politics. There was the opportunity of combining his accession to the leadership with a party reconstruction that could capitalize on the vitality of the All For Australia movement. By March Latham was under pressure from the leading members of the National Union to step aside. He resisted for a fortnight, telling them that it was for the members of parliament to decide the leadership, but on 17 April he surrendered his post to Lyons and in the following month a new party, the United Australia Party, swallowed up the Nationalists, the All For Australia Leagues and other citizens' groups. Another casualty was Tom Bavin, the ailing leader of the Nationalists in New South Wales, and indeed it was a notable irony that the principal victims on the conservative side of the call to 'clean up' politics were both men of exceptional probity. The machine politicians survived unscathed. The conservative reconstruction was a remarkable feat of legerdemain which, as one Nationalist veteran put it, began as a 'movement which set out to replace incompetent politicians by capable business and professional men, and finished by placing those same politicians more firmly in the saddle'.[41]

Faced with these ominous developments in the early months of 1931, the government had little room to manoeuvre. Theodore pressed ahead with his scheme of reflation and introduced legislation to take Australia off the gold standard and to authorize the printing of a special issue of £18 million

(£12m to finance public works, £6m to assist wheatgrowers). After listening to the advice of Gibson, the Senate rejected the government's Bills. A meeting of the Loan Council was due and Gibson found it his 'unpleasant duty' on 2 April to advise that 'a point is being reached beyond which it would be impossible for the bank to provide further assistance for the Government'. The Loan Council took the hint and established a sub-committee to review all expenditure in order to identify potential economies. The sub-committee, with the assistance of four leading economists, formulated proposals for cuts which were taken to a premiers' conference the following month.[42]

The composition and indeed the very establishment of the sub-committee revealed the disarray in Labor ranks. New South Wales, having boycotted the proceedings, took no part. Even then, the ALP representatives of the Commonwealth, South Australia and Victoria could have outvoted the conservative representatives of the remaining states. But the two Labor premiers were scarcely the men to take a stand. Ned Hogan, the Victorian, had already fallen out with his party executive; a man of the soil with an unshakeable attachment to the values of thrift, hard work and financial probity, he brusquely dismissed the plans of Theodore and Lang as the work of 'bilkers' and 'welshers'. Lionel ('Slogger') Hill, the South Australian, was dominated by advisers, notably the managing director of the Elder Smith pastoral company and the editor of the major Adelaide newspaper, the *Advertiser*. Their counsel, he insisted, was 'purely non-political'. Hill's infuriating combination of pusillanimity and vanity soon exhausted Theodore's patience. 'I can't really make up my mind,' said the premier. 'You bloody old woman, you haven't got a mind to make up,' snapped the federal treasurer.[43]

With these two die-hards holding the balance of power when the premiers' conference assembled in Melbourne in May 1931, it was apparent that cuts were inevitable. Taking as its basis the proposals of the sub-committee, the meeting adopted what became known as the Premiers' Plan. First, all adjustable government expenditure was to be cut by 20 per cent (with the exception that the old-age pension would be cut by 12½ per cent, from 20s to 17s 6d). Second, taxes would be increased. Third, internal interest rates would be reduced by 22½ per cent, including interest on existing loans which would be converted 'voluntarily'. Altogether, it was estimated that these measures would reduce the public deficit by £25 million and leave a manageable deficit of about £15 million in the next financial year. The signatories acknowledged the hardships that the plan would impose but took comfort in the claim that they were distributed in the best possible way both to restore equilibrium and to satisfy requirements of equity: 'The Conference has adopted a plan which combines all possible remedies in such a way that the burden falls as equally as possible on everyone.'[44]

The principle of 'equality of sacrifice' was crucial to the acceptance of the Premiers' Plan. Since cuts were unavoidable, the Labor representatives could take cover behind the fact that bond-holders as well as pensioners were to make their contribution to national recovery. Hogan and Hill assented readily, Scullin saw no alternative and even Lang agreed to sign on condition that the reduction of interest rates preceded curtailment of his government's expenditure. That part of the plan, however, proved more contentious. The Melbourne business community was openly critical of the conversion of existing loans, which they regarded as tantamount to repudiation of lawful contracts. Staniforth Ricketson said the proposal to enforce conversion 'outLangs Lang'; Robert Menzies, the rising young lawyer with good connections in Collins Street, called it 'a very clever and well-considered scheme for bilking the public creditor' and added that he would rather see every Australian citizen die of starvation than fail to honour contractual debts in their entirety. If either of these indignant worthies had ever gone to bed on an empty stomach, their protests would have carried more weight. As it was, they were disowned by the majority conservative opinion for which Frederic Eggleston spoke when he warned, 'I do not like to contemplate the position of the bondholders in a community like Australia if they were the only class that refused to make the sacrifice.' In the event, 97 per cent of the bond-holders made a voluntary conversion.[45]

The Scullin government's capitulation to the Premiers' Plan brought to an end more than twelve months of pre-

Joseph Lyons and supporters, election day 1931

varication. Unable to control the economic slide, unwilling to impose its authority on the financial institutions, the government had twisted and turned in an effort to put off the day of reckoning, and in doing so had suffered defections from its right and left wings. Now it accepted the final indignity—a reduction in the income support of the dependent members of the community, the pensioners and recipients of benefits. It met a storm of protest. There were fresh resignations from the ministry and Scullin depended on opposition support to pass the enabling legislation through the House of Representatives.[46] State executives of the ALP denounced the plan. So did an emergency meeting of the ACTU and the Labor Federal Conference. But none of these bodies was prepared to call on the government to resign. It limped on, bereft of coherence or credit, until the Lang group of MHRs voted with the opposition on a motion of confidence in November.[47] With more than a quarter of the workforce unemployed, the result of the election that followed was a foregone conclusion. When Scullin told a meeting that 'for the first time since I have been Prime Minister, I can see daylight ahead', a man in the audience observed, 'You must have very good eyesight.' Campaigning on a platform of sound, honest finance which would restore the honour and credit of the country, backed by the treasury chest of the National Union, drawing on the vast numbers of the All For Australia Leagues, Lyons swept the polls. The United Australia Party won 40 seats in the House of Representatives, the Country Party 16, Labor 13 and Lang Labor 4.[48]

The Hogan government remained in office until April 1932 when, in the premier's absence overseas, it was brought down for weakening its adherence to the Premiers' Plan. Only then was Hogan expelled from the party because of his defiance of policy. The Hill government carried on despite its expulsion from the Labor Party until in 1933, his usefulness exhausted, the conservatives cast him aside. Lang was brought down after dramatic confrontations with Canberra and Government House. From March 1931, when he first defaulted on interest payments to London, he preserved his reputation with calculated acts of defiance. In April 1932 the Lyons federal government used a Financial Agreements Enforcement Act to take over the revenues of New South Wales. Lang, who had already withdrawn more than £1 million in cash from the Sydney banks, responded by ordering his public servants not to pay money to the Commonwealth. The governor held that the instruction was illegal and dismissed him. Contrary to widespread expectations, Lang accepted his dismissal. 'We were prepared to go to the people,' he recalled, 'knowing that in a very little more time they will realise that our cause was their cause, and that only a Labor movement can give them back the self-governing rights that have been filched from them.' He was defeated at the ensuing election and Labor remained out of power in New South Wales for a decade.[49]

The electoral reverses suffered by Labor in these years were scarcely surprising—between 1929 and 1933 there was only one federal or state government, that of Tasmania, that survived an election.[50] The economic crisis served to emphasize that the electorate was principally concerned with the protection of living standards by customary means. Radical ideologies of both left and right were short lived in their appeal. Awesome in their sudden mobilization, the extra-parliamentary organizations lasted no longer than the immediate political crisis and were satisfied by the election of an eminently conventional government. The set-back to the labour movement, however, went further than this. For the past thirty years it had worked on the expectation that capitalism could be civilized, that regulation of the labour market could ensure employment at good wages for men and that Labor governments could protect Australian living standards. By 1931 the unions were powerless, the ALP was routed and a large proportion of the population reduced to helpless indigence. The Depression therefore marked a watershed in the course of social reform, a graphic demonstration of the inadequacy of the old ways and the need for new.

12

WINNERS AND LOSERS

THE HUMAN DIMENSIONS of the Depression defy precise statistical measurement. Unemployment figures were based on the returns of trade unions and these now covered less than a third of the workforce. If, as seems likely, their figures did not overstate the severity of unemployment, then from the middle of 1930 until the last months of 1934, more than a fifth of wage- and salary-earners were out of work. To these must be added the school-leavers who failed to find a place in the workforce; others, mostly women, who withdrew from it in despair; and a further group of employees working reduced hours. In the second quarter of 1932, when unemployment among trade unionists reached a peak of 30 per cent, it would therefore seem likely that as many as one million people in a total workforce of a little over two million lacked full-time employment. The census of the following year revealed that two-thirds of all breadwinners had received an income of less than the basic wage in the year 1932–33.[1]

'Unemployed at last!' Thus begins that extraordinary novel of the last great depression of the 1890s, *Such is Life*, whose author derived a 'grim fakeer-like pleasure' from his inversion of the natural order.[2] 'Unemployed' was the way the victims of this Depression understood their plight also, and at times the term seemed to define a caste of pariahs. No matter that the catastrophe engulfed previously secure employees, white-collar as well as blue, and swallowed up so many small proprietors that in the 1933 census the appellation 'self-employed' became a common euphemism for unemployed. No matter that the unemployed person was in practice likely to cross and recross the boundary that separated the idle from the active, here picking up a few days' labouring, there a harvesting job, and constantly exchanging labour and commodities in the burgeoning informal economy. In popular estimation it was the longevity and seeming finality of their exclusion from the regular workforce that set the 'dolies' apart. Nor is this surprising. The chief energies of the organized labour movement over the past three decades had aimed at securing an availability of work at living wages for all breadwinners. The performance of wage labour was at once the source of a workingman's income, the basis of his standing among family and friends, and an affirmation of his very identity. Small wonder, then, that the unemployed understood their plight in these terms. 'What do we want? Work!' 'We want work, not charity!' These were the slogans of demonstrators in the late 1920s and 1930s.[3]

The unavailability of loan funds meant that the traditional standby, public works, would no longer be offered to unemployed men. Sheer weight of numbers swamped the charities and government relief agencies that had provided for indigent women and children up to this time. From 1930 the states were therefore forced to establish unemployment funds, financed by emergency income taxes in most cases, which they eked out in two forms of public assistance: sustenance and special relief work.

Sustenance was distributed to the unemployed and their dependants, principally in the form of ration orders that could be exchanged for a limited range of foodstuffs—meat, bread, vegetables, tea and sugar. While each state devised its own scale and procedures, the monetary value of sustenance, which was fixed according to family size, varied from 5s to 7s per week for an adult and the same or less for a child. A family of five was thus forced to make do with less than half of a basic wage that supposedly measured its essential requirements. All applicants for sustenance had to have been registered as unemployed for a period before becoming eligible, all assets with the exception of the roof over their heads had to have been realized, and there were draconian limits on other sources of income. For clothing, the unemployed had to make do with cast-offs distributed by charities, though boots, tunics and greatcoats were made available by the Commonwealth Defence Department and dyed black to fit out 'Scullin's Army'. Scarcely any assistance was given with housing, though there were sporadic attempts to provide some legal protection to mortgage-holders and tenants. Sustenance was not usually available to single women, and single men were subject to special conditions: at first they were offered meal tickets but later, to get them off the streets, they were brought into camps or pushed out to the country and forced to tramp from town to town to collect their track rations. Yet 'susso', with its attendant measures, was the principal means of meeting the immediate emergency and by 1931 it was costing the states £9 million a year.

Subsequently, with the easing of financial stringency, there was greater resort to various forms of unemployed relief work. Unemployed men were required to labour on construction or improvement projects of the sort that had previously generated employment on public works. But those who now swung picks and shovels were no longer earning a wage. Indeed, the more niggardly states of South Australia, Tasmania and Victoria made such tasks a condition of sustenance. Elsewhere the relief worker was given a small increment on sustenance, but still well below the basic wage, and a man's acceptance of such an offer, no matter how far afield it took him, became a condition of his family's further assistance. This forced labour was thought preferable to idleness and dependency, so strong was the attachment to the work ethic, and wherever possible the unemployed were forced to perform it in the later Depression years.[4]

The inadequacy of these provisions was obvious. Even if ration orders were used to maximum nutritional advantage, they were barely able to satisfy bodily needs. It was common for men who had been out of work, upon eventually resuming employment, to collapse after trying desperately to keep going on an empty stomach—and there were plenty of others to take their places. Teachers in working-class areas reported a high incidence of malnutrition among their pupils. The weakened state of mothers was reflected in an increase in infant mortality. Every town and city had its shanty camps, with primitive shelters made of packing cases, hessian and corrugated iron, lacking sanitation and infested by vermin.

Beyond such basic hardships, mass unemployment reshaped the lives of its victims. An immediate effect of losing a job was the breaking of a daily routine of work and leisure. Not only did the unemployed lose the companionship of workmates, but they were deprived of access to commercial leisure pursuits (cinema attendance halved between 1930 and 1932) and the friendships that leisure activities supported. Coupled with this severance of established habits was a sense of inadequacy. Some found ways to give meaning and purpose to their lives either by putting their idleness to good account (the use of libraries doubled during the same period) or by turning their hands to whatever was going (such as growing vegetables); others found the loss of earning capacity a crippling blow to their self-esteem. In this respect the previously secure suffered most. Casual labourers already had techniques of survival and were adept in scavenging, shifting their accommodation by night and practising all the other lurks that enabled them to make ends meet. Clerks or salesmen or secretaries, on the other hand, had no such experience and were more likely to hide their new poverty behind the drawn curtains of their suburban villas. Again, working-class neighbourhoods supported forms of mutual support that did not exist in the more privatized middle-class suburbs. The unfortunates who were turned out onto the street suffered a sharper contraction of their social network, which was all the more traumatic if they took to the track. Such nomads might form friendships, but with thousands constantly on the move such associations were likely to be transitory. They might receive assistance from householders and shopkeepers, but sympathy was a fragile quality and constantly frayed by the pressure of numbers. Thus for some time the people of Cairns gave hospitality to the small army of unemployed that came each autumn and camped at the showground to escape the rigours of the southern winter, but in July 1932 this hospitality was withdrawn and 500 vigilantes organized by leading businessmen evicted 150 'hobos' in a pitched battle fought with clubs, cane knives and jam-tin bombs. An ostensible willingness to support the members of the community meant, in practice, an unwillingness to help outsiders. Hardship evoked both generosity and selfishness.[5]

The family was the chief refuge from the ravages of unemployment and the principal solace against its isolating effects. But the Depression put the family, and its gender roles, under immense strain. While the husband's loss of a job threatened his standing as breadwinner, the scrimping and stretching of meagre resources imposed an added burden on his wife. The fact that female employment held up better than male employment (principally because women worked for lower wages and in industries that were less severely affected) did not protect women from allegations that they were taking men's jobs, nor did it alter the gender division of domestic labour. Furthermore, the very use of the family as the unit of relief exacerbated its vulnerability. Public policy was still permeated by the traditional belief that while it might be appropriate to assist dependent women and children, it was necessary to guard against the importunities of those who ought to be working—the obscene term 'dole bludger' derives from moral opprobrium against a man living off a woman. Thus there were instances where a family received an allowance only in the absence of the father; and since children who had passed the school-leaving age might no longer be included in sustenance calculations, many of them went onto the track rather than remain a burden on their parents. Above all, it was easier for an individual to chase after the opportunities to pick up a few shillings. 'The only way to survive in those days was to be on your own—married people, oh, married people . . .!' was how one who lived through the period put it. The postponement of marriage during the early 1930s, the decline of the birth rate to an historic low in 1934, and the increased incidence of desertion all lend weight to his statement. Yet it is also evident that the family offered an infinitely precious emotional resource to those in need. Subsequent testimony from husbands ('It was harder on the women') and wives ('The men suffered most') suggests that while married couples may have remained trapped in established gender roles, there was indeed a mutual concern and an appreciation of each other's sacrifices.[6]

In responding to their predicament, the unemployed lacked one obvious resource that was available to those in work. Bereft of the ultimate sanction, withdrawal of their labour, they found it hard to bargain with the state, much less the employer, for a better deal. Given the organized labour movement's strong attachment to such methods, and indeed its complete imbrication in their institutional procedures, it would have been difficult for the unions and the Labor Party to give an effective lead to the workless even if these bodies had maintained their strength. An unemployed worker was more likely to remain part of the local labour movement in communities with a strong occupational identity, such as those on the minefields or around the ports. Elsewhere the loss of a job was followed by loss of trade union membership, and his interests then came second to those of former workmates. Organizations for the unemployed, most notably the Communist-led Unemployed Workers' Movement, therefore maintained an uneasy relationship with the labour movement. During the worst years of the Depression, in fact, the Communists concentrated their attack against the 'Labor fascists' whom they accused of betraying the working class, but even the 'official' unemployed organizations that were established by trades and labour councils were wont to throw off restraints and condemn an unresponsive Labor government or municipal council.

Imprisoned by chains that they themselves had forged, the federal and state governments were unable to offer any solution to the crisis. Politicians and businessmen, economists and clergymen, all expressed sympathy for the unemployed while insisting that they were powerless to alter the circumstances that afflicted them. The very perception of the Depression as a catastrophe beyond the control of human agency made it difficult for the unemployed to find a focus for their discontent. In 1930 a number of Perth's homeless unemployed announced their intention of going to Canberra to put their plight before the Commonwealth authorities. The state government gave them a free railway pass as far as the edge of the Nullarbor, thereby ensuring that it would be some time before they would return, and there they disappeared from the historical record.[7] With this propensity to shift the blame to remote scapegoats went a particular interpretation of responsibility. As individual supplicants, the unemployed were tolerated and offered whatever charity was available; collectively, or when they asserted rights, they were denied recognition.

What were the rights of the unemployed? It had long been the custom in times of seasonal distress for the workless to gather and ventilate their grievances. A meeting would assemble, march to the premier's office and there wait while a deputation presented details of their plight and requested that public works be established to enable the men to support their families. This ritual was re-enacted in the early years of the Depression in all the state capitals. But since there were no funds in the state treasuries to finance public works, the deputations were turned away empty-handed. The frequency and increasing turbulence of these public gatherings alarmed the authorities who, by 1930, were refusing to receive deputations and even banning street marches altogether. Some of the most violent affrays occurred during the nation-wide day of protest on 6 March 1931 when the Victorian countryside was put on alert against an expected uprising. Yet a police officer in Perth, where mounted police and batons were used to break up a crowd of 2000, appreciated the actual significance of what was happening: 'men do not rampage up and down the streets waving the Hammer and Sickle, demanding food and relief, for no reason at all'.[8] Partly as a result of the savage repression, partly because of subsequent efforts to drive the single unemployed from the cities and partly because of the shift from sustenance to relief work, this form of protest by the unemployed declined after 1931.

In its place the unemployed devised multiple forms of protest and resistance.[9] The most spectacular and probably the most significant action was opposition to evictions, since it was here, over the struggle for possession of accommodation, that the dominant code of property rights was directly contested. Where tenants were threatened with eviction for arrears of rent, neighbouring residents gathered to remonstrate with the landlord and agent, and, if necessary, prevent them from emptying the tenants' furniture onto the street. Since landlords were unable to let accommodation in the depths of the Depression and since a house was an easy target for sabotage (if smashing windows was too easy, a bag of cement down the toilet was even more persuasive), these sanctions were often sufficient. A topical joke conveys the spirit of defiance:

> Does that fellow collect your rent?
> No!
> Who is he?
> Our rent collector.[10]

If the landlord called in the police, there could be more trouble. Some of the most celebrated battles took place in Sydney in mid-1931, culminating in the storming of a house in Newtown by forty police. The attackers fired a fusillade of pistol shots, the defenders replied with stones and other weapons until, upon their submission, they were batoned and kicked senseless. That night demonstrators broke the windows of city stores as well as the offices of Lang's *Labor Daily* since it was his government that authorized the action. Lang then conceded new legislation that delayed evictions.[11] In Melbourne, similarly, a concerted attack on the premises of estate agents tempered their zeal.

The unemployed even managed to adapt that most familiar of strategies, the strike. In May 1931 the mining families of Bulli, on the south coast of New South Wales, declared the dole 'black' when the Lang government turned over the administration of the dole to the police. In October 1932 those on the northern coalfield refused to fill in a questionnaire designed to catch out imposters—'Are you residing with your wife?' was the query that caused particular offence. Elsewhere there were attempts to boycott acceptance of reduced rations.[12] Such desperate expedients were manifestly difficult to sustain. With the subsequent shift to relief work, however, gangs of men were more inclined to use the tactic. Most relief workers were in fact forced to join the AWU, which regarded them as little more than dues fodder and even extracted the dues in such a way that the men were denied a voice in the union. But in response to speed-ups and harassment, and more generally to protest against the principle of working for less than the basic wage, they formed their own relief workers' unions and could sometimes win the support of other unions. In this fashion the breach between the working and workless sections of the labour movement began slowly to close. Even then, the authorities usually treated the unemployed with disdain. Nearly 300 relief workers marched from the Frankland River in the far south-west of Western Australia to protest against impossible contract rates; they were met in Perth by mounted police and their leaders were arrested. A ganger in the middle-class Melbourne suburb of Hawthorn abused his workers ('I want some bloody work from you two-faced fucking bastards') and lunged at one of them with a spanner; his victim was imprisoned for defending himself with a shovel.[13]

Those who fought were those most likely to cope. But it was not easy to maintain energy and morale in the face of frequent victimization and unremitting hardship. The prevalent expectation that you should grin and bear it, and a tendency among contributors to oral history to romanticize youthful tribulations, should not obscure the depths to which many were reduced. If we marvel at the ingenuity with which the unemployed kicked against the pricks, we may well wonder at the absence of a more explosive upheaval against the indignities to which they were subjected. For every stalwart who looked the relief officer in the eye, there was an anonymous letter-writer offering the information that his or her neighbour was double-doling and perhaps another of the workless who retreated into an almost catatonic passivity. An examination of activism can reveal a great deal about unemployed responses, but a fuller understanding requires consideration of the web of social relations in which they were enmeshed.

Few households escaped altogether from the loss of income caused by the economic crisis. A drop in the number of boarders at the wealthy private schools bore testimony to the reduced circumstances of pastoralists; empty first-class lounges on passenger liners were a sign of belt-tightening among the *rentier* class. But if some residents of the dress-circle suburbs were forced to economize, others remained affluent. Leading companies kept up their dividends, even if from reserves, and those that traded in essential commodities maintained profitability. The usual round of entertainments certainly continued unabated. The major event in Melbourne's social calendar for 1933 was the wedding of Lord Mayor Harold Gengoult Smith to a daughter of Norman and

Mabel Brookes. The consular corps, leading politicians and a roll-call of the pastoral and urban establishment headed a guest list of a thousand—it was said that never before had so many morning suits gathered in the same place. Upwards of 40 000 more Melbournians joined vicariously in the celebrations. They began taking up vantage points around St Paul's Cathedral six hours before the ceremony, spilling up Swanston Street to the town hall where the reception was to be held and filling every window in the office blocks opposite. The bride and groom emerged to a forest of handkerchiefs and a cacophony of cheers and klaxons. The more enthusiastic spectators even followed the bridal party to the Brookes's house in Domain Road where there was an evening ball. Towards midnight the hosts took out pieces of the wedding cake and distributed them among the well-wishers 'to sleep on for luck'.[14]

How many of the unemployed were offered crumbs from the tables of the rich? The level of subscriptions to charities was low. Some prominent individuals, notably self-made businessmen with a strong paternalist streak, gave generously, but others, like BHP's Essington Lewis, regarded the Depression as an almost salutory retribution for Australians' want of industry and thrift. The principal relieving organizations were unable to raise the funds they sought. There was in any case a belief that indiscriminate benevolence of this sort would merely encourage the wrong sort of recipient. Newspapers frequently reported cases of imposition and Melbourne's influential Charity Organisation Society warned in 1931 that 'increased mendicancy is invariably associated with hard times which are exploited by the poorest types of socially disabled'. The stigma attached to charity was strong. Case histories compiled by relieving organizations confirm that applicants came to them only when they had exhausted all other sources of support from family and friends, and usually after they were deeply in debt to shopkeepers and landlord.[15]

The more common form of assistance was direct and personal. An employer would keep on a superfluous employee, a housewife would find a few shillings for a man down on his luck who offered to chop the firewood, a pedestrian would drop a coin in a busker's hat. It is tempting to construe these

Distributing Christmas gifts to the children of the unemployed, Footscray, 1932

myriad acts of generosity as a safety-valve whereby the haves took the edge off the hunger of the have-nots. Certainly the gift relationship served often as a process of social affirmation, one that eased the conscience of the giver and won the gratitude of the recipient. But the relationship was seldom as neat as this. A party of itinerant unemployed was as likely to find a bed and a meal with a struggling cocky farmer as at a pastoral station, and an unemployed family in an inner-city suburb had more chance of obtaining help from immediate neighbours than from any Lady Bountiful. Moreover, unpredictability was an essential characteristic of this, the most basic expression of a common humanity. The public conscience, on the other hand, was something more than the aggregate of these individual responses. Strongly influenced by the work ethic, slow to accept the unpalatable implications of the fact that work was unavailable, it took comfort from the fictional 'equality of sacrifice' and was quick to seize on the shortcomings of recipients of sustenance. Particular cases of misfortune pricked it. Albert Jacka, Australia's most celebrated winner of the Victoria Cross, who was reduced to selling soap from door to door until his death in January 1932, personified the betrayal of the digger. So too did the discovery in the following year that John ('Barney') Hines, an Aboriginal soldier of great renown, had been living in a dilapidated shack, unemployed for the previous four years.[16] But respectable opinion was untroubled by the fact that ex-servicemen toiled among the relief workers beautifying the approaches to the Melbourne Shrine, and one conservative journal even rekindled memories of 'Gibbit Backsheesh', the beggars' cry in the streets of Cairo, in order to deride the dole complex.[17] Such attitudes survived the Depression and even strengthened as economic recovery began.

The timing of the recovery was by no means obvious. Unemployment, having reached a peak in the winter of 1932, began slowly to recede. National output climbed less slowly from its deep trough, though company profits for the financial year 1933/34 were close to their pre-Depression level. By 1933 the economist Douglas Copland could lecture to an English audience on how Australia had achieved its recovery. Yet when his colleague Giblin reviewed the state of the national economy at this time, the prospects seemed bleak. Export earnings were still below those of the 1920s, the British money market was not yet prepared to risk new investment, unemployment remained over 20 per cent. 'In general,' he concluded, 'things are going from bad to worse.'[18] From their respective vantage points both judgements were valid. The worst of the Depression was over by the end of 1933, and recovery was clearly under way. Giblin's pessimism can be understood in part as an expression of impatience with the undue delay caused by those who controlled economic policy (he had described Sir Robert Gibson a year earlier as 'a menace to Australia').[19] More fundamentally, Giblin was worried that Australia would fail to take advantage of the unique opportunity, so painfully secured, to place its convalescent industries on a healthier regimen. If the producer organizations reverted to their racketty habits, if they were permitted to exact from the pub-

Table 12.1: Gross domestic product, 1931/32 and 1938/39 (£m at 1910/11 prices)[20]

Industry	1931/32	1938/39
Pastoral	55.9	55.4
Agricultural	35.9	22.9
Dairying, forestry, fisheries	24.0	31.7
Mining	5.9	11.6
Manufacturing	51.3	79.2
Construction	17.3	29.6
Distribution	60.0	97.4
Finance	7.7	11.8
Railways, other public undertakings and government services	40.5	46.6
Other services	44.2	50.8
Rents	43.0	47.1
Other	1.2	1.8
Total	386.9	485.9

lic purse what they could not obtain from the market-place, then Australia would fail to achieve its full potential. The national accounts for the 1930s lent support to these fears. While output resumed an impressive upward course, real domestic product per head of population was but a few shillings greater in 1938–39 (£70.12) than it had been in 1920–21 (£70.04).

The difficulties of the rural industries are readily apparent from table 12.1. While the harvest of 1938–39 was unusually bad, the earnings of pastoralists and farmers had scarcely advanced on the Depression. And these statistics tell only part of the story. The volume of wheat, wool, butter and other primary production increased significantly; it was the prices that tumbled and, although there was some recovery on world markets after 1933, these items failed to regain their monetary value of the 1920s. For wheatfarmers especially, the consequences were disastrous. From 1930, when Scullin had exhorted them to 'Grow More Wheat', until 1935, half of the 60 000 growers were unable to recover production costs from the sale of their crops. Heavily encumbered by debt, their plight was grim. Wheat cheques went straight to the creditor, preventing the farmer from renewing his plant

Table 12.2: Workforce by industry, 1931/32 and 1938/39 ('000s)[21]

Industry	1931/32	1938/39
Rural	531.0	550.3
Mining	50.4	64.3
Manufacturing	380.6	615.4
Gas, electricity, water	21.1	28.4
Construction	141.8	252.4
Transport	164.4	186.5
Commerce	301.3	396.7
Community and business services	148.1	169.0
Finance and property	50.4	56.2
Other	276.0	325.8
Totals:		
Workforce	2065.1	2645.0
Population (excluding Aboriginals)	6552.6	6929.7

('tractors don't have foals', observed one) and denying his family access to the commodities that would lighten their hardship.[22] Some walked off the land, others were evicted. But in general they clung to their farms with a resolution comparable to that of the unemployed tenants whose methods they emulated. Out on the Western Australian wheatbelt an outsider attended a foreclosure auction at his peril. The locals controlled the bidding and repurchased all livestock and machinery on behalf of their neighbour for a few pounds. Governments eventually acceded to the farmers' demands with debt moratorium legislation.[23]

The 1933 census disclosed a marked increase in the rural workforce, one whose lingering effects are still apparent in the second column of table 12.2. It made obvious sense in the depths of the Depression for teen-age children to stay on the family farm, as it did for city relatives to head for an uncle or aunt's holding. You might get heartily sick of rabbit and boiled wheat, but at least they filled the stomach. Moreover, farming lent itself to the practice of a subsistence economy around which an elaborate system of barter and exchange developed—more than one country doctor took his fee in bags of wheat. But as the decade wore on, with still no improvement in farmers' fortunes, the economists became more perplexed by the 'irrational' willingness to plant crops or run livestock on which there was not even the possibility of a profitable return. It was not that farmers were wholly satisfied with life on the land. A Victorian survey conducted at the end of the decade found that while most expressed a favourable attitude to farming, a significant minority had neutral or negative attitudes. Women especially harboured reservations. 'I've gotter like it. The old man thinks farming's Christmas. I don't.' Yet very few were prepared to leave.[24]

Giblin could only conclude that

> The farmer has been taught for so many generations that he is the backbone of the country, and that his activities are more of the nature of pious devotions than of sordid commercial ventures, that it is hard for him to believe that his country will ever fail to maintain him in the production of increased quantities of wheat and butter, even if nobody wants them.[25]

This was a bitter jest for those wheatfarmers who in 1930 had been promised a national pool with a guaranteed price of 4s a bushel and who then received, more than a year later, a bounty of 4d a bushel. The farmers certainly expected assistance but subsidies of this sort—amounting to £10 million over the next four years—hardly maintained those in greatest need. It was not until 1935 that the Commonwealth struck at the root of the problem, debt, with the apportionment of funds to restructure farmers' obligations. Even then, the £12 million spent on debt adjustment hardly dented accumulated debts of over £150 million. The more common practice, applied in every branch of rural industry except wool, was to secure the Australian market against overseas produce and make local consumers subsidize the price at which the commodity was offered to overseas customers. Such a wholesale extension of protection all round was hardly conducive to efficiency, but it did at least reorient state activity away from the earlier mania for development at all costs. The Depression therefore marked the end of a long phase of extending rural settlements, with its associated programmes of direct and indirect assistance to the settler: henceforth growth would depend on improved farming of existing holdings.[26]

Where were the primary products to be sold? It was a buyer's market in the 1930s and low commodity prices reflected an overcapacity among the producer countries in an international economy that was choked by economic nationalism. These circumstances increased Australia's reliance on its principal customer, the United Kingdom, and led to new trade agreements struck at an Imperial Economic Conference held at Ottawa in July 1932. Australia undertook to review its protective arrangements and to ensure that the British producer enjoyed 'full opportunity of reasonable competition', which was interpreted to mean a more comprehensive tariff preference over other imports; Britain, in return, gave preferential treatment to Australian meat, butter, fruit and other farm products, and lesser privileges in respect of wheat and minerals, though not wool. Consequently Australia was able to place more than half its exports on the British market and to maintain a favourable balance of trade. But imperial preference was not without disadvantages. Britain's competitors, now penalized in the Australian market, retaliated by purchasing wool elsewhere. Japan and the United States were angered particularly by a supplementary policy of trade diversion, announced in 1936, whereby certain of their products which outsold British lines were subjected to additional penalties. An expensive trade war with Japan lasted six months.[27]

There were other British concessions. The accumulated overseas debt hung like a millstone on the Australian economy during the early 1930s, much of it in short-term, high-interest loans that were all the more onerous because of reduced foreign earnings—by 1933 more than a third of Australia's export income was needed to service this debt. The Lyons government therefore sought to convert maturing loans in order to reduce interest charges. Bruce, who had returned to the ministry with special responsibility for economic affairs, conducted the negotiations in London with vigour, though neither the chancellor of the exchequer nor the London bankers were quite as compliant as he later claimed. But in their own good time they arranged for the conversion over several years of loans worth £160 million, which brought annual savings in interest and exchange of £3 million.[28] One consideration that weighed favourably with the British was Australia's restraint in raising new loans. No one again wanted to play host to Niemeyer. Having felt the consequences of their earlier profligacy, the Commonwealth and states were doubly wary of incurring fixed interest obligations, so the Loan Council exercised a strict control over public projects. The external public debt amounted to £621 million in 1939, an increase of less than £30 million over the position reached at the end of the previous decade.[29]

The absence of the two traditional stimuli, high export prices and large capital inflows, made the economic recovery all the more remarkable. As tables 12.1 and 12.2 indicate, manufacturing achieved the most pronounced gain in output during the 1930s and made the greatest contribution to new employment. Conditions for growth had been established during the Depression when wage costs were lowered, while devaluation and higher tariffs increased the effective level of protection by as much as 80 per cent.[30] Such circumstances enabled numerous backyard workshops to start up (comparison of the tables reveals that output per worker in this sector failed to increase), but the Depression also brought about takeovers and mergers that increased the market share of the big companies. With the recovery, the engineering, chemical and electrical industries all broke fresh ground. Production of iron and steel expanded with the construction of a new blast furnace at Port Kembla, which was acquired in 1935 by BHP and developed along with its Newcastle works into a highly efficient operation. Output passed 1 million tons in 1937. BHP's ability to produce steel as cheaply as foreign competitors brought lower costs of production for Australian manufacturers and a broader range of locally processed and fabricated goods.[31]

These were important achievements. The domestic orientation of the revival gave evidence of a reduced dependence on overseas conditions and this increased self-sufficiency would be vital to Australia's survival in the years to come. But it was small comfort to the army of men and women—never less than 200 000—who remained out of work even as the tempo of economic activity increased. The unskilled suffered most. Wage fixation had kept the earnings of untrained men close to those of tradesmen over the past two decades. Now, in an economy with less need for casual navvies, wage differentials increased. Even at a reduced rate, there was less demand for their services. The changes in the manufacturing

Men and their bicycles outside the employment office of the Newcastle steelworks

sector, the lingering rural malaise, the contraction of public works and the continued decline of domestic service all reduced employment opportunities for untrained and unqualified labour, so that in 1934 engineering workers were attracting above-award payments while men and women still queued for casual vacancies.

Neither arbitration nor direct negotiation could win a restoration of earnings for the majority of wage-earners. In 1932 the Commonwealth Arbitration Court rejected the unions' application that they be given back the 10 per cent cut from the basic wage the previous year. The court based its 1933 award on a new price index that yielded a small increase, and in the following year it again conjured a further tiny increment from a new basis of calculation. Even in 1937, a 'prosperity loading' of 6s a week fell far short of union claims. Though the court would not acknowledge the fact, the old 'needs' basis of wage determination had long since given way to the court's assessment of employers' ability to pay.[32] Awards were commonly circumvented in any case. Employers flouted manning agreements, demarcation rules and apprenticeship provisions; some forced their workers to hand back part of their wages, other did not even bother to disguise their violation of the law. Much of the increased profitability and growth of the 1930s was sweated out of a weakened workforce.

The unions were scarcely able to resist. In many cases a foreman merely had to point to those waiting for a job outside the gates to quell protests against speed-ups. With the crushing defeat of those once-powerful battalions of the labour movement, the miners, timber-workers and wharf labourers (who again lost preference as soon as the Lyons government took office), few other groups of wage-earners were likely to make a stand. Acts of defiance among those whose conditions of work came under concentrated assault, such as the meatworkers and pastoral workers, were quickly snuffed out. The cutters in the Queensland sugar industry struck against unsafe working conditions but were bludgeoned into submission. Union membership fell back from more than 900 000 in 1928 to 739 000 by 1933 and known militants made up a high proportion of the casualties. Such was the hostility towards unionism in the steel industry that the Federated Ironworkers' Association took on the character of a secret army, with branch officers putting prospective members through a third degree to ensure that they were not management spies. An attempt to revive militancy among the seamen in 1935 collapsed when the Lyons government applied the Transport Workers Act and licensed 6977 volunteers to take the jobs of 2300 strikers.[33]

Slowly and with immense difficulty, the shattered ranks reformed. The residents of Wonthaggi showed in 1934 that it was possible to withstand an economic siege. Theirs was a coal-town and the miners struck in support of colleagues sacked for resisting a speed-up. They held out because they were able to tap the resources of the community, women as well as men, shopkeepers and even local farmers. The strike committee collected and distributed food, it organized a boot repair depot, sewing centres and a hairdresser, and it sent representatives to draw support from further afield. After five months the state government reinstated the dismissed miners. It was the first battle since the late 1920s from which a union had emerged strengthened rather than weakened by conflict. Miners elsewhere were heartened by the victory and adapted the same organizational forms:

> For it's no longer Jack and Joan, it is united 'we',
> Oh comrades, toast the Ladies of the Strike Auxiliary.

Combining new forms of industrial action, including the stay-down strike, with calculated use of the arbitration system, they won improved pay and conditions during the later 1930s.[34] Similar advances were achieved among transport and heavy industry unions.

In the course of the 1934 strike the Wonthaggi miners elected as their new president Idris Williams, a Welsh-born Communist who best captured their spirit of resolution. So too, other unions threw up new leaders during the recovery. An older generation, all too often broken or dispirited, gave way to younger activists with fire in their bellies. Among the Communists who rose to prominence were Bill Orr and Charlie Nelson, who were elected to national leadership of the miners in 1934; Ernie Thornton, who became general secretary of the Federated Ironworkers' Association in 1936; Jim Healy, who took the same position in the Waterside

Workers' Federation in the following year; Eliot Elliott, who rebuilt the Queensland branch of the Seamen's Union before becoming national secretary in 1940; and a host of officers of smaller unions as well as branch officials and delegates who made their successes possible. The growth of the revolutionary left during the Depression had been slow (indeed, the social credit message of Major Douglas exerted greater electoral appeal than that of Marx). In 1934 the Communist Party had 3000 members, most of them out of work. But they comprised an outstanding generation of activists, and their experience of hunger and hardship hardened into a fierce hatred of capitalism. They brought to the struggle an almost military appreciation of strategic possibilities. A Leninist vocabulary of class warfare, with its associated conceptions of party vanguard and class rank and file, and a corresponding tendency to substitute the wisdom of the former for the frailties of the latter, may have encouraged an unduly instrumental understanding of politics but in the desperate circumstances it was undoubtedly effective. Abandoning the sectarian denunciation of the Labor Party in favour of the united front, they catalysed the grievances of their workmates. In place of the all-purpose general stoppage, they made use of selective bans, work-to-rules, arbitration applications and whatever else would bring improvements for their members, while always keeping their larger purpose intact. Nor was their scope merely industrial. Their respect for Marxist doctrine and veneration for the Soviet Union gave them a wider perspective on world events. They would therefore be prominent in the progressive campaigns that stirred Australia from its torpor.[35]

The growth of the left, however, was confined for the most part to an industrial base, and a restricted base at that. In mining, transport and heavy industry, large groups of men worked for remote employers at dangerous and physically demanding jobs yielding low pay and very little security. Such conditions bred a particular brand of militancy: robust, forthright and aggressively masculine. The objectives of these workers—decasualization, better pay, safer working conditions—sprang from their working conditions, as did their friendships, interests and values. To this extent, the wage relationship was at once a source of discontent and the principal solace against that discontent. Other sectors of the workforce were more divided and less responsive to unionization. Moreover, it was far more difficult to maintain a sense of camaraderie and common purpose outside the workplace. Within the political arena, a large part of the electorate regarded industrial militancy merely as a form of anti-social behaviour. The absolute prerogatives of capital seemed natural, the limited powers of labour unnatural. How could the social order be contested when its central mechanism was taken for granted?

13

DEVIL'S DECADE

FEW OF THOSE who took office under Joseph Lyons at the beginning of 1932 expected him to remain their leader for the next seven years. He seemed merely the man for the moment, judged by his backers as the one best able to serve their needs. Bruce and Latham, the senior statesmen, liked to think that they were essential to the success of the ministry. 'Good as the Prime Minister is', Bruce wrote to Latham in evident surprise after working with him for six months, 'it would not be fair to leave him without either you or me when Parliament is sitting.' Yet Bruce departed Canberra for London in 1933, and Lyons confided to his wife that 'It is strange but I feel more capable of handling the job when poor old Latham is not here – he fidgets and worries himself into ill health.' Latham retired to the High Court after the government completed its first term.[1] Lyons's Melbourne sponsors might have expected that they would be able to call the tune but found in practice that there were limits to their influence over the prime minister.[2] The Country Party, eager from the outset to reassert its influence over tariff policy, received short shrift. Immediately after the election, Lyons offered to take just three Country members into the new ministry, the three and their portfolios to be chosen by him, and he made it clear that customs would not be available. Page discovered

The UAP ministry formed in January 1932 is here presented as a thirteen-man cricket team: Hughes, with the drinks, makes an unlikely fourteenth man

that the offer was not negotiable and took his followers to the cross-benches with all the dignity he could muster.[3]

The UAP therefore governed in its own right with a majority of five. The 1934 election reduced its strength in the House of Representatives to 33, with Labor advancing to 27, and 14 members of the Country Party holding the balance of power. Negotiations between the two non-Labor parties again broke down, but Page's threat to cross the floor forced Lyons's hand. Page became deputy prime minister and his party was allocated two full portfolios and two assistant portfolios. While the urban and rural alliance was not always harmonious, that between the two leaders prospered and they remained on good terms until Lyons's death.

The prime minister was neither innovative nor assertive.

His genial reputation as 'Honest Joe' belied a shrewdness and, at times, a toughness without which he could hardly have survived as long as he did in the cut-throat world of conservative politics. But his instinctive caution and dependence on the expertise of senior ministers allowed a familiar array of pressure groups to reassert their influence. Charles Hawker, the only UAP minister from a rural background, resigned late in 1932 against the sacrifice of primary producers to the bad old habits of indiscriminate protection and regulation. It was, he said, 'a sort of government of the feeble for the greedy'.[4] This was a jaundiced description of an administration that was trying to juggle the claims of its clients while holding expenditure within tight limits. The overriding objective, to maintain national solvency, meant that assistance would have to come from consumers rather than from the treasury. Manufacturers sought and obtained the retention of a high tariff schedule. But even before the Country Party entered the ministry, the Lyons government was trying to offset the effects of industrial protection by means of the Ottawa Agreement, which gave farmers privileged access to the British market. After 1934 there was an increase in support for primary producers as well as an extension of marketing regulation.

The vagaries of Commonwealth activity in this field illustrate the way the administration approached the task of reconciling divergent interests. Its intention was to make high domestic prices subsidize export prices for farm produce. The Commonwealth still lacked a clear jurisdiction over the domestic market (extra powers were sought by referendum in 1937 but the voters failed to provide them), so it relied on state marketing authorities to police operations within their own boundaries. But since growers were naturally tempted to dispose of their output in the more profitable home market, it is necessary to fall back on ever more complex forms of compulsory acquisition, production quotas and other expedients of shaky legality and dubious efficiency. If any rationale can be discerned in this regulatory labyrinth, it was the government's commitment to maintaining at least the bare livelihood of all producers while clinging to the shibboleths of self-sufficiency and individual initiative.

As with the producers, so with the states. The operation of

a national customs union weighed heavily on the outlying regions, whose dependence on primary industry remained as pronounced in the 1930s as it had been on the eve of the First World War. Western Australia, Tasmania and South Australia had long received financial assistance as compensation for their disabilities, but the calculation of these grants was a persistent cause of disputation. What was needed, the new government recognized, was a 'competent, impartial and permanent body' to assess claims. In 1933 Lyons announced the creation of a Commonwealth Grants Commission, chaired by Frederic Eggleston. In establishing its guiding principles, the commission decided that it could not be expected to take account of the alleged consequences of federation, nor should it compensate a state for the paucity of its economic resources. Its task was simply to ensure that poorer states were able to 'function at a standard not appreciably below that of other States'. In practice the commission increased the level of support to the claimant states. Giblin, who was a member, confessed in 1938 that he and his colleagues often used their judgement to arrive at a 'reasonable' figure which was then 'dressed up in arithmetical terms'. And on most occasions the states were satisfied, despite their protests. Meetings of the Loan Council and allocations of special purpose grants for road building produced similar histrionics, but a rough-and-ready consensus usually emerged.[5]

It was fortunate that this was so. The disadvantaged states felt a keen discontent in the early part of the decade when Western Australia even sought to secede. Separatist feeling had grown there during the Depression until Mitchell's Nationalist government conceded a referendum on the question in 1933; 138 653 voted in favour of secession, 70 706 against. Canberra blundered badly when it despatched Lyons, Hughes, Pearce and even the South Australian Hill to put the case for the Commonwealth—support for federation was greatest in Labor circles and these emissaries were Labor renegades. Among conservatives these 'wise men from the east' fared no better. They faced an unruly meeting in the Perth town hall and when the playing of 'Advance Australia Fair' failed to quell the disorder, the organizers had to fall back on 'God Save the King'. The new state government—Labor having defeated the Nationalists in the 1933 election—

showed more finesse. It kept faith with the result of the plebiscite by appointing a committee to prepare a detailed *Case of the People of Western Australia* for submission to the imperial parliament, but drew out the exercise for more than a year. It also released William Somerville from the state Arbitration Court so that he could help the Commonwealth formulate a rebuttal, *The Case for Union*, copies of which were distributed to all voters and those in South Australia and Tasmania for good measure. The leading secessionists departed for London late in 1934 with a petition 26 feet long; in the following year they were told that the Commonwealth could not be dissolved except with the consent of the Commonwealth parliament, and by then the strength of the secession movement was spent.[6]

The UAP's best years were its early ones. It began with the support of the mass-based citizens' organizations, the resources of the business community and a simple but attractive programme of putting the country back on sound business lines. The circumstances of its creation presented the promoters with an opportunity to build an effective national party, one that could span the propertied classes and channel their energies into a coherent unity. The opportunity was botched. Instead, the party faithful were enlisted into what one organizer conceded was a 'low voltage organisation', largely dormant between elections, denied any real voice in the formulation of policy and dependent for its revenue on the old finance committees of Melbourne and Sydney. The state branches of Tasmania and Western Australia retained the old Nationalist title and organization, Queensland and South Australia operated as a hybrid UAP and Country Party, Victoria was split between three party units. Since there was no federal secretariat, indeed no formal federal structure of any kind, it was left to Lyons as party leader to co-ordinate activities of the state branches. An explanation for this failure probably lies in the inability of different fractions of capital to compose their differences—the patronage exercised by the Melbourne paymasters, who were identified most closely with finance capital, certainly antagonized producer groups. Be that as it may, the UAP soon fell back into the tergiversating habits of its Nationalist predecessor.

This did not matter while the electoral popularity of the

With the accession of Lyons to national leadership, the Prime Minister's Lodge in Canberra became a cynosure of family values. Here, Joe, Enid and the children line up for a portrait

Lyons government was high. Electioneering had passed from the street corner and town hall to the radio and cinema, and conservatives were more capable than their Labor opponents of speaking to this mass electorate. The Lyons family—homely Joe, his formidable wife Enid and their large brood—made good advertising copy. But the UAP was losing momentum by the mid-1930s. Representation in the House fell from 40 in 1931 to 33 in 1934 and 28 in 1937, when UAP candidates won only 34.4 per cent of the vote. Party discipline was never strong and the loss of leading ministers—first Hawker, Bruce and Latham, later Gullett and Hughes—weakened the calibre of the administration. Younger men began to anticipate Lyons's retirement and manoeuvre for the succession. Supporters of Menzies, who had replaced Latham as attorney-general, made up one faction; Parkhill, the senior New South Wales figure, headed another. Legislative and administrative programmes fell prey to constant dissension and Lyons himself is reported by a boyhood friend to have lamented that he had ever left

Tasmania—'I had good mates there, and was happy, but this situation is killing me.' Under these circumstances the government drifted into inanition.[7]

A point of complete paralysis was reached in the drawn-out argument over social policy. As early as 1932, the government declared its intention to put existing pensions onto a contributory basis, both as an economy measure and to encourage greater self-reliance. Ministers shrank from the electoral consequences of such a drastic step, though they did reduce the level of pensions, and one wing of the UAP then pressed for a system of social insurance covering unemployment and sickness as well as old age. Reports and recommendations gathered dust for several years. Richard Casey (who had entered politics in 1932 and risen swiftly to the office of treasurer) guided a scheme onto the statute book that covered pensions and health benefits only, but even this fell victim to the doctors and financial institutions, while constant revision to meet their objections delayed implementation of the Act. Cabinet split into warring factions. Finally, in March 1939, Menzies resigned in protest against the failure 'to go on with an Act which represents two years of labour, a vast amount of organisation and considerable expenditure of public and private funds'.[8] Menzies's sincerity was questionable—he himself was prime minister when the last rites were performed—but his exasperation was genuine. The UAP was paralysed by its subordination to its supporters.

This failure has a more general significance, one that provides an insight into the class order and the way it had hardened since the turn of the century. The basis of that order was a minority's ownership of the means of production and the majority's dependence on wage labour. While the class configuration was modified by rural producers, shopkeepers and others working for themselves, the wage relationship remained the crucial determinant of the destiny of most Australians. Thirty years of state activity threw up economic and social institutions that were meant to ensure the benign operation of the labour market, but they could not alter its essential character—and the Depression showed the ultimate futility of the elaborate regulatory edifice. From the capitalist viewpoint, the problem was one of high wages and inefficient rigidities in the allocation of economic resources. If the problem was to be solved within the confines of political democracy, then it was necessary to create some alternative mechanism of welfare, one that would address citizens' needs directly and free the labour market from the burden of doing so. State insurance was not the only possibility but there were impressive foreign precedents and it possessed the undeniable attraction of leaving the power of private capital intact. This much the Nationalists discerned. Yet the objections of particular fractions of capital had once again overcome larger collective interests. Why was this? The protracted struggle over social insurance suggests a fundamental incoherence of the propertied class. The disjunction of domestic and export producers remained as sharp in the 1930s as it had been at the beginning of the century, the dependent and derivative character of the Australian ruling class just as marked. The majority preferred the familiar to the unknown, short-term advantage to the longer-term possibilities of a thorough transformation of the political economy.

The UAP was more united but no more convincing in the field of foreign policy. From the moment it assumed office, the government placed an exaggerated emphasis on links with Britain, reviving even those ceremonial trappings that the Scullin government had discarded in a gesture of independence. Imperial honours were restored. The speaker of the House of Representatives once more donned wig and gown. Incidents that strained Australian loyalty, notably the success of Harold Larwood in the Test cricket series of 1932–33 and the infuriating arrogance of his captain Jardine were of great concern. The home country took on a talismanic importance for conservatives such as Menzies, who wrote in his diary when he first landed on the sacred soil:

> At last we are in England. Our journey to Mecca has ended and our minds abandoned to reflections which can so strangely (unless you remember our traditions and upbringing) move the sense of those who go 'home' to a land they have never seen.[9]

Given this attachment, it was hardly surprising that Australia followed closely on the heels of Britain as it wove a diplomatic path through the international crises. The difficulty was that Britain's movements were so unpredictable. One moment it was holding fast to Geneva and offering lip-service to the principles of collective security through the League of Nations; the next moment it was off to Rome or Berlin to conduct direct negotiations with the aggressive powers. Denied forewarning, let alone consultation, the Australian government clung all the more tightly to Britain's coat-tails. Alarmed in their own region by the growing strength of Japan, the Australian ministers declared repeatedly that our defence arrangements 'dovetailed' into Imperial defence plans, but they could not get a satisfactory account of what those plans were. External affairs were separated from the Prime Minister's Department in 1935, and made a department in their own right, but still the government betrayed halting uncertainty as it recited apologies for appeasement.[10]

Meanwhile Labor rebuilt its shattered ranks. The split between Lang's followers and ALP loyalists affected not just New South Wales, where Labor remained out of office for a decade, but the federal scene, where the Langites retained their separate identity until 1936. In those states in which Labor governments had implemented the Premiers' Plan, South Australia and Victoria, bitter recriminations left the ALP in permanent opposition. Thus New South Wales was governed by a UAP–Country Party coalition, the South Australian Liberal Country League began thirty-two years of continuous rule, while Victoria even found itself under a Country Party administration.[11] The three remaining states returned to Labor control: Queensland in 1932, Western Australia in 1933 and Tasmania in 1934. This allowed a resumption of the old patterns of economic development with an emphasis on generating employment by state enterprise and public works. As before, the AWU was the chief beneficiary of such projects (a clear majority of the Queensland and Western Australian cabinets had formal ties with that union) and dissidents were crushed. By preference, if not by necessity, these regimes pursued rural development rather than heavy industry. The premier of Queensland insisted in 1932 that 'this state will continue for all time to be a primary producing state. It is desirable that it should be so. Primary production is the natural occupation of mankind.'[12]

The political advantages of such a programme were undeniable. Knowing that it could count on the working-class electorates of the metropolis, the Forgan Smith government in Queensland and the Collier (from 1936, the Willcock) government in Western Australia could concentrate resources on crucial rural areas and provide voters with the considerable benefits at their disposal. When all else failed, it was not unknown for a gang of navvies to be dispatched to a marginal electorate—and no one doubted who received their votes. The economic rationale was that rural growth would increase demand for the goods and services of the city. But there were limits to the strategy, especially during a time of stringent controls on public expenditure. Moreover, the opportunities for further development were no longer obvious. Tasmania could embark on the hydro-electric projects that were to skew its politics for the next half century, but elsewhere, as Premier Willcock acknowledged, 'All railways needed had been built.' As for urban workers, they could expect little more than union preference and marginal improvements in pay and working conditions. The unions themselves, preferring earnings to leisure, lacked enthusiasm for their declared objective of a forty-hour week.[13]

In short, the political efficacy of the old Laborism was all but spent. The mobilization of the working class around programmes of state intervention for the protection of jobs and wages had faltered at the federal level by the end of the First World War. The momentum slowed in the industrially advanced states during the 1920s. Now, in the outlying regions where such methods retained their appeal, electoral success could hardly conceal a bankruptcy of purpose. Indeed, Labor's very durability as an electoral organization enabled it to postpone a reappraisal of its policies. John Curtin, who replaced Scullin as federal leader in 1935, brought a new awareness of the importance of social welfare; Ben Chifley, who served as Labor's representative on the Royal Commission on Banking from 1936 to 1937, appreciated that mere hostility to Money Power hardly solved the problems of economic management. But these were scarcely more than presentiments of a new approach that would congeal in the wartime emergency and be beaten into shape on the anvil of Keynesian economics.[14]

Already an alternative way forward was being explored. South Australia had a non-Labor government that was no less concerned with fostering economic growth but was more appreciative of the greater potential of secondary industry. Here the state did not attempt to construct or operate new undertakings; rather it attracted private capital to do so by means of concessions and other inducements. General Motors-Holden was given special rates of wharfage and taxation; Imperial Chemical Industries received planning assistance and a generous provision of ancillary services; BHP obtained a favourable lease to situate a blast furnace at Whyalla.[15] It would be some time before this new style of courting national and international capital eclipsed the old patterns of state enterprise; New South Wales and Victoria did not yet have to engage in the Dutch auction of public resources. Even so, the threatening implications for the old Laborism were already apparent.

The success of the conservatives, both federally and in the leading states, and the lack-lustre performance of Labor were accompanied by a repressive conformity. As soon as it assumed office, the Lyons government mobilized against the communist menace. It banned the importation of seditious material (three men were convicted on a charge of smuggling lantern slides of the Soviet Union). It prevented the transmission of local publications through the post. It put pressure on employers to dismiss communist employees and on proprietors to deny the use of meeting halls and party rooms. It amended the Crimes Act and prosecuted the editor of the *Workers' Weekly* for soliciting funds for a revolutionary organization, though his conviction was quashed by the High Court. State authorities, similarly, restricted or forbade street gatherings, demonstrations and public meetings. In 1933 there was the epic battle for free speech in Sydney Road, Brunswick, where the young painter Noel Counihan addressed a crowd from an iron cage.[16]

The intolerance spread to all forms of dissent, allowing any troublemaker to be labelled a 'Commo'. Late at night in Brisbane, towards the end of 1932, two unemployed activists with posters and pots of glue were apprehended. One complained that the arresting officer had called him a 'communist bastard', though he withdrew the complaint when an official investigation followed. Nevertheless, the secretary of the Police Union complained to the minister that the constable was only obeying instructions to suppress communism and insisted that he be exonerated from any criticism. The minister, a Labor man, wrote across the secretary's letter: 'I am getting the impression that the secretary of the union is a communist.'[17] The Labor premier of Western Australia used the same ploy in more sinister circumstances. In Kalgoorlie, on the Australia Day weekend of 1934, an abusive customer died after being put out of a hotel by an Italian barman. Full-scale race riots followed for several days as mobs combed the hotels, boarding-house and residential areas of the Italian and Yugoslav miners, causing two deaths and damage estimated at £100 000. The trouble had been brewing for some time and the tension between the Anglo-Celtic majority and the Mediterranean minority was exacerbated by allegations that the 'Dings' curried favour with the mining companies. Yet two days of violence were allowed to pass before the government sent police reinforcements to restore order. The evidence suggests that the premier, Philip Collier, allowed the delay because of intelligence reports that Kalgoorlie was a hotbed of communism. There was in fact a strong anti-fascist feeling among Italians on the goldfield as well as a large branch of Yugoslav members of the Communist Party and a smaller group of 'Australian' party members, and these radicals risked their safety by standing up at public meetings to call for peace and condemn racial hate. But this did not prevent Collier from blaming the whole affair on 'the machinations of a few Communists'.[18]

As the spectre of an unemployed uprising receded in the conservative imagination, it was replaced by two equally urgent fears: disloyalty and immorality. On the left there was a growing international awareness of the need to contain fascism and defend democracy, but the right regarded such cosmopolitan concerns as inimical to patriotism and imperial loyalty. Even then, there was considerable sympathy on the right for the fascist dictators in their assertion of firm authority and cultivation of vigorous national pride. Harry Lawson, the conservative premier of Victoria, had greeted Mussolini on his accession to power as 'the man whom Providence wanted to lead Italy'. Menzies visited Germany in 1938 and

Egon Kisch, leg in plaster, is carried into court to appeal against his prohibition

found there was a 'really spiritual quality in the willingness of young Germans to devote themselves to the service and well-being of the State'.[19] In Australia the Commonwealth Investigation Bureau harassed Italian anti-fascists and tolerated the thuggery of the local *fascisti*. Similarly, the German consul-general was able to prevail on the federal government to ban Clifford Odets's anti-Nazi play *Till the Day I Die*, and Lyons rebuked H.G. Wells for describing Hitler as a 'certifiable lunatic'. Victims of Hitler and Mussolini who sought refuge in Australia had to pay a landing fee of £200, and £1000 if they were Jewish, on the grounds that it was necessary to guard against anti-Semitism! The mounting international tension only intensified these attitudes. From 1938 radio stations had to submit scripts of any talks on international affairs before they were broadcast.[20]

Yet the government was unable to stifle revulsion against the methods of right-wing dictatorships and their military aggression in Abyssinia, the Rhineland, Spain, Czechoslovakia and China. A Movement Against War and Fascism was launched in 1933, on communist initiative and with support from religious, civic and other political organizations. Moreover, the government's ham-fisted response to the movement's inaugural Peace Congress in 1934 only increased that support. Two overseas delegates, one Czech and the other a New Zealander, were denied entry. The first, Egon Kisch, jumped ship in Melbourne and broke a leg. He was bundled back on board and taken to Sydney, but released there on the order of the High Court. The Commonwealth then administered a language test under the Immigration Act of 1901, and, since Kisch was proficient in a number of European languages, put the test in Gaelic. Kisch failed the test and was sentenced to six months' imprisonment and deportation. On the next day the New Zealander received the same sentence for being 'so British that he could not speak Dutch'. The absurdity of such procedures was readily apparent. 'What would the government do if Signor Mussolini wished to come to Australia—give him a dictation test in Gaelic?' asked Giblin, who had recently helped to establish an anti-censorship league. The High Court ruled that Gaelic was not a language under the Act and Kisch was set free to address the large crowds attracted by the publicity.[21]

As an exercise in mismanagement the Kisch affair was exceeded only by the pig-iron dispute of 1938. Here the maritime unions imposed an embargo on the export of Australian iron to Japan in protest against that country's invasion of China. The federal government had itself stopped the sale of iron ore in 1937, but insisted that the sale of scrap and pig-iron should not be interrupted. In November 1938 the waterside workers of Port Kembla refused to load a ship bound for Japan. Menzies, who acted for the government in the dispute, replied by invoking the Transport Workers Act and offering licences to volunteer labour. None were taken out. Further excesses such as the temporary closure of a Labor-owned radio station which criticized the government's

Attorney-General Robert Menzies ('Pig-Iron Bob') is escorted into Wollongong, January 1939

handling of the dispute merely increased support for the unions. The workers held out for two months and abandoned their stand only because BHP closed its works to starve the community into submission. Henceforth Menzies carried the derisive nickname 'Pig-Iron Bob', and many who bore the brunt of Japanese shells and bullets just three years later had good cause to remember how he had earned it.[22]

The government exercised an equal vigilance over literature, art and public morality. In 1930 a customs officer laid down the principle that work of literature should not be permitted into Australia unless 'the average householder would accept the book in question as reading matter for his family'. By 1936 the list of prohibited publications included works by Joyce, Defoe, Hemingway and Orwell, and amounted to some 5000 titles.[23] Australian writers also found great difficulty in finding publishers for works that violated the canons of respectable taste. Katherine Susannah Prichard's *Coonardoo*, which dealt with Aboriginal–white relations, and Leonard Mann's anti-war novel *Flesh in Armour* were both rejected by Angus and Robertson, the leading publisher of Australian fiction. Even Jock Neilson, as unpolitical a writer as ever put pen to paper, was worried in 1934 that the publication of his collected poems might somehow be interpreted as seditious and jeopardize his job as a messenger in the Victorian Country Roads Department. 'You may laugh at this', he told his editor, 'but some departmental heads take things like this very seriously. Some of them regard any slight attack on capitalism as a clergyman would regard an attack on his faith.'[24]

It was not that the censors flouted public opinion. For every delegation from the Book Censorship Abolition League that waited on the minister, there were a dozen calls for greater vigilance from church bodies, women's clubs, parents' associations, the RSSILA and similar organizations. Even the Fellowship of Australian Writers demanded a ban on American comics. Every time the Council for Civil Liberties drew attention to an abuse, it was drowned out by newspaper editorials and correspondence. The government was more vulnerable when it showed ineptitude, as in the Kisch case, than when it simply clamped down on dissent. There was in fact a calculated intent in its use of bluster and threat, as in the boast of the minister responsible for the Australian Broadcasting Commission that he knew nothing of broadcasting, was not interested in it, and wished only that he could bring it under the Vermin Act to poison the people responsible for radio talks and commentaries.[25] Such ukases served as a substitute for the poverty of conservative ideology. Lacking any clear vision of national purpose, unable to sustain the altruistic ideals of duty and service and unwilling to repudiate the ethic of competitive self-satisfaction, caught between loyalty to Empire and national interest at a time when they threatened to fly apart, the ruling class fell back on symbols. Yet it was among the writers, artists and intellectuals that dissent was greatest. A characteristic episode was Menzies's promotion in 1937 of an Academy of Art to enforce a conventional 'healthy' aesthetic and proscribe the 'decadent' modernism of a younger generation of painters. Menzies believed that in art, as in politics, unifying institutions must prevail if proper standards were to be maintained. The academy enlisted only part of the establishment and even stimulated the formation of a far more influential Contemporary Art Society.[26]

The divisions in the art world also reveal the gulf that separated Anglophiles like Menzies from the more cosmopolitan members of the intellectual community. Critics were quick to exploit the irony of taking from Britain the model, even the name, of an institution that was meant to bring coherence and unity to the nation's art. That argument between Empire loyalists and nativists was, of course, well established. Indeed, it had long since exhausted its creative vitality. Once nationalism took on a conservative valency, as it surely had by the end of the First World War, its conformist demands became stultifying and oppressive—the rabid excesses of the *Bulletin* were as inimical to cultural advance as the *Bulletin* of the 1890s had been favourable. Creative and progressive minds recoiled from the drab mediocrity of their native land.

The crises of the epoch transcended this impasse. Communism, fascism, the failure of capitalism—these were global forces encouraging a new kind of national sentiment, no longer pro- or anti-colonial but post-colonial. It could be seen in the views of P.R. Stephensen, who had abandoned his youthful communism and was lurching towards the right. His essay on *The Foundations of Culture in Australia* (1936) was a forceful statement of the need for a distinctively Australian culture:

> In what, at present, can an Australian take pride? In our cricketers, merino sheep, soldiers, vast open spaces—and what then? Until we have a culture, a quiet strength of intellectual achievement, we have nothing except our soldiers to be proud of.

At the same time, Stephensen appreciated the inadequacy of the 'gum and wattle' school: 'cultures must remain local in creation and universal in appreciation'.[27] On the left, there was something analagous—a conscious exploration of an identity that was at once distinctively Australian and part of a larger movement. Perhaps the most seminal figure here was Brian Fitzpatrick, the civil libertarian and freelance writer who published *British Imperialism in Australia 1788–1833* in 1939, and its sequel, *The British Empire in Australia*, in 1941, along with a popular *Short History of the Australian Labor Movement* in 1940. These works traced a line of continuity from the convicts, the diggers and bush-workers to the unionists and radicals of his day, a tradition of popular resistance and struggle that he helped to maintain even as he wrote. As the titles suggest, Fitzpatrick situated the Australian experience within its imperial context, but his insight into the dynamics of capitalist exploitation enabled him to present this, in turn, within a class framework. While the Australian people could draw on deep native roots in their pursuit of freedom and equality, the values were universal.[28]

The historian R.M. Crawford, who was himself involved in the intellectual and political controversies of the late 1930s, has claimed that this period saw 'a new maturity' in Australian life. He cites as evidence the decision to recruit graduates to the Commonwealth Public Service, the expansion of the Council for Scientific and Industrial Research, the revival of the universities and the consolidation of the Australian Broadcasting Commission. 'The evidence crowds in on the historian as he looks for signs of growing maturity that a corner was turned in Australian history at that time.' He was conscious also, when he returned from England to Australia in 1935, of a more informed discussion of current affairs, a greater willingness among writers to give up the tired clichés of the bush and explore life in the cities, and the beginning of an architectural renaissance. Nor was this all. To Crawford's list can be added the increased attention to secondary education and public libraries, renewed attempts to improve housing standards and social services generally, and even some rethinking of Aboriginal policy.[29]

Such an inventory suggests a new mentality. The difference is partly one of timing. As Australia emerged from the stringencies of the Depression, it was possible to relax the stultifying parsimony and breathe life into public institutions. It is also a matter of perspective. The bodies in Crawford's roll-call are those concerned with the training and professional practice of administrators, policy-makers, scientists, teachers, artists and intellectuals. Together they comprise a new middle class whose skills were of increasing importance in the post-Depression years. By this time an institutional base existed for bourgeois professionalism in Australia, and the earlier disappointments could be left behind.[30] Critical though many of these professional men and women were of the injustice, the drab mediocrity and philistinism that they saw all about them, struggle though they might against entrenched conservatism, they could at least find reassurance in the conviction that the future lay with their knowledge and expertise. That comfort was denied to the unemployed activists, the Port Kembla strikers, those who gave their lives fighting fascism in Spain, and others bearing the brunt of the struggle. In short, a distinction should be drawn between the progressive middle-class reformers, who had an acknowledged place in their society, and the embattled left-wing activists, who did not. It is not merely a question of class origins and loyalties. There was an acute sense during this decade, unprecedented in its intensity and cutting through the political spectrum, of the shallowness of capitalist civilization in Australia. Estrangement took different forms, from the savage nationalism of P.R. Stephensen or Xavier Herbert to the high-minded humanism of Nettie and Vance Palmer and their circle. Never before had so many of the country's creative minds been so fundamentally alienated from their society. The distinction is important, nevertheless, because the middle-class reformers were so closely involved in reshaping and administering that society's institutions.

The process is apparent in the housing campaigns of the late 1930s. A philanthropic concern with urban conditions went back to the beginning of the century and drew added force from environmentalist beliefs that slums bred social problems. The reformers' attention shifted after the First World War to patterns of behaviour, especially reproduction, child-rearing and socialization, and away from the physical fabric of working-class lives. Given the interruption of house construction during the Depression years, the deterioration of the existing stock and the movement of evicted families into makeshift alternatives, it was hardly surprising that the issue revived. In Melbourne a Methodist accountant, Oswald Barnett, took the lead. He was able to interest the premier and conduct him, with an entourage of eighteen cars, through the inner suburbs of Fitzroy, Collingwood, West and Port

Children playing in an inner-suburban street: a photograph published in the Report of the Housing Investigation and Slum Abolition Board, 1937

Melbourne—the inhabitants of the improvised shanties of Dudley Flats turned out respectfully because they thought a funeral was passing through. Following Barnett's disclosures, a Housing Investigation and Slum Abolition Board was created to assess needs, and a permanent Housing Commission followed in 1938.[31] Norman Dick, a manufacturer, played a similar role in mobilizing public opinion in Sydney, leading to a Housing Conditions Investigation Committee in 1936 and a Housing Improvement Board later in the same year.[32]

It was not until after the war that these permanent instrumentalities came into their own, but the story of their creation is instructive. First, the language and approach of the reformers proceeded from middle-class assumptions. For them the problem was not just overcrowding, leaking roofs and poor sanitation; it was also children playing in streets and family members congregating around mother in the kitchen.

The terrace house was condemned by its very nature and the superiority of a house and garden in the suburbs taken for granted. All the more surprising, therefore, was the readiness of some Labor politicians to collaborate in the project of demolishing their own working-class neighbourhoods.[33] Second, the reforming impulse found its justification in the authority of the expert. Concerned Christians though Barnett and Dick were, the era of charity had passed. Oswald Barnett laid the foundation for his campaign in a post-graduate thesis, 'The Economics of the Slums', written in Copland and Giblin's School of Commerce at Melbourne University. The Victorian survey was based on an extensive statistical inquiry into the condition of some 7000 homes and the circumstances of their inhabitants. The residents themselves played little part in the whole process, beyond answering questions and lining up to be photographed in their squalor. (These methods were duplicated in other states and extended in a massive social survey of Melbourne conducted between 1941 and 1943.) Dick also was convinced that the slum clearance required 'expert town planners' with 'a scientific approach'; and both campaigns called on the services of academics, architects and public servants, the very groups who would take charge of this new area of social policy.[34]

Social reform usually involved a greater contest of purpose between the administrators and the administered. Take the case of Aboriginals. Their earlier confinement on reserves had been based on an expectation of dwindling numbers and eventual extinction. Hence the work of missionaries such as Daisy Bates, with all her sympathy for the 'primitive, lawless creatures', sprang from the belief that they were assisting in 'the wonderful easing of their inevitable passing'.[35] This approach relied on strict segregation. It removed Aboriginals to designated reserves where they were kept in absolute dependence and virtual isolation from the outside world, including dealings with light-skinned Aboriginals. These so-called 'half-castes' were deemed capable of supporting themselves, as long as they were prevented from falling back into old ways, and it was hoped that several generations of intermarriage with the rest of the community would expunge their Aboriginality. In the words of the chief protector of Aboriginals in the Northern Territory, the idea was to 'outbreed colour altogether'. This perspective provided an additional reason for vesting the administrators of the reserves with absolute authority over their charges, even the power to separate husband and wife and parent and child. By the 1930s, however, it was becoming apparent that the Aboriginals were not going to die out. Their numbers, as recorded in the official census (though they were still not included in the national population), had fallen from 93 000 in 1901 to 71 000 in 1921, but recovered in 1933 to 80 000. During the 1930s, also, researchers who worked among Aboriginals drew attention to the atrocious conditions on the reserves. These anthropologists were largely responsible for the 'New Deal' that was announced by the federal minister for the interior at the beginning of 1939, involving the establishment of a Native Affairs Department staffed by trained personnel with the welfare of Aboriginals as their chief concern. Henceforth, the minister said, the Commonwealth's intention was to qualify Aboriginals 'to accept the privileges and responsibilities of citizenship'.[36]

The shift from segregation to assimilation was more easily announced than achieved. The state reserves continued to work on the old lines. Furthermore, even as the minister talked of raising Aboriginals to citizenship, those outside the reserves were denied the most elementary rights of employment, housing and education. Far from merging into the general community, those like George Dutton who had previously enjoyed rough acceptance, were excluded in an increasingly systematic fashion. White administrators did little to combat such discrimination; rather, it was the Aboriginals themselves who mobilized and campaigned. An Australian Aborigines League, formed in 1932, petitioned for representation in the federal parliament. An Aboriginals' Progressive League, formed in 1937, declared the sesquicentennial anniversary of white settlement a Day of Mourning. The showpiece of the official celebrations on 26 January 1938 was a re-enactment of Governor Phillip's landing, including the putting to flight of a party of Aboriginals. On the same day that this grotesque pageant was performed, the Aboriginals' Progressive Association pointed out that the white settlers had taken their lives, land and freedom. The Aboriginal demands were simple. 'We ask only for justice, decency and fair play.' They have still to win it.[37]

Advocates of reform in Aboriginal administration looked with approval on the Australian record in the territories of Papua and New Guinea. Here, they claimed, was an exemplary model of humane concern for an indigenous people. The colonial administration brought the benefits of peace, justice, medicine, education and material progress while protecting the native against the destructive consequences of uncontrolled exploitation by white settlers. Certainly the lieutenant-governor of the Territory of Papua and the administrator of the Mandated Territory of New Guinea tried to restrain the zeal of the missionaries, the cupidity of gold prospectors and the rapacity of plantation owners. Sir Hubert Murray, the benevolent autocrat who served as lieutenant-governor of Papua from 1906 until his death in 1940, was accused repeatedly of molly-coddling the natives.[38]

The basic instrument of government was the foot patrol. The white officer, with a handful of native police, could spend only a brief period in each of the villages within his district. He would hoist the flag, parade the inhabitants, instruct them to keep the peace and dispense summary justice to wrongdoers. Several hundred of these patrol officers had to maintain authority over some 2 million inhabitants, many of whom lived in regions that were still unexplored. Port Moresby and Rabaul, the seats of government in the two territories, had but skeleton administrative staffs working with meagre budgets. Canberra took only sporadic interest in the territories; it was enough that they should balance their books and be conducted in such a way as to avoid embarrassing questions from the Mandates Commission of the League of Nations.[39]

If only to raise revenue, the administrators had to allow white settlers to recruit native labour into the cash economy. Both gold-miners and plantation owners used indentured labour for which they paid—or so the labour laws stipulated—from 5s to 10s a month at the completion of a contract lasting a year or more. There were 50 000 indentured labourers in the late 1930s. The effects of the institution were momentous on black and white alike. Villages suffered a disruptive imbalance when denuded of their young men; upon their return (and there was an annual death rate of 2 per cent) these same men brought wealth, prestige and attitudes that upset established ways. On the white side, while the planters treated their 'boys' no differently than white masters have treated bonded labour ever since Europeans first established foreign empires—'A stern father but a loving father, this is what we liked to think of ourselves anyhow', was the way one put it—for Australian prospectors who fossicked in Papua, New Guinea and the great arc of islands fringing the Australian continent, there was a significant change. The battler was now a *masta*, with a team of boys to carry his equipment and perform his labour.[40]

There remained a crucial difference between Australian treatment of Aboriginals and Melanesians. The thrust of domestic policy during the 1920s was towards the destruction of the Aboriginal social unit; the aim was to drive Aboriginals off their tribal lands and onto reserves, to split groups into families and families into individuals, so as to promote assimilation of the fragmented remnants. In Papua and New Guinea, on the other hand, the village remained the basis of administration. Murray, whose influence is difficult to exaggerate, had a strong sense of the cultural effects of change. He took from functionalist anthropology the notion of the social organism as an interconnected whole such that interference with one element would lead to its disintegration. In his view the Papuan was 'crossing the gulf which divides the Stone Age from the twentieth Century, and it would be foolish to ignore the perils that surround his path'.[41] Hence he sent members of his staff for training at the department of anthropology at the University of Sydney, which he had helped to establish. This did not make Murray a progressive. Far from it: his insistence on preserving distance between the Papuan and the European went to the lengths of a prohibition on local men wearing clothing above the waist. Indeed, the extraordinary paternalism of Murray's regime produced a crippling lack of self-confidence in the Papuan community. But at least it retained its land, its social structure and a core of its culture.

Finally there was the disadvantaged section of Australian society that was not far short of the majority—women. The whole thrust of social policy since the turn of the century had been to reinforce their domestic responsibilities within the family. Here too experts had been at work, turning household management into a domestic science, regulating sexuality, reducing fertility, institutionalizing childbirth and child-rearing practices. Although professional women played a prominent part in these enterprises, there was no intention to upset gender boundaries. When the first female member of an Australian legislature proposed (for polemical effect rather than with serious intent) that housewives be brought within the scope of industrial arbitration, the National Council of Women replied in shocked disbelief: 'The woman is queen in the home and the man is king outside.'[42] Mass bodies like the Country Women's Association and church auxiliaries were pillars of conservatism, and even the smaller feminist groups sought to expand the public rather than repudiate the domestic sphere. Even so, the substitution of reason and science for older practices sanctioned by the ascription of innate qualities to the female gender altered the position of women. As portrayed in the enormously successful *Women's Weekly*—it achieved a circulation of 400 000 within six years of first publication in 1933—the modern woman was a home-maker and an incurable romantic with a vicarious interest in the lives of the famous; but she was also concerned with social issues such as Aborigines and housing, she took an interest in women's careers and she wanted to know her legal rights.[43]

While the struggle for emancipation took place across a wide front, it concentrated around the question of women's employment. A woman's right to work was hedged by her exclusion from areas designated as men's jobs and by rates of pay set at about half the male rate. Trade union resistance to female workers was therefore reinforced by male workers' determination to protect wage standards. In this they were helped by industrial tribunals, which continued to follow Higgins in calculating male wages on the basis of a family's needs and female wages on the needs of a single person. 'The whole spirit of the Anglo-Saxon race', insisted Jethro Brown of the South Australian Industrial Court, 'supports the view that the woman should be supported by the man.'[44] Muriel Heagney campaigned against these disabilities in her work among unions in Melbourne during the 1930s; upon moving to Sydney she concentrated on the wage issue and became president of a Council of Action for Equal Pay, established with union sponsorship in 1937. The council drew also on the support of the United Associations of Women, led by the redoubtable Jessie Street, though she favoured a more gradual approach than Heagney and later withdrew from the council. In any case, the Commonwealth Arbitration Court rejected the council's application for equal pay. Further progress in this area would not come until women moved into new areas of the workforce during the war.[45]

In 1938 an American scholar, who had spent eighteen months studying in Australia, offered his hosts a candid statement of his findings. Unlike the investigators who had come at the turn of the century, intrigued by the precocious antipodean democracies and their social experiments, Hartley Grattan was less enthusiastic. 'Australia *was* an advanced country forty years ago, but, perverse though the world may be, it has moved forward and overtaken Australia in many directions.' The past forty years had been taken up with filling in the outlines of a scheme of development that had been laid down then, one that entrenched the interests of what he called 'the owning-producers'. Manufacturers, farmers, wage-earners—they all shared the spoils, and Grattan was 'not deeply impressed with compensating and ameliorative legislation in Australia'.[46] Similar views were expressed by American diplomats. Australians were found to be a free-and-easy people, fond of outdoor leisure and sport. (This was certainly true. Football, cricket and racing were among the few interests that spanned the classes and excited a common enthusiasm. 'What is the use of winning a High Court decision [against Lang's policy of repudiating debts] and losing Phar Lap? asked Lyons in 1932.[32]) Australian businessmen were judged to lack energy and imagination. The much-vaunted protective institutions merely reduced ambition to a minimum and exacerbated, rather than resolved, class conflict.[48]

Such observations probably tell us as much about American preoccupations as about Australia, but they do draw attention to the persistence of important continuities. A narrative history is inclined to dramatize change and neglect the slow, almost imperceptible alteration in everyday conditions of life. The people who worked and played in the late 1930s did so in circumstances that were different from those of their parents and grandparents, but not overwhelmingly so. They enjoyed somewhat better living standards. They worked shorter hours; they were more likely to retire at sixty-five and to live longer after retirement; family sizes had become smaller so there were fewer dependants to support. Advances in medicine, industrial and domestic technology, transport and communications lightened the burden of toil. Yet average incomes in real terms had progressed very little, if at all, and while inequalities of wealth would seem to have lessened in the first two decades of the country, the next two decades saw little change.[49]

Conditions in a coal-mining town were recorded in a wartime survey. The usual house had four rooms: one bedroom for the parents, another for the children, a kitchen which doubled as the living-room, and the 'best room'. Floors were covered with linoleum and mats, except for the carpeted front room which also boasted a lounge suite and probably a wedding photo and other family mementoes. The older and poorer households still used iron bedsteads. In the kitchen the stove burned solid fuel, an ice-chest held meat and dairy products, while the sink had only a cold-water tap. The privy was out the back. A survey of wheatfarms found their housing standards were no better. The investigator was struck by the bleak, unattractive appearance of many of the houses, a third of which were in need of repair. The majority depended on kerosene lamps for lighting and lacked a kitchen sink. Five out of every six relied on a pan latrine. Lack of sanitation was also a feature of country towns. The decline of these towns as trading centres, the disappearance of shops and service industries, the deterioration of the housing stock and the high proportion of aged residents underscored the continuing drift to the cities. Yet investigations of inner-city conditions revealed overcrowding and physical decay.[50] The suburbs offered varying degrees of comfort to small proprietors, salary-earners and skilled workers. But far from lifting the material conditions of Australians as a whole, the passage of forty years had increased the distance between well-to-do home-owners and the unskilled labourers, single mothers, unemployed and other battlers.

What of those who commanded and benefited from the labour of others? The story told here suggests that this class retained its affluence and authority. It had met the political challenge of the organized working class during the early part of the century and demonstrated conclusively in 1931 the resources at its command. It had created a federal compact and maintained its checks and balances. It had closed off Australia to the non-British world and maintained the imperial link. Albeit reluctantly, it had accepted measures designed to tailor capitalism to the needs of wage-earners, and adapted protectionist principles for its own purposes. Grattan is surely correct, therefore, in suggesting that forty years had been spent elaborating that scheme of development. But was he right to emphasize its limitations? By entrenching themselves, the producers had certainly closed off more adventurous lines of development. Their domestic compact precluded pursuit of more openly competitive, harshly exploitative strategies that, if they had been pursued, might have yielded richer rewards. This was a choice that confronted other white settler societies. Australia, New Zealand, southern Africa, the Americas—all had been colonized by Europeans to become successful producers of raw materials for the world economy during the nineteenth century. All exhibited that distinctive configuration of sparsely populated but immensely productive hinterlands and prosperous but parasitic cities. The white settler societies were precocious democracies, secular, acquisitive, unrestrained by hereditary privilege, rapacious in the use of nature as well as in their treatment of the original inhabitants. The pressure was strong for high wages, industrialization and self-sufficiency. But only one of them, and that the most bountiful, managed to throw off dependency. Grattan generalized too readily from the particular experience of his American homeland. In Australia the buccaneers were in a minority. The majority of employers were content to work with the existing institutional framework, knowing that it preserved their income and class power. This was their choice and in the main they were content with it.

14

'FREE OF ANY PANGS'

AT 8 P.M. EASTERN AUSTRALIAN time on the evening of 3 September 1939, a despondent Neville Chamberlain broadcast the news that Britain's attempts to preserve peace had failed. Seventy-five minutes later, Australian listeners heard their own prime minister declare it his melancholy duty to confirm officially that Britain was at war with Germany and 'as a result, Australia is also at war'.[1]

The speaker was Robert Menzies, who had become prime minister when Lyons died five months earlier. The choice of words was characteristic of his fulsome loyalty. Menzies meant to establish in law and in sentiment that there was never any doubt where Australia stood: 'where Great Britain stands there stand the people of the entire British world.'[2] The dominion had delayed establishing its own diplomatic representation, even though the United States pressed it to open a legation in Washington. It had followed Britain uncritically down the path of appeasement, even though it received scant warning of the twists and turns. It had supported Britain in the final, belated guarantee of Polish independence, even though it had not been consulted and in spite of Menzies's private conviction that 'nobody really cares a damn about Poland as such'.[3] It even made its declaration of war seven hours before the Dominions Office got round to sending official notification that Britain had done so.

Thus far Australia followed Britain into the Second World War as unquestioningly as it had rushed into the first. Volunteers were enlisted into a second AIF, vessels were assigned to the Royal Navy, air squadrons trained in Canada and were sent on to Britain. But Australia held back from sending an expeditionary force. 'This is not the last war', Menzies explained, 'when there was no real problem of Australia's security from attack.'[4] He was concerned principally about the threat from the north. In 1914 the Japanese fleet had escorted Anzac troopships across the Indian Ocean; in 1939 the greatly strengthened Japanese forces were arrayed menacingly, poised for action. The only real obstacle to their ambitions in South East Asia was the British base at Singapore, and since Britain's navy was likely to be fully occupied in the Atlantic, Australia wanted to know that Singapore was safe. Britain gave the necessary assurances—even if Japan entered the war, an assault on Singapore would require a siege of at least four months, plenty of time for British warships to reach the base and drive off the attackers. As a further inducement, the Australians were invited to make their contribution in the Middle East where they would serve their own interests by safeguarding the Suez Canal. Australia sent a division to the Middle East at the beginning of 1940 and another two divisions subsequently. They played an important part in the campaigns in North Africa, the defence of Greece and Crete, and the occupation of Syria. But from the beginning there was friction between the British and the Australian generals over the status of the dominion troops. Blamey, the Australian commander, had constantly to insist on his right to control the deployment of his forces and to communicate directly with his government. He was not consulted over the ill-fated expedition to Greece where 6000 British soldiers were sacrificed on the altar of British strategy. He was unable to obtain relief for Australian units under siege at Tobruk until the Australian prime minister insisted point-blank that they be withdrawn. Blamey asked the British commander, 'If I were a French or an American commander making this demand, what would you say about it?' 'But you're not' was the reply.[5]

Such was the cost of security. Menzies, like Deakin, Hughes and Bruce before him, was not a blind imperialist oblivious of Australian interests, but he was convinced that those interests were best served by tying Australian foreign and defence policy to Britain's. Without a British base at Singapore to safeguard South East Asia, without a British navy protecting the trade routes, without Britain's survival against the Axis offence, Australia was in peril. Menzies also recognized that the perspectives of the home country and the dominion did not always coincide: 'What Great Britain calls the Far East is to us the near north.' Hence Australia was at once more alarmed by Japan and more anxious to appease it. Moreover, Menzies sometimes found it necessary to insist that Australia should not be taken for granted: in responding to the initial request for military assistance, he complained that there was 'a quite perceptible disposition to treat Australia as a Colony'.[6] Such slights did not weaken the Australian allegiance, indeed they strengthened the determination to maintain it. Bruce, as high commissioner in London with unique access to the corridors of power, was therefore the principal agent of Australian diplomacy in 1939–41. Casey led a mission to London shortly after the war began; Menzies himself spent four months there in 1941 and even tried to revive the precedent of an Imperial War Cabinet. Yet nothing the Australians said or did during the grim European struggle could still their nagging fear—was Singapore secure?

It could hardly be said that Australia itself was braced for the emergency. Upon the outbreak of war the government assumed powers under the Defence Act to call up the militia for home service and raise volunteers for overseas service. But there were no brass bands and banners to lead men to the recruiting tables in the summer of 1939, and the niggardly decision to cut infantrymen's pay from 8s a day to 5s hardly increased enthusiasm. Many volunteers were turned away on the grounds that they worked in skilled occupations. In 1941, after two years of fighting, there were no more Australians in uniform than there had been in 1918.[7] A National Security Act (corresponding to the War Precautions Act of the previous war) was used with greater vigour and less discretion. German-born residents were rounded up in 1939 and Italians in 1940 when Mussolini entered the war, though many of the Italian internees were known anti-fascists. The Communist Party, which opposed the war prior to Hitler's attack on the Soviet Union in 1941, was made an illegal organization and a number of its members were imprisoned. An ultra-nationalist Australia First Movement, including its leader, P.R. Stephensen, met the same fate. Others to be placed behind barbed wire included a shipload of mainly Jewish refugees from Germany who were declared aliens and kept in custody, in some cases until 1945. As well as entrusting the army with censorship powers, the government appointed the newspaper magnate Keith Murdoch as Director-General of Information and allowed him not just to suppress information but to force newspapers to publish the official viewpoint.[8]

There was plenty to suppress. The government had not attempted to strengthen Australia's feeble defences until the very eve of war—in 1938/39 it spent just £13 million on the armed forces—and years of parsimony could not be remedied overnight. Australia had no mercantile fleet and little more than the embryo of an aircraft industry; its munitions factories had to be built along with the various support industries and administrative machinery needed to sustain the war effort. Spending increased rapidly: £50 million in 1939/40, £170 million in 1940/41, £308 million in 1941/42. Even so, personal consumption levels remained as high in 1941 as they had been in 1939. The formation of the armed forces and the construction of the war economy proceeded not by curtailing living standards but by drawing into use those resources that had remained idle in peace (including the unemployed who made up 10 per cent of the workforce in September 1939). Few commodities were rationed and a Manpower Directorate had still to assume powers to allocate labour where it was needed. Civilians had not yet been called upon to make sacrifices.[9] Menzies had suggested shortly after the commencement of the war that they should get on with 'business as usual', and they were happy to do so—a record amount was bet on the 1940 Melbourne Cup. Small wonder that when Blamey returned to Australia in November 1941, he was astounded by the complacency he saw and likened his compatriots to a herd of gazelles in a dell on the edge of a

"Cripes, I wonder if the jockey was hurt!"

A comment on domestic indifference to the war in the Middle East, 1941

jungle. Yet on the following Saturday he himself went to Flemington to present a cup to the winner of the principal event.[10]

There was little sense of urgency or common purpose. Workers in key industries, where the boss had held the whip hand for the past decade, saw little reason why they should not make the most of their improved bargaining position, especially when profits and dividends were so high. Disputes were frequent, especially on the coalfields. Farmers, who had only recently recovered from years of hardship, insisted that the war should not deprive them of a good price for their produce. Wheatfarmers had to be placated with a guaranteed price for the grain that could not be shipped to overseas markets. Young men and women who had known only intermittent employment since leaving school were understandably cynical when told that their country now needed them. And why should they respond to the rhetoric of Empire? A survey of public attitudes in 1941 warned of 'a sense of disillusionment, disappointment, futility, distrust, disgust, diffidence and indifference which so many possess with regard to politics and society in general and the war in particular'.[11]

The governing coalition was unable to provide effective leadership because it was itself rent by internal divisions.

Menzies had exposed the rifts in the UAP when he resigned from the Lyons ministry in March 1939. He won the contest for the succession when Lyons died in the following month, even though the influential Melbourne businessmen of the National Union opposed him (and even explored the possibility of restoring Bruce). There was a further consequence of Menzies's accession. Page, who had still not forgiven his betrayal of Lyons, launched an emotional attack on the floor of parliament where he even revived the allegation that Menzies shirked his duty in the First World War. There could be no chance of continuing the coalition after this, so the UAP formed a minority government. This in turn opened divisions in the Country Party. A section of that party disowned Page, and the peppery doctor surrendered his leadership shortly after. While a fresh coalition was then negotiated, Menzies's sharp tongue and lordly manner made many enemies. His relations with the Melbourne establishment were strained and a public row with the *Sydney Morning Herald* eroded his position in that city. The ministry was weakened further in August by the deaths in an air crash of three ministers.[12]

Beyond the debilitating divisions, there was a further reason for Menzies's failure to direct a stronger war effort. He hinted at it in a remarkable letter written to Bruce shortly after the war began where he set down 'in a rambling and personal way' something of what was on his mind. He could see no good coming from the war and every prospect of disaster. This did not mean that he shrank from support for Britain—that was beyond question. But he confessed to a 'horrible feeling that by the time we have sustained three years of carnage and ruin, law and order will tend to be at a discount in every combatant country, and our last state may be worse than our first'.[13] At the back of his mind he seems to have been thinking of the outcome of the last world conflict, when revolutionary unrest fed on hardship and destruction until it brought down most of the European monarchies and threw up the Soviet Union. Might not the impending war precipitate another period of turbulent instability? Might not the Soviet Union again be the beneficiary of a fight between the non-communist powers? Menzies was prescient about these and other implications of the course on which they had embarked. If only to influence neutral opinion, he favoured the adoption of a statement of Allied war aims that would embrace a pledge to build a new post-war order free of insecurity and hunger.[14] Other members of the cabinet held back from so radical a step, as indeed they hesitated before an all-out prosecution of the war effort. Full mobilization would bring about sweeping changes in the organization of the economy; it would expand the power of central government in ways that alarmed conservatives, and promote a national self-sufficiency that might endanger imperial links. Above all, it would require sacrifices from the populace that might in turn give rise to an irresistible demand for social reform. In short, the ruling parties shrank from the prospect of a People's War.

An election in September 1940 left the government in a precarious position in the House of Representatives. Labor won 36 seats, the UAP 23, the Country Party 14, and the balance of power was held by 2 independents. Menzies sought to overcome his weakness by inviting the ALP to join in an all-party war administration. Curtin rejected the offer—he had yet to secure a whole-hearted support in Labor ranks for the war effort—but agreed to the formation of an Advisory War Council, consisting of representatives of all the parties. For the time being Menzies could therefore expect co-operation in the parliamentary transaction of government business, and he was able to go to London, where Australian priorities urgently needed to be pressed. In this he was unsuccessful. 'What irresponsible rubbish these Antipodeans talk!', the permanent head of the Foreign Office noted after listening to Menzies and Bruce.[15] But publicly, like Hughes before him, Menzies enjoyed considerable success as an eloquent speaker, relished the role of statesman, and even toyed with the possibility of transferring to Westminster. He returned to Australia in May 1941 with a distinctly Churchillian manner and a corresponding distaste for what he was unwise enough to call the 'diabolical game of politics'.[16] The malcontents had certainly been busy during his absence and the coalition, despite a cabinet reshuffle, was falling apart. Menzies again called on Curtin to join in an all-party government; Curtin again refused. Dispirited and weary, Menzies resigned. In his place a joint meeting of the United Australia and Country parties elected Arthur Fadden, the Country Party leader, but he lasted only forty days. On 7 October 1941 the two independents crossed the floor to defeat the government. Curtin then formed a Labor government.[17]

Labor took office just as the storm was about to break. For the past two years Australia had been fighting a distant war which did not demand its whole-hearted support and permitted a dilatory mobilization of resources. Within these limits, the government had paid the premium on imperial defence and was slow to recognize that the insurer might default. A sober appraisal of Britain's beleaguered position in the later months of 1940, when massed squadrons of the German air force flew daily over the English Channel, would have suggested that pledges east of Suez were unlikely to be redeemed. The appointment of John Latham as Australian representative in Tokyo signalled Australia's unease, yet it was not until Menzies reached London in February 1941 and heard the British equivocations with his own ears that he began to grasp how much Australia was taken for granted. Shocked though he was, Menzies preserved the façade of imperial unity. Curtin suffered from no such vestigial loyalty. As soon as Japan entered the war, striking simultaneously on 8 December at the American base on Pearl Harbor and the British position on the Malay Peninsula, he appreciated that Australia would depend on the United States. 'Without any inhibitions of any kind', he declared on 27 December 1941, 'I make it quite clear that Australia looks to America, free of any pangs as to our traditional links or kinship with the United Kingdom.'[18]

Even as Curtin looked across the Pacific, Churchill was in conference with Roosevelt. Their confirmation of an earlier agreement to 'beat Hitler first' relegated the war in Asia to secondary importance, though Australia did not stumble upon the full significance of this fact for a further five months. Moreover, the Australians were denied any effective voice in the conduct of the Asian war, and Curtin had to insist that the divisions returning from the Middle East defend Australia, rather than India as Churchill desired.[19]

The course of events in Malaya and Singapore fulfilled Australia's worst fears. The Japanese landed in force immediately to the north of Malaya in present-day Thailand. When the news was brought to the governor of Singapore, he replied, 'Well, I suppose you'll shove the little men off.' Four hours later, bombs were falling round Government House. Two days later the Japanese sank the *Prince of Wales* and the *Repulse*, giving them complete control of the sea and the air. Their army made rapid progress down the Malay Peninsula, outflanking the defenders and inflicting heavy losses on those British and Indian units that stood their ground. Lines of defence were no sooner formed than circumvented and cut to ribbons, when the survivors would fight their way south to regroup in fresh positions and repeat the same grim routine. The failure of command became farcical. On one occasion when the British commander was conferring with a subordinate, the operator broke in to inform him, 'Your three minutes have expired, sir,' and broke the connection. Major-General Gordon Bennett, the acerbic leader of the Australian Eighth Division, denounced the 'effete conservatism and arrogance' of his British superiors, yet his men were no more successful when they went into action in the new year. In just fifty days the Japanese progressed 500 miles, and by the end of January 1942 they were looking across the Strait of Johore to the island of Singapore.[20]

On Singapore the 8000 Europeans who controlled the lives of 600 000 Malays and Chinese maintained their customary round of activities. They still dressed for dinner, gathered for cocktails, danced at Raffles and excused themselves from air raid duty on the grounds that they had to play tennis. Protected by censorship, they had little appreciation of the shambles to the north. Even if the Japanese reached Singapore, they reasoned, there was a garrison of 90 000 troops and the massive fortifications of the naval base overlooking the strait. When it was pointed out that the base's batteries were designed to repulse an assault from sea rather than land, and indeed could not be brought to bear on attackers across the strait, the British commander still refused to build new fortifications on the grounds that to do so might cause panic among civilians. (Perhaps the full extent of the predicament only came home to the whites when shopkeep-

The fall of Singapore: A stick of bombs bursts on the aerodrome five days before the final Japanese assault

ers began demanding cash.) The battle for Singapore began on 8 February. The Japanese expended their dwindling supplies in a massive bombardment, then slipped across the narrow waterway and overran the perimeter. As they closed in on the city, panic set in. Some newly arrived Australians, untrained reinforcements, were among the troops who aban-

doned their posts to pillage or push their way onto ships in the last exodus. Amid the devastation of shelling and bombing, fires raged and water supplies were lost. Faced with the complete destruction of Singapore and the death of many of its civilians, the commander surrendered on 15 February. Altogether, 130 000 fighting men were taken into captivity, including 15 000 Australians.[21]

The fall of Singapore was only one disaster among many in the early months of 1942. Four days after this capitulation, Darwin was bombed. Rabaul had fallen in January; there were Japanese landings on Timor in February and on New Guinea in March. Even so, the loss of the principal British base in South East Asia had a special significance. This was the linchpin of imperial defence, the physical expression of an identity and loyalty that Australians had maintained since federation. Warned at the end of January that the British were considering evacuating Singapore, Curtin told Churchill that this would be regarded in Australia as an 'inexcusable betrayal'. The subsequent surrender was therefore understood as something more than a tragedy in which Australia was implicated along with Britain. It was a final demonstration that Australia could no longer rely on an Empire whose power was broken and pledges were worthless. Mary Gilmore expressed the feeling of outrage in the poem she wrote for the *Women's Weekly* a few weeks later:

> They grouped together about the chief,
> And each one looked at his mate
> Ashamed to think that Australian men
> Should meet such a bitter fate!
> And black was the wrath in each hot heart
> And savage oaths they swore
> As they thought of how they had all been ditched
> By 'Impregnable' Singapore.[22]

So much for the past. What of the future? The immediate threat to Australia caused Vance Palmer to assess the meaning of its civilization. He anticipated that the next few months would decide not only whether Australia would survive as a nation but whether it deserved to survive. If it should fall, what evidence would there be to show a future historian that these people had made it their homeland? The country had been merely exploited, the settlements had a fragile look. There were 'decadent' elements—whisperers, faint-hearts and near fascists, those grown rotten with easy living who had held power but now felt it slipping away from them. They would have to be pushed into the background and, if necessary, every yard of Australian earth would have to be made a battle-station. For Palmer believed that what was significant would survive and that the people would come out of the struggle stronger than they went in. He felt there was another Australia, 'an Australia of the spirit, submerged and not very articulate', that was quite different from these 'bubbles of old-world imperialism'.[23] And he hoped that from this Australia a new mood would arise and an authentic democracy, sound and egalitarian, would be built. Time would show the outcome of his hopes.

NOTES

ABBREVIATIONS

ABC	Australian Broadcasting Commission
ADB	*Australian Dictionary of Biography*
AEHR	*Australian Economic History Review*
AGPS	Australian Government Publishing Service
AJPH	*Australian Journal of Politics and History*
ANU	Australian National University
ANZ	Australian and New Zealand Book Company
APCOL	Alternative Publishing Cooperative Limited
AQ	*Australian Quarterly*
BPP	*British Parliamentary Papers*
CAR	*Commonwealth Arbitration Reports*
CLR	*Commonwealth Law Reports*
CPD	*Commonwealth Parliamentary Debates*
CPP	*Commonwealth Parliamentary Papers*
CSIR	Council for Scientific and Industrial Research
CUP	Cambridge University Press
JAS	*Journal of Australian Studies*
JIR	*Journal of Industrial Relations*
JRH	*Journal of Religious History*
JRAHS	*Journal of the Royal Australian Historical Society*
MUP	Melbourne University Press
NLA	National Library of Australia
NSWPD	*New South Wales Parliamentary Debates*
NSWPP	*New South Wales Parliamentary Papers*
NZJH	*New Zealand Journal of History*
OUP	Oxford University Press
QPD	*Queensland Parliamentary Debates*
RHSVJ	*Royal Historical Society of Victoria Journal*
RSSS	Research School of Social Sciences
SAPD	*South Australian Parliamentary Debates*

SAPP	*South Australian Parliamentary Papers*
SUP	Sydney University Press
SWAH	*Studies in Western Australian History*
THRAPP	*Tasmanian Historical Research Association Papers and Proceedings*
UQP	University of Queensland Press
UWA	University of Western Australia
VPD	*Victorian Parliamentary Debates*
VPP	*Victorian Parliamentary Papers*
WAPD	*Western Australian Parliamentary Debates*
WAVP	*Western Australian Votes and Proceedings*
WEA	Workers' Educational Association

PROLOGUE

1 Boulder *Evening Star*, 1 January 1901; *Kalgoorlie Miner*, 28 December 1900–3 January 1901; Kalgoorlie *Sun*, 6 January 1901; *Western Argus*, 8, 15 January 1901; *Westralian Worker*, 3 January 1901. The events at Kalgoorlie are summarized and compared with those elsewhere in Australia by Gavin Souter, *Lion and Kangaroo: The Initiation of Australia 1901–19*, Collins, Sydney, 1976, especially pp. 32–3.

CHAPTER 1: SOME AUSTRALIANS

1 Lord Casey, *Australian Father and Son*, Collins, London, 1966; Goldsbrough Mort papers, ANU Archives (especially the memoirs of Niall, 2/151, and correspondence books of Casey and Niall, 2/152–3); profile in Melbourne *Punch*, 7 November 1912; Casey's will in the Victorian probate office; information on Shipping House from Paul de Serville. J.W. McCarty, 'British Investment in Western Australian Gold Mining 1894–1914', *University Studies in History*, 4, 1(1961–62), pp. 7–23; Neville Cain, 'Financial Reconstruction in Australia 1893–1900', *Business Archives and History*, 6 (1966), pp. 166–83; Geoffrey Blainey, *The Rush that Never Ended: A History of Australian Mining*, 3rd edn, MUP, Melbourne, 1978, pp. 187–9, 232–47; W.S. Robinson, *If I Remember Rightly*, ed. Geoffrey Blainey, Cheshire, Melbourne, 1967, pp. 71–5.

2 John Shaw Neilson, *Collected Poems*, ed. R.H. Croll, Lothian, Melbourne, 1934: note that 'The Poor Country' was actually written on an earlier selection; A.R. Chisholm (ed.), *The Poems of John Shaw Neilson*, Angus and Robertson, Sydney, 1965; Hugh Anderson and L.J. Blake (eds), *Green Days and Cherries. The Early Verses of Shaw Neilson*, Red Rooster Press, Ascot Vale, 1981; Nancy Keesing (ed.), *The Autobiography of John Shaw Neilson*, NLA, Canberra, 1978; James Devaney, *Shaw Neilson*, Angus and Robertson, Sydney, 1944; Hugh Anderson and L.J. Blake, *John Shaw Neilson*, Rigby, Adelaide, 1972; Edgar Dunsdorfs, *The Australian Wheat Growing Industry 1788–1948*, MUP, Melbourne, 1956, pp. 154–61.

3 Somerville papers in the Battye Library (the gardening journal is Acc. 458A, the letterbook from which comes his rebuke of the Governor, 19 May 1920, is Acc. 460). His university papers are in the Reid Library, University of Western Australia. I have drawn also on family information from Mrs Margaret Hicks and from two unpublished typescripts, Ann Clover, 'Biography of William Somerville', 1958, and Bill Latter, 'Dr William Somerville', 1980, both in the Battye

Library. *Monthly Reports* of the ASE; K.D. Buckley, *The Amalgamated Engineers in Australia, 1852–1920*, Dept of Economic History, RSSS, ANU, 1970, especially, chap. 8, and pp. 10–14; 'The Arbitrator from Albert Street', *Mosman Park Review*, 2, 7 (February 1955); Fred Alexander, *Campus at Crawley*, Cheshire, Melbourne, 1963.

4 Jocelyn Treasure, 'Deborah Watt's Narrative, 1897–1914', *SWAH*, 7 (December 1983), pp. 92–100; Watt narrative in Battye Library, and further material from Jocelyn Treasure. Living conditions on the goldfields are described in Margaret Bull, *White Feather: The Story of Kanowna*, Fremantle Arts Centre Press, Fremantle, 1981, and Gavin Casey and Ted Mayman, *The Mile that Midas Touched*, Rigby, Adelaide, 1964. The background to agricultural settlement is given in Sean Glynn, *Government Policy and Agricultural Development*, UWA Press, Nedlands, 1975.

5 This account of George Dutton is based on the work of Jeremy Beckett: 'Marginal Men: A Study of two Half-Caste Aborigines', *Oceania* 29 (1958), pp. 91–108; 'Kinship, Mobility and Community among Part-Aborigines in Rural Australia', *International Journal of Comparative Sociology*, 6 (1965), pp. 7–23; 'Marriage, Circumcision and Avoidance among the Maljangaba of North-west New South Wales', *Mankind*, 6 (1967), pp. 456–64; 'George Dutton's country: Portrait of an Aboriginal Drover', *Aboriginal History*, 2 (1978), pp. 3–31. I have also drawn on C.E.W. Bean, *On the Wool Track*, Angus and Robertson, Sydney, 1956 and *The 'Dreadnought' of the Darling*, Alston Rivers, London, 1911; Bobbie Hardy, *West of the Darling*, Jacaranda, Brisbane, 1969 and *Lament for the Barkindji*, Rigby, Adelaide, 1976; Luise Hercus, 'George Dutton in South Australia', manuscript, 1970; Jenny Lee, 'A Black Past, A Black Prospect: Squatting in Western New South Wales 1879–1902', MA thesis, ANU, 1980.

CHAPTER 2: GETTING AND SPENDING

1 Alfred Deakin, *Federated Australia: Selections from Letters to the Morning Post 1900–1910*, ed. J.A. La Nauze, MUP, Melbourne, 1968, p. 19.

2 J.C. Foley, *Droughts in Australia*, Department of Meteorology, Melbourne, 1957, especially pp. 61, 208–11; observation from Queensland quoted in Lorna McDonald, *Rockhampton: History of City and District*, UQP, St Lucia, 1981, p. 243; *Gillen's Diary: The Camp Jottings of F.J. Gillen on the Spencer and Gillen Expedition Across Australia, 1901–02*, Library Board of South Australia, Adelaide, 1968, p. 7.

3 See generally R.L. Heathcote, *Back of Bourke: a Study of Land Appraisal and Settlement in Semi-arid Australia*, MUP, Melbourne, 1965; Geoffrey Bolton, *Spoils and Spoilers: Australians Make their Environment 1788–1980*, Allen and Unwin, Sydney, 1981, chaps 8, 12.

4 N.G. Butlin, *Australian Domestic Product, Investment and Foreign Borrowing 1861–1938/9*, CUP, Cambridge, 1962, pp. 460–1.

5 N.G. Butlin and J.A. Dowie, 'Estimates of Australian Workforce and Employment, 1861–1961', *AEHR*, 9 (1969), p. 144; M. Keating, *The Australian Workforce 1910–11 to 1960–61*, Department of Economic History, RSSS, ANU, Canberra, 1973, pp. 356–7.

6 See Alan Barnard (ed.), *The Simple Fleece: Studies in the Australian Wool Industry*, MUP, Melbourne, 1962.

7 See A.R. Callaghan and A.J. Millington, *The Wheat Industry in Australia*, Angus and Robertson, Sydney, 1956, and Edgar Dunsdorfs, *The*

Australian Wheat-Growing Industry, 1788–1948, MUP, Melbourne, 1956.

8 See S.M. Wadham and G.L. Wood, *Land Utilization in Australia*, MUP, Melbourne, 2nd edn, 1950, chaps 2, 8, 9.

9 G. Blainey, *The Rush that Never Ended: A History of Australian Mining*, 3rd edn, MUP, Melbourne, 1978, parts II and III.

10 A comprehensive survey of manufacturing, with statistics up to 1900, is provided by G.J.R. Linge, *Industrial Awakening: A Geography of Australian Manufacturing, 1788 to 1890*, ANU Press, Canberra, 1979. Thereafter see Butlin, *Australian Domestic Product, Investment and Foreign Borrowing*, chap. 8, and B.D. Haig, 'Manufacturing Output and Productivity, 1910 to 1948/9', *AEHR*, 15 (1975), pp. 136–61.

11 C.B. Schedvin, 'Rabbits and Industrial Development: Lysaght Brothers and Co. Pty Ltd, 1884 to 1929', *AEHR*, 10 (1970), pp. 27–55.

12 'First Progress Report of the Royal Commission on the Shortage of Labour and the Clothing and Boot Trades', *SAPP*, 1912, vol. 2, p. 70.

13 'Report on the Working of the Factories and Shops Act . . . during the Year 1911', *NSWPP*, 1912, vol. 2, p. 44; Thomas Cherry, *Victorian Agriculture*, D.W. Paterson, Melbourne, 1913, pp. 51, 120; K.M. Dallas, *Horse Power*, Fullers Bookshop, Hobart, 1968; Donald F. Dixon, 'Origins of the Australian Petrol Distribution System', *AEHR*, 12 (1972), pp. 39, 43.

14 Alfred Guy, 'One Man's Journey', manuscript autobiography in family possession, pp. 57–8; 'Royal Commission of Inquiry into the Hours and General Conditions of Employment of Female and Juvenile Labour', *NSWPP*, 1911–12, vol. 2, pp. 24–35; R.F. Holder, *Bank of New South Wales: A History*, vol. 2, Angus and Robertson, Sydney, 1970, p. 504; J.R. Poynter, *Russell Grimwade*, MUP, Melbourne, 1967, p. 67; Beverley Kingston, *My Wife, My Daughter and Poor Mary-Ann: Women and Work in Australia*, Nelson, Melbourne, 1975, chap. 2.

15 As suggested by Graeme Davison in his survey, 'The Australian Energy System in 1888', *Australia 1888*, 10 (September 1982), p. 14.

16 L.R. Smith, *The Aboriginal Population of Australia*, ANU Press, Canberra, 1980, pp. 199–202.

17 Neville Hicks, *'This Sin and Scandal': Australia's Population Debate 1891–1911*, ANU Press, Canberra, 1978, p. 157; *Commonwealth Yearbook 1901–14*, pp. 93, 1085. See also Hicks, 'Demographic Transition in the Antipodes: Australian Population Growth and Structure, 1891–1911', *AEHR*, 14 (1974), pp. 123–42.

18 A.C. Kelley, 'International Migration and Economic Growth in Australia, 1865–1935', *Journal of Economic History*, 25 (1965), pp. 333–54; D. Pope, 'Empire Migration to Canada, Australia and New Zealand, 1910–1929', *Australian Economic Papers*, 7 (1968), pp. 167–88; D. Pope, 'Contours of Australian Migration, 1901–30', *AEHR*, 21 (1981), pp. 29–52.

19 Edward Dyson, 'In Town', in Leon Cantrell (ed.), *The 1890s: Stories, Verse and Essays*, UQP, St Lucia, 1977, p. 151; Immigration League president in *Australia To-Day*, 1911, p. 41; *Bulletin*, 8 December 1900, reprinted in *A.G. Stephen: Selected Writings*, ed. Leon Cantrell, Angus and Robertson, Sydney, 1978, pp. 404–5.

20 ibid., pp. 21–2.

21 J.W. McCarty in McCarty and C.B. Schedvin (eds), *Australian Capital*

Cities, Historical Essays, SUP, Sydney, 1978, pp. 6–7.

22 C.T. Stannage, *The People of Perth*, Perth City Council, Perth, 1979, p. 292; 'Report of the Board upon the Stock Markets, Stockyards, Abbatoirs ...', *VPP*, 1911, vol. 2, p. 402; 'Report of the Select Committee on the Abbatoir', *NSWPP*, 1903, vol. 3, pp. 665–803.

23 Urban history is better served for the latter part of the nineteenth century than it is for the twentieth. Apart from the essays in McCarty and Schedvin, the following are useful: Stannage, *The People of Perth*; Ronald Lawson, *Brisbane in the 1890s*, UQP, St Lucia, 1973; Michael Williams, *The Making of the South Australian Landscape*, Academic Press, London, 1974, chap. 9.

24 C.P. Trevelyan, *Letters from North America and the Pacific 1898*, Chatto and Windus, London, 1969, p. 199; John Shaw Neilson, *Selected Poems*, ed. R.H. Croll, Lothian, Melbourne, 1934, p. 177.

25 Robin Boyd, *Australia's Home*, 2nd edn, Penguin, Ringwood, 1968, chaps 5–6.

26 N.G. Butlin, *Investment in Australian Economic Development 1861–1900*, 2nd edn, Dept of Economic History, RSSS, ANU, Canberra, 1971, p. 235; McDonald, *Rockhampton*, pp. 116–22.

27 Susan Priestley, *Echuca: A Centenary History*, Jacaranda, Brisbane, 1965, p. 126; Donald S. Garden, *Albany: A Panorama of the Sound*, Nelson, Melbourne, 1977, p. 259.

28 *Farmer's Advocate*, 14 September 1917, quoted in L. Lomas, 'Graziers and Farmers in the Western District, 1890–1914', *RHSVJ*, 46 (1975), p. 255.

29 This paragraph draws on the ideas of Donald Denoon, 'Understanding Settler Societies', *Historical Studies*, 18 (1978–79), pp. 511–27; Figures derived from Butlin, *Australian Domestic Product, Investment and Foreign Borrowing*, p. 11; D.C.M. Platt, *Latin America and British Trade 1806–1914*, A. and C. Black, London, 1972, p. 111; S.B. Saul, *Studies in British Overseas Trade 1870–1914*; Liverpool University Press, Liverpool, 1960, p. 67.

30 A.R. Hall, *The London Capital Market and Australia 1870–1914*, ANU Press, Canberra, 1963, p. 92.

31 Australian statistics come from the *Pocket Compendium of Australian Statistics*; 'Royal Commission of Inquiry as to Food Supplies and Prices', *NSWPP*, 1913, first session, vol. 4, pp. 284–6, second session, vol. 3, p. 414; Butlin, *Investment in Australian Economic Development*, pp. 221, 259. For Britain see John Burnett, *Plenty and Want: A Social History of Diet in England from 1815 to the Present Day*, Nelson, London, 1966, pp. 96–9.

32 N.G. Butlin, 'Long-run Trends in Australian Per Capita Consumption', in K. Hancock (ed.), *The National Income and Social Welfare*, Cheshire, Melbourne, 1965, pp. 1–19; Ian McLean and Jonathan Pincus, *Living Standards in Australia 1890–1940: Evidence and Conjectures*, Working Papers in Economic History, ANU, Canberra, 1982.

33 The principal Australian evidence is a census of income and wealth undertaken by the Commonwealth in 1915 and published as G.H. Knibbs, *The Private Wealth of Australia and Its Growth*, Government Printer, Melbourne, 1918. See F. Lancaster Jones, 'The Changing Shape of Australian Income Distribution, 1914–15 and 1968–9', *AEHR*, 15 (1975), pp. 21–34.

34 This budget is based on an inquiry undertaken by the Commonwealth Bureau of Census and Statistics for the period June 1910–July 1911.

Of the 1500 budget books distributed, only 212 were returned and of these 49 were for families of more than four members with an income of less than £200 p.a., so the findings are merely illustrative. *Commonwealth Yearbook 1901–11*, pp. 1167–84. It may be compared with the household budgets based on the expenditure of £106 p.a. by eleven 'housekeeping women' cited in the Harvester Judgement, 2 *CAR*, 5–7.

CHAPTER 3: CLASS AND SOCIETY

1 P.G. Macarthy, 'Wages for Unskilled Work and Margins for Skill, Australia, 1901–21', *AEHR*, 12 (1972), pp. 142–60.
2 J. Hagan and C. Fisher, 'Piece Work and Some of its Consequences in the Printing and Coal Mining Industries in Australia, 1850–1930', *Labour History*, 25 (November 1973), p. 26.
3 Beverley Kingston, *My Wife, My Daughter and Poor Mary-Ann: Women and Work in Australia*, Nelson, Melbourne, 1975, chap. 4.
4 E.J. Holloway, 'From Labour Council to Privy Council', typescript memoirs, p. 48; NLA 2098; Simon Hickey, *Travelled Roads*, Cheshire, Melbourne, 1951, p. 7.
5 Adelaide *Weekly Herald*, 23 May 1908, quoted in Pavla Cook, 'Education and the Labour Movement in South Australia, 1890–1910' (unpublished paper, 1980). The estimate of earnings is calculated from G.H. Knibbs, *The Private Wealth of Australia and its Growth*, Government Printer, Melbourne, 1918.
6 See for example, 'Royal Commission on the Operation of the Factories and Shop Laws in Victoria', *VPP*, 1902–03, vol. 2, pp. 422–4.
7 Kerry Wimshurst, 'Child Labour and School Attendance in South Australia, 1890–1915', *Historical Studies*, 19 (1980–81), pp. 388–411; H.L. 'Duke' Tritton, *Time Means Tucker*, Shakespeare Head Press, Sydney, 1964, pp. 19–20.
8 Margaret Bridson Cribb, 'The A.R.U. in Queensland: Some Oral History', *Labour History*, 22 (May 1972), p. 18. The patterns of poverty are elaborated in Anne O'Brien, 'The Poor in New South Wales, 1880–1918', PhD thesis, University of Sydney, 1982, especially chap. 2.
9 Janet McCalman, 'Class and Respectability in a Working-Class Suburb: Richmond, Victoria, Before the Great War', *Historical Studies*, 20 (1982–83), pp. 90–103.
10 R.K. Nobbs, 'Ventures in Providence: the Development of Friendly Societies and Life Assurance in Nineteenth Century Australia', PhD thesis, Macquarie University, 1978, p. xxiii; Jill Roe (ed.), *Social Policy in Australia. Some Perspectives 1901–75*, Cassell, Stanmore, 1976, p. 17.
11 R.V. Jackson, 'Owner Occupation of Houses in Sydney, 1871–91', *AEHR*, 10 (1970), pp. 138–54; A.E. Dingle and D.T. Merrett, 'Home Owners and Tenants in Melbourne, 1891–1911', *AEHR*, 12 (1972), pp. 21–35, and 'Landlords in Suburban Melbourne, 1881–1911', *AEHR*, 17 (1977), pp. 1–24.
12 Geoffrey Serle, 'John Monash's Description of the Navvy, 1891', *Labour History*, 40 (May 1981), p. 94.
13 N.G. Butlin, A. Barnard and J.J. Pincus, *Government and Capitalism: Public and Private Choice in Twentieth Century Australia*, Allen and Unwin, Sydney, 1982, p. 72.
14 R.F. Holder, *Bank of New South Wales: A History*, vol. 2, Angus and Robertson, Sydney, 1970, pp. 502–3.

15 *Australian Journal of Education*, 3 (March 1906), p. 15, quoted in R.M. Pike, '"The Cinderella Profession": The State School Teacher of New South Wales, 1880–1963', PhD thesis, ANU, 1965, p. 40.
16 Rev. T.E. Clouston, writing in the Presbyterian *Messenger*, 3 April 1903, quoted in Richard Broome, *Treasure in Earthen Vessels: Protestant Christianity in New South Wales Society 1900–14*, UQP, St Lucia, 1980, p. 21.
17 *Australia To-Day*, 1911, p. 23; P.F. Rowland, *The New Nation*, Smith Elder, London, 1903, p. 119; 'Judge Stretton's Reminiscences', *La Trobe Library Journal*, 5, 17 (April 1976), p. 7.
18 H.L. Wilkinson, *The Trust Movement in Australia*, Critchley Parker, Melbourne, 1914.
19 Bernard Barrett, *The Inner Suburbs*, MUP, Melbourne, 1971, pp. 159–61. A.G. Lowndes (ed.), *South Pacific Enterprise. The Colonial Sugar Refining Company Limited*, Angus and Robertson, Sydney, 1956, pp. 43–5, 301; Geoffrey Blainey, *The Politics of Big Business: A History*, Academy of Social Sciences Lecture, ANU, Canberra, 1976, pp. 1–11.
20 Calculated from Knibbs, *The Private Wealth of Australia*.
21 William Epps, *Anderson Stuart M.D.*, Angus and Robertson, Sydney, 1922, p. 112; A.J. Hannan, *The Life of Chief Justice Way*, Angus and Robertson, Sydney, 1960, pp. 221–2; Roger B. Joyce, *Samuel Walker Griffith*, UQP, St Lucia, 1984), p. 184. Salomons in A.B. Piddington, *Worshipful Masters*, Angus and Robertson, Sydney, 1929, p. 204; Geoffrey Serle, *John Monash*, MUP, Melbourne, 1982, p. 167. For architecture and its opportunities see J.M. Freeland, *The Making of a Profession*, Angus and Robertson, Sydney, 1971.
22 N.G. Butlin, *Australian Domestic Product; Investment and Foreign Borrowing 1861–1938/9*, CUP, Cambridge, 1962, p. 231; T.S. Pensabene, *The Rise of the Medical Practitioner in Victoria*, ANU, Canberra, 1980, pp. 83–4.
23 *QPD*, vol. 104, p. 672 (6 Dec. 1909); Epps, *Anderson Stuart M.D.*, p. 124.
24 A.J. McLachlan, *McLachlan: An F.A.Q. Australian*, Lothian, Melbourne, 1948, p. 46; Warren Perry, 'The Late Sir John Latham', *RHSVJ*, 35 (1964), p. 95 and Latham Papers, NLA 6409; J.M. Bennett (ed.), *A History of the New South Wales Bar*, Law Book Company, Sydney, 1969, p. 102.
25 Norman Verschuer Wallace, *Bush Lawyer*, Rigby, Adelaide, 1976, p. 50; Arthur Dean, *A Multitude of Counsellors: A History of the Bar of Victoria*, Butterworths, Sydney, 1962, chap. 10.
26 Stevan Eldred-Grigg, 'Moral Capitalism in the Hunter Valley, 1880–1914', *JRAHS*, 66 (1980–81), pp. 132–46.
27 Quoted from a government inquiry into farm settlement by Marilyn Lake, '"Building Themselves Up With Aspros": Pioneer Women Re-assessed', *Hecate*, 7 (1981), p. 18. The problems confronting the settler in a comparatively favourable region are discussed in D.B. Waterson, *Squatter, Selector and Storekeeper: A History of the Darling Downs 1859–93*, SUP, Sydney, 1968, chap. 7 where a minimum outlay of £221 is calculated; but it is based on rental of only 160 acres and makes no allowance for clearing or living expenses in the initial period.
28 G.H. Gibson, *Ironbark Splinters from the Australian Bush*, T. Werner Laurie, London, 1912, p. 123. The standard account of pastoral expansion and contraction is that of N.G. Butlin, *Investment in Australian*

Economic Development 1861–1900, 2nd edn, Dept of Economic History, RSSS, ANU, Canberra, 1971, chaps 2 and 6.

29 S.H. Roberts, *History of Australian Land Settlement (1788–1920)*, MUP, Melbourne, 1924, chaps 26, 28.

30 All three writers spent their early years in the country and settled in cities, with the exception of Paterson who spent the period from 1908 to 1912 as a pastoralist. The contradictions are explored by Graeme Davison, 'Sydney and the Bush: an urban context for the Australian Legend', *Historical Studies*, 18 (1978–79), pp. 191–209. See generally Edgar Waters, 'Some Aspects of the Popular Arts in Australia, 1880–1915', PhD thesis, ANU, 1962.

31 Thomas Cherry, *Victorian Agriculture*, D.W. Paterson, Melbourne, 1913, p. 279; Charles Fahey, 'The Wealth of Farmers: A Victorian regional study, 1879–1901', *Historical Studies*, 21 (1984–85), pp. 29–51.

32 M.E. Robinson, *The New South Wales Wheat Frontier 1851–1911*, ANU Press, Canberra, 1969, pp. 146–54.

33 J.W. McCarty, 'The Inland Corridor', *Australia 1888*, 5 (1980), pp. 33–48, who draws on the workforce statistics of Butlin and Dowie, cited in the previous chapter; see generally Harriet Friedmann, 'World Market, State and Family Farm: Social Bases of Household Production in the Era of Wage Labour', *Comparative Studies in Society and History*, 20 (1978), pp. 545–86.

34 A.P. Elkin (ed.), *Marriage and the Family in Australia*, Angus and Robertson, Sydney, 1957, p. 2; Patricia Grimshaw, 'Women and the Family in Australian History', *Historical Studies*, 18 (1978–79), p. 415.

35 Ailsa Burns and Jacqueline Goodnow (eds), *Children and Families in Australia. Contemporary Issues and Problems*, Allen and Unwin, Sydney, 1979, chap. 2.

36 'Royal Commission on the Decline of the Birth Rate', *NSWPP*, 1904, vol. 4.

37 'Judge Stretton's Reminiscences', p. 9; Jan Carter, *Nothing to Spare. Recollections of Australian Pioneering Women*, Penguin, Ringwood, 1981, p. 166. See also Stephen Murray-Smith's material on children's pastimes in Guy Featherstone (ed.), *The Colonial Child*, Royal Historical Society of Victoria, Melbourne, 1981, pp. 73–87.

38 'Royal Commission of Inquiry into Hours and General Conditions of Employment of Female and Juvenile Labour', *NSWPP*, 1911–12, vol. 2, pp. 30–3. The onerous conditions of domestic service are elaborated in Kingston, *My Wife, My Daughter and Poor Mary-Ann*, chaps 2 and 3.

39 Evidence of a South Australian employer to 'Royal Commission on the Shortage of Labour', *SAPP*, 1912, vol. 2, p. 12.

40 See, for example, the 'Report on the Working of the Factories and Shops Act', *NSWPP*, 1912, vol. 2, p. 31. And see generally Edna Ryan and Ann Conlon, *Gentle Invaders: Australian Women at Work 1788–1974*, Nelson, Melbourne, 1975, chaps 2 and 3.

41 Jill Reekie, 'Female Office Workers in Western Australia, 1895–1920', *Time Remembered*, 5 (1982), pp. 1–35; Jennifer MacCulloch, '"This Store is Our World": Female Shop Assistants in Sydney to 1930', in Jill Roe (ed.), *Twentieth Century Sydney: Studies in Urban and Social History*, Hale and Iremonger, Sydney, 1980, p. 168; Claire McCluskey, 'Women in the Victorian Post Office', in Margaret Bevege et al. (eds), *Worth Her Salt: Women at Work in Australia*, Hale and Iremonger,

Sydney, 1982, pp. 60–1. See generally Margaret Power, 'The Making of a Woman's Occupation', *Hecate*, 1, 2 (1975), pp. 25–34.

42 'Report of Minister Controlling Education', *SAPP*, 1901, vol. 2; Richard Selleck, 'State Education and Culture', *Papers on Australian Cultural History*, 1 (1981), pp. 29–42; Alan Barcan, *A History of Australian Education*, OUP, Melbourne, 1980, chaps 10 and 11.

43 Ronald Fogarty, *Catholic Education in Australia 1806–1950*, vol. 2, MUP, Melbourne, 1968, especially pp. 306–11, 356–70.

44 Rupert Goodman, *Secondary Education in Queensland 1860–1960*, ANU Press, Canberra, 1968, pp. 115–20; A.E. Pratt, *Life of Dr W.S. Littlejohn*, Lothian, Melbourne, 1934, pp. 215, 231.

45 C.P. Trevelyan, *Letters*, p. 199.

46 Ernest Scott, *Historical Memoir of the Melbourne Club*, Specialty Press, Melbourne, 1936 and *Rules and Regulations of the Melbourne Club*; Alfred Hart, *History of the Wallaby Club*, Anderson Gowan, Melbourne, 1944, pp. 19, 67–8; David M. Dow, *Melbourne Savages: A History of the First Fifty Years of the Melbourne Savage Club*, Savage Club, Melbourne, 1947; Minutes of the Boobooks in Sir William Harrison Moore Collection, University of Melbourne Archives; Joan M. Gillison, *A History of the Lyceum Club*, Lyceum Club, Melbourne, 1975.

47 Vic Courtney, *All I May Tell*, Shakespeare Head, London, 1939, p. 181; R.H. Goddard, *The Union Club 1857–1957*, Halstead, Sydney, 1957, p. 72.

48 *SAPD* (Legislative Council, 1904), p. 295; James Murray, *Larrikins: Nineteenth Century Outrage*, Lansdowne, Melbourne, 1973, includes useful information in a generally sensationalized treatment; see also Ambrose Pratt, 'Push Larrikinism in Australia', *Blackwood's Magazine*, 170 (1901), pp. 27–40.

49 C.E. Jacomb, *'God's Own Country': An Appreciation of Australia*, Max Goschen, London, 1914, pp. 206–7.

50 'Royal Commission on Tramway Employes' Grievances', p. 124, *VPP*, 1898, vol. 3; quoted in Margaret Indian, 'Leisure in City and Suburb: Melbourne 1880–1900', PhD thesis, ANU, 1980, chap. 2; J.F. Archibald in *Lone Hand*, 2 December 1907, quoted in Ian Turner, *The Australian Dream*, Sun Books, Melbourne, 1968, p. 267; Patrick O'Farrell, *The Catholic Church and Community in Australia: A History*, Nelson, Melbourne, 1977, p. 232.

51 Gordon Inglis, *Sport and Pastimes in Australia*, Methuen, London, 1912, p. 19; Richard Broome, 'The Australian Reaction to Jack Jones, Black Pugilist, 1907–1909', in Richard Cashman and Chris McKernan (eds), *Sport in History*, UQP, St Lucia, 1979, p. 347.

52 Brian Stoddart, 'Sport and Society 1890–1940: A Foray', in C.T. Stannage (ed.), *A New History of Western Australia*, UWA Press, Nedlands, 1981, pp. 652–74; Chris Cunneen, 'The Rugby War: the Early History of Rugby League in New South Wales, 1907–15', in Cashman and McKernan (eds), *Sport in History*, pp. 293–306.

53 The quotations, one from 1888 and the other from 1913, are taken from K.S. Inglis, 'Religious Behaviour', in A.F. Davies and S. Encel (eds), *Australian Society: A Sociological Introduction*, 2nd edn, Cheshire, Melbourne, 1970, p. 472.

54 These are estimates of religious practice; the number of nominal adherents to the Church of England was higher than one-third and the number of Roman Catholics lower. See Walter Phillips, 'Religious

Profession and Practice in N.S.W., 1851–1901: the Statistical Evidence', *Historical Studies*, 15 (1971–73), pp. 378–400, who draws Victorian and South Australian comparisons; and *Commonwealth Year Book*, 1, 1901–07, pp. 172–4.

55 Broome, *Treasure in Earthen Vessels*, p. 11, quoting a report of the Anglican Synod of Sydney in 1901. For similar pictures in other states see Renate Howe, 'The Response of Protestant Churches to Urbanism in Melbourne and Chicago, 1875–1914', PhD thesis, University of Melbourne, 1971; and Ronald Lawson, 'The Political Influence of Churches in Brisbane in the 1890s', *JRH*, 7 (1972), pp. 144–62.

56 Patrick and Deidre O'Farrell, *Documents in Australian Catholic History*, vol. 2, Geoffrey Chapman, London, 1969, p. 21.

57 Quoted in O'Farrell, *The Catholic Church and Community in Australia*, p. 295.

58 ibid., p. 211.

59 See for example H.M. Moran, *Viewless Winds*, Peter Davies, London, 1939, p. 10; Dame Enid Lyons, *So We Take Comfort*, Heinemann, London, 1965, p. 63.

60 Broome, *Treasure in Earthen Vessels*, chap. 4; Cyril Pearl, *Wild Men of Sydney*, W.H. Allen, London, 1958, chap. 9.

61 Patrick O'Farrell, 'The Cultural Ambivalence of Australian Religion', *Papers on Australian Cultural History*, 1 (1981), pp. 3–8.

62 Alfred Guy, 'One Man's Journey', pp. 64–8.

63 Anthony Splivalo, *The Home Fires*, Fremantle Arts Centre Press, Fremantle, 1980, pp. 18–19.

64 Geoffrey Dutton, *Snow on the Saltbush*, Penguin, Ringwood, 1984, pp. 224–5.

65 James Devaney, *Shaw Neilson*, Angus and Robertson, Sydney, 1944, p. 32. For an introduction to the cultural history of the period see Geoffrey Serle, *From Deserts the Prophets Come: The Creative Spirit in Australia*, Heinemann, Melbourne, 1973, chaps 5 and 6.

CHAPTER 4: PATTERNS OF POLITICS

1 Holman and Reid quoted in H.V. Evatt, *William Holman: Australian Labour Leader*, Angus and Robertson, Sydney, 1940, chap. 21, and F.K. Crowley, *Modern Australia in Documents*, vol. 1, Wren, Melbourne, 1973, p. 70.

2 F.W. Eggleston in E.H. Sugden and Eggleston, *George Swinburne*, Angus and Robertson, Sydney, 1931, p. 52.

3 Quoted in P. Weller, 'Preselection and Local Support: An Explanation of the Methods and Development of Early Non-Labor Parties in New South Wales', *AJPH*, 21 (1975), p. 24.

4 See Gerald A. Caiden, *Career Service*, MUP, Melbourne, 1965, chap. 1.

5 'Royal Commission on Administration of the Lands Department', *NSWPP*, 1906, vol. 2; Frank Clune, *Scandals of Sydney Town*, Angus and Robertson, Sydney, 1957, pp. 132– 57; Cyril Pearl, *Wild Men of Sydney*, W.H. Allen, London, 1958, chap. 12.

6 Sugden and Eggleston, *George Swinburne*, pp. 169–70; W.G. Spence, *Australia's Awakening: Thirty Years in the Life of an Australian Agitator*, Worker Trustees, Sydney, 1909, pp. 332–3.

7 'Royal Commission on the Victorian Police Force', especially pp. 205–8, *VPP*, 1906, vol. 3; Pearl, *Wild Men of Sydney*, chap. 14.

8 'Royal Commission on Administration of the Lands Department', p. 815.
9 P. Loveday, A.W. Martin and R.S. Parker (eds), *The Emergence of the Australian Party System*, Hale and Iremonger, Sydney, 1977, p. 18.
10 Weller, Preselection and Local Support', p. 23, and *ADB*, vol. 9, p. 10.
11 *QPD*, vol. 93, p. 124 (29 September 1904); Colin A. Hughes and B.D. Graham, *A Handbook of Australian Government and Politics 1890–1964*, ANU Press, Canberra, 1968, p. 286.
12 D.R. Hall, 'The Historical Development of Australian Political Parties', in W.G.K. Duncan (ed.), *Trends in Australian Politics*, Angus and Robertson, Sydney, 1935, pp. 33–4. George Meudell, *The Pleasant Career of a Spendthrift*, Routledge, London, 1929, p. 35.
13 W.M. Hughes, *Crusts and Crusades: Tales of Bygone Days*, Angus and Robertson, Sydney, 1948, pp. 207–14; L.F. Fitzhardinge, *William Morris Hughes: A Political Biography. That Fiery Particle 1862–1914*, Angus and Robertson, Sydney, 1964, p. 155.
14 Sir Ronald Munro Ferguson, Memorandum of Interview with Joseph Cook, 2 June 1914, Novar Papers, NLA 696/557. King O'Malley in *CPD*, vol. 63, p. 3636 (4 December 1911); 'Report of Royal Commission upon the Commonwealth Electoral Law and Administration', *CPP*, 1914–17, vol. 2, pp. 437–9.
15 A.R. Chisholm, *Men Were my Milestones: Australian Portraits and Sketches*, MUP, Melbourne, 1958, chap. 4.
16 Quoted in R. Norris, *The Emergent Commonwealth. Australian Federation: Expectations and Fulfilment 1889–1910*, MUP, Melbourne, 1975, p. 15.
17 Geoffrey Sawer, 'The Australian Constitution and the Australian Aborigine', *Federal Law Review*, 2 (1966), pp. 17–36.
18 Geoffrey Sawer, *Australian Federal Politics and Law 1901–29*, MUP, Melbourne, 1956, pp. 56–7, 82–4, 107–9; Leslie Zines, *The High Court and the Constitution*, Butterworths, Sydney, 1981, especially chap. 1.
19 Quoted in L.F. Crisp, *Australian National Government*, 3rd edn, Longman, Melbourne, 1973, pp. 59–60.
20 D.I. Wright, 'The Political Significance of "Implied Immunities", 1901–10', *AJPH*, 5 (1959), p. 394; Conrad Joyce, 'Attempts to Extend Commonwealth Powers 1908–19', *Historical Studies*, 9 (1959–61), p. 296.
21 Alfred Deakin, *Federated Australia*, p. 97. See also J.A. La Nauze, *Alfred Deakin: A Biography*, MUP, Melbourne, 1965, vol. 2, pp. 587–92, and D.I. Wright, 'The Politics of Federal Finance: The First Decade', *Historical Studies*, 13 (1967–69), pp. 460–76.
22 J.A. La Nauze, *The Hopetoun Blunder: The Appointment of First Prime Minister to the Commonwealth of Australia December 1900*, MUP, Melbourne, 1957.
23 *Maitland Daily Mercury*, 18 January 1901, quoted in Crowley, *Modern Australia in Documents*, vol. 1, pp. 2–3.
24 Henry Gyles Turner, *A History of the Colony of Victoria*, Longmans Green, London, 1904, p. 347.
25 A.G. Austin (ed.), *The Webbs' Australian Diary, 1898*, Pitman, Melbourne, 1965, p. 26.
26 The Free Trade side of national politics is in need of attention. I have found useful the suggestions of L.F. Crisp in his *George Houston Reid:*

Federation Father, Federal Failure?, ANU, Canberra, 1979.

27 See D.J. Murphy (ed.), *Labor in Politics: The State Labor Parties in Australia 1880–1920*, UQP, St Lucia, 1975. The Australian Labor Party began as a federation of pre-existing state organizations which retained their various titles (Political Labour League, etc.) until after the First World War. The expression and spelling 'Labor Party' is therefore strictly an anachronism but is used here for convenience.

28 *Worker*, 3 February 1900, quoted in Ian Turner, *Industrial Labour and Politics: The Dynamics of the Labour Movement in Eastern Australia 1900–21*, ANU Press, Canberra, 1962, p. 26.

29 The leader of the Labor Party in the Senate, reported in *CPD*, vol. 1, p. 131 (22 May 1901). The organisation of the federal party followed at the 1902 conference.

30 See P. Loveday, 'Support in Return for Concessions', *Historical Studies*, 14 (1969–71), pp. 376–405, which demonstrates Labor's failure to vote as a bloc.

31 In this extensive literature the more important works are V.S. Clark, *The Labour Movement in Australasia: A Study in Social Democracy*, Holt, New York, 1906; Albert Métin, *Socialism without Doctrine*, trans. Russel Ward, APCOL, Sydney, 1977, and the selection of German writings edited by Jurgen Tampke, *Wunderbar Country: Germans Look at Australia 1850–1914*, Hale and Iremonger, Sydney, 1982. Then there were the British Labour leaders Ben Tillett, Keir Hardie, Ramsay and Margaret MacDonald, and Tom Mann, who all wrote books about Australia.

32 Quoted in D.W. Rawson, 'Labour, Socialism and the Working Class', *AJPH*, 7 (1961), pp. 83–4.

33 Address to the Sixth Queensland Labour-in-Politics Convention, quoted in D.J. Murphy et al. (eds), *Prelude to Power: The Rise of the Labour Party in Queensland 1885–1915*, Jacaranda, Milton, Qld, 1970, p. 285. See also R.W. Connell and T.H. Irving, *Class Structure in Australian History: Documents, Narrative and Argument*, Longman Cheshire, Melbourne, 1980, pp. 197–8.

34 A.E. Cahill, 'Catholics and Socialism: the 1905 Controversy in Australia', *JRH*, 1 (1960), pp. 88–101; Evatt, *William Holman*, p. 122; Turner, *Industrial Labour and Politics*, p. 53. The best survey is Verity Burgmann, *'In Our Time': Socialism and the Rise of Labor, 1885–1905*, Allen and Unwin, Sydney, 1985.

35 Matt Reid, quoted in Murphy et al. (eds), *Prelude to Power*, p. 281.

36 H.S. Broadhead, 'J.C. Watson and the Caucus Crisis of 1905', *AJPH*, 8 (1962), p. 96. See generally L.F. Crisp, *The Australian Federal Labor Party 1901–51*, 2nd edn, Hale and Iremonger, Sydney, 1978, chaps 7, 9.

37 Quoted in La Nauze, *Alfred Deakin*, vol. 2, p. 363.

38 W.M. Hughes, quoted in Fitzhardinge, *William Morris Hughes*, p. 157; Deakin quoted in La Nauze, *Alfred Deakin*, vol. 2, p. 378.

39 Crowley (ed.), *Modern Australia in Documents*, vol. 1, pp. 87–8.

40 Quoted in Sawer, *Australian Federal Politics and Law 1901–29*, p. 60.

41 Voting figures for the House of Representatives taken from Hughes and Graham *A Handbook of Australian Government and Politics*, pp. 286–301.

42 Loveday, Martin and Parker (eds), *The Emergence of the Australian Party System*, pp. 413–42; John Rickard, *Class and Politics: New South Wales, Victoria and the Early Commonwealth 1890–1910*, ANU Press, Canberra, 1976, p. 26.

43 *CPD*, vol. 49, pp. 1156–7 (14 July 1909).
44 Joan Rydon and R.N. Spann, *New South Wales Politics 1901–10*, Cheshire, Melbourne, 1962, p. 42; J.B. Hirst, *Adelaide and the Country 1870–1917: Their Social and Political Relationships*, MUP, Melbourne, 1973, pp. 207–11. Loveday, Martin and Parker (eds), *The Emergence of the Australian Party System*, p. 293; B.D. Graham, 'The Place of Finance Committees in non-Labor Politics, 1910–1930', *AJPH*, 6 (1960), pp. 50–1.
45 H.L. Nielsen, *The Voice of the People, or the History of the Kyabram Reform Movement*, Arbuckle, Waddell and Fawkner, Melbourne, 1902; Sugden and Eggleston, *George Swinburne*, pp. 78–107.
46 Rickard, *Class and Politics*, pp. 177–9.
47 Rydon and Spann, *New South Wales Politics*, pp. 4, 32; J.D. Bollen, *Protestantism and Social Reform in New South Wales 1890–1910*, MUP, Melbourne, 1972, pp. 139, 156.

CHAPTER 5: MAKING A COMMONWEALTH

1 *Official Report of the National Australasian Convention Debates*, Government Printer, Sydney, 1891, pp. 590–2. See also W.K. Hancock, 'A Veray and True Comyn Weal', in *Politics in Pitcairn, and Other Essays*, Macmillan, London, 1947, pp. 94–109, and J.A. La Nauze, 'The Name of the Commonwealth of Australia', *Historical Studies*, 15 (1971–73), pp. 59–71.
2 Dancey in the Melbourne *Punch*, quoted in Marguerite Mahood, *The Loaded Line: Australian Political Caricature 1788–1901*, MUP, Melbourne, 1973, p. 278.
3 See R.A. Gollan, *Radical and Working Class Politics: A Study of Eastern Australia 1850–1910*, MUP, Melbourne, 1967, ed., pp. 155–60, 183–9; J.H. Portus, *The Development of Australian Trade Union Law*, MUP, Melbourne, 1958, chap. 7; Stuart Macintyre, 'Labour, Capital and Arbitration, 1890–1920', in Brian Head (ed.) *State and Economy in Australia*, OUP, Melbourne, 1983, pp. 98–114.
4 W. Pember Reeves, *State Experiments in Australia and New Zealand* (1902), reissued with an introduction by John Child, Macmillan, London, 1968, vol. 2, p. 96.
5 Wise in *NSWPD*, vol. 103, p. 651 (4 July 1900), p. 45; Deakin in *CPD*, vol. 15, p. 286, 2865 (30 July 1903) quoted in Keith Hancock, 'The Wages of the Workers', *JIR*, 11 (1969), p. 17.
6 P.G. Macarthy, 'Wage Determination in New South Wales, 1890–1920', *JIR*, 10 (1968), pp. 192–6.
7 'Explanatory Memorandum in Regard to New Protection', *CPP*, 1907–08, vol. 2, p. 1187. The best treatment of New Protection is that of John Rickard in *Class and Politics: New South Wales, Victoria, and the Early Commonwealth 1890–1910*, ANU Press, Canberra, 1976, chap. 7.
8 2 *CAR* 3–17; P.G. Macarthy, 'Justice Higgins and the Harvester Judgement', *AEHR*, 11 (1969), pp. 17–38.
9 Peter Macarthy, 'The Harvester Judgement—An Historical Assessment', PhD thesis, ANU, 1967; Colin Forster, 'Indexation and the Commonwealth Basic Wage, 1907–22', *AEHR*, 20 (1980), pp. 99–118.
10 While Macarthy's study of the Harvester Judgement considers wage relativities in Commonwealth awards, we still lack a comprehensive study of the effects of arbitration on workforce hierarchies. That it

benefited unskilled workers soon became the convention; see, for example, H.M. Murphy, *Wages and Prices in Australia: Our Labour Laws and their Effects*, George Robertson, Melbourne, 1917, especially pp. 13–14. On the effects of the Commonwealth tariff see C. Forster, 'Federation and the Tariff', *AEHR*, 17 (1977), pp. 95–116.

11 The fragile equilibrium is discussed in J.W. McCarty, 'Australian Regional History', *Historical Studies*, 18 (1978–79), pp. 88–105.

12 Ralph Shlomowitz, 'The Search for Institutional Equilibrium in Queensland's Sugar Industry, 1884–1913', *AEHR*, 19 (1979), pp. 91–122.

13 R.S. Parker, 'Public Enterprise in New South Wales', *AJPH*, 4 (1958), pp. 208–23; J.R. Robertson, 'The Foundation of State Socialism in Western Australia: 1911–16', *Historical Studies*, 10 (1961–63), pp. 309–26; D.J. Murphy, 'The Establishment of State Enterprises in Queensland', *Labour History*, 14 (May 1968), pp. 13–22. Theodore quoted in K.H. Kennedy, *The Mungana Affair. State Mining and Political Corruption in the 1920s*, UQP, St Lucia, 1978, p. 2.

14 *Australasian Insurance and Banking Record*, 10 July 1905, quoted in R.F. Holder, *Bank of New South Wales: A History*, Angus and Robertson, Sydney, 1970, vol. 2, p. 555.

15 Robin Gollan, *The Commonwealth Bank of Australia: Origins and Early History*, ANU Press, Canberra, 1968; see also Geoffrey Blainey, *Gold and Paper: A History of the National Bank of Australasia*, Georgian House, Melbourne, 1958, pp. 266–71, and S.J. Butlin, *Australia and New Zealand Bank: The Bank of Australasia and the Bank of Australia Limited 1828–1951*, Longmans, London, 1961, p. 341–52.

16 T.H. Kewley, *Social Security in Australia 1900–72*, 2nd edn, SUP, Sydney, 1973, part I.

17 Brisbane Crêche and Kindergarten Association, quoted in Kay Daniels and Mary Murnane (eds), *Uphill All the Way: A Documentary History of Women in Australia*, UQP, St Lucia, 1980, p. 127. See generally Graeme Davison, 'The City-Bred Child and Urban Reform in Melbourne, 1900–40', Peter Williams (ed.), *Social Process and the City*, Allen and Unwin, Sydney, 1983, pp. 143–74, and Leonie Sandercock, *Cities for Sale*, MUP, Melbourne, 1977, chap. 1. Some prominent progressives are described in Michael Roe, *Nine Australian Progressives: Vitalism in Bourgeois Social Thought*, UQP, St Lucia, 1984.

18 Rupert Goodman, *Secondary Education in Queensland, 1860–1960*, ANU Press, Canberra, 1968, p. 206; A.G. Austin and R.J.W. Selleck, *The Australian Government School 1830–1914*, Pitman, Carlton, 1975, p. 226.

19 R.J.W. Selleck, *Frank Tate*, MUP, Melbourne, 1982, p. 186.

20 *Our First Half-Century: A Review of Queensland's Progress*, Government Printer, Brisbane, 1909, p. 245; 'Report of the Royal Commission on the Establishment of a University, *WAVP*, 1910–11, vol. 2, pp. 11, 14.

21 Dianne Barwick, 'Coranderrk and Cumeroogunga: Pioneers and Policy', in T. Scarlett Epstein and David H. Penny (eds), *Opportunity and Response: Case Studies in Economic Development*, C. Hurst, London, 1972, pp. 11–68. C.D. Rowley, *The Destruction of Aboriginal Society*, Penguin, Ringwood, 1972, chaps 10–13.

22 Women's Association quoted in Daniels and Murnane (eds), *Uphill All the Way*, p. 286; Derham, quoted in Sarah Stephen, 'A Quest for Collegiate Identity: Women in Ormond 1885 to 1910', in Stuart

Macintyre (ed.), *Ormond College Centenary Essays*, MUP, Melbourne, 1983, pp. 62–78.

23 Murphy et al. (eds), *Prelude to Power*, p. 84. See, more generally, Anne Summers, *Damned Whores and God's Police: The Colonization of Women in Australia*, Penguin, Ringwood, 1975, chap. 11; Kerreen Reiger, *The Disenchantment of the Home: Modernizing the Australian Family 1880–1940*, OUP, Melbourne, 1985, chaps 1 and 2, and Judith Allen, 'Breaking into the Public Sphere', in Judy Mackinolty and Heather Radi (eds), *In Pursuit of Justice: Australian Women and the Law 1788–1979*, Hale and Iremonger, Sydney, 1979, pp. 107–17.

24 6 *CAR* 72, quoted in Edna Ryan and Ann Conlon, *Gentle Invaders: Australian Women at Work 1788–1974*, Nelson, Melbourne, 1975, p. 95. See also Edna Ryan, *Two-thirds of a Man: Women and Arbitration in New South Wales, 1902–08*, Hale and Iremonger, Sydney, 1984.

25 Judith Allen, 'The Making of a Prostitute Proletariat in early Twentieth-Century New South Wales', in Kay Daniels (ed.), *So Much Hard Work: Women and Prostitution in Australian History*, Fontana, Sydney, 1984, pp. 192–232.

26 Cyril Pearl, *Wild Men of Sydney*, W.H. Allen, London, 1958, pp. 113–14, and Michael Cannon, *That Damned Democrat: John Norton, an Australian Populist, 1859–1916*, MUP, Melbourne, 1981. See also Keith Dunstan, *Wowsers*, Cassell, Melbourne, 1968.

27 Richard Broome, *Treasure in Earthen Vessels: Protestant Christianity in New South Wales Society 1900–14*, UQP, St Lucia, 1980, pp. 131, 155.

28 ibid., chap. 7; Renate Howe, 'The Response of Protestant Churches to Urbanism in Melbourne and Chicago, 1875–1914', PhD thesis, University of Melbourne, 1971; Michael McKernan, 'An Incident of Social Reform, Melbourne 1906', *JRH*, 10 (1978), pp. 70–85.

29 Peter Coleman, *Obscenity, Blasphemy, Sedition: 100 Years of Censorship in Australia*, 2nd edn, Angus and Robertson, Sydney, 1974, chap. 1.

30 J.A. La Nauze, *Alfred Deakin: A Biography*, MUP, Melbourne, 1965, vol. 2, p. 604.

31 Graeme Osborne, 'Town and Company: the Broken Hill Industrial Dispute of 1908–09', in John Iremonger et al. (eds), *Strikes: Studies in Twentieth Century Australian Social History*, Angus and Robertson, Sydney, 1973, pp. 26–50; *Tom Mann's Memoirs*, 2nd edn, Macgibbon and Kee, London, 1967, p. 190; Paquita Mawson, *A Vision of Steel: The Life of G.D. Delprat*, Cheshire, Melbourne, 1958, pp. 152–9. See generally Geoffrey Blainey, *The Rise of Broken Hill*, Macmillan, Melbourne, 1968, pp. 114–23; and Brian Kennedy, *Silver, Sin and Sixpenny Ale: A Social History of Broken Hill 1883–1921*, MUP, Melbourne, 1978, chap. 8.

32 Edgar Ross, *A History of the Miners' Federation of Australia*, Australasian Coal and Shale Employees' Federation, Sydney, 1970, p. 176.

33 L.F. Fitzhardinge, *William Morris Hughes: A Political Biography. That Fiery Particle, 1862–1914*, Angus and Robertson, Sydney, 1964, p. 228 and W.M. Hughes, *The Case for Labour*, 2nd edn, SUP, 1970, p. 29. See generally V.G. Childe, *How Labour Governs: A Study of Workers' Representation in Australia*, Labour Publishing Company, London, 1923, chap. 8; and Ian Turner, *Industrial Labour and Politics: The Dynamics of the Labour Movement in Eastern Australia, 1900–21*, ANU Press, Canberra, 1962, pp. 55–67.

34 *NSWPD*, vol. 36, p. 4582 (16 December 1909); Ross, *A History of the*

Miners' Federation, chap. 10; Robin Gollan, *The Coalminers of New South Wales*, MUP, Melbourne, 1963, chap. 6.

35 Quoted in M.H. Ellis, *A Saga of Coal: The Newcastle Wallsend Coal Company's Centenary Volume*, Angus and Robertson, Sydney, 1969, p. 162.

36 H.L. Wilkinson, *The Trust Movement in Australia*, Critchley Parker, Melbourne, 1914.

37 8 *CLR* 330; 15 *CLR* 92–103; Fitzhardinge, *William Morris Hughes*, pp. 267–70; Andrew Hopkins, 'Anti-Trust and the Bourgeoisie, 1906 and 1964', in E.L. Wheelwright and Ken Buckley (eds), *Essays in the Political Economy of Australian Capitalism*, vol. 2, ANZ, Sydney, 1978, pp. 87–109. The owners' case is put by N.L. McKellar, *From Derby Round to Burketown: The A.U.S.N. Story*, UQP, St Lucia, 1977, chap. 17.

38 Trevor Matthews, 'Business Associations and Politics: Chambers of Manufacturers and Employers' Federations in New South Wales, Victoria and Commonwealth Politics to 1939', PhD thesis, University of Sydney, 1971, especially pp. 184–5.

39 A.A. Morrison, 'The Brisbane General Strike of 1912', in D.J. Murphy et al. (eds), *Prelude to Power: The Rise of the Labour Party in Queensland 1885–1915*, Jacaranda, Milton, 1970, pp. 127–40.

40 K.H. Kennedy, 'Theodore, McCormack and the Amalgamated Workers' Association', *Labour History*, 33 (November 1977), pp. 14–28.

41 See 'Report of Royal Commission on the Sugar Industry', *CPP*, 1913, vol. 3, especially pp. 107–89.

42 W. Jethro Brown, *The Prevention and Control of Monopolies*, John Murray, London, 1914, pp. 14, 21–2. See Michael Roe, 'Jethro Brown: the First Teacher of Law and History in the University of Tasmania', *University of Tasmania Law Review*, 5 (1977), pp. 209–47.

CHAPTER 6: AUSTRALIANS IN THEIR WORLD

1 Ian Turner (ed.), *The Australian Dream*, Sun Books, Melbourne, 1968, p. 160; Bernard Smith, *Australian Painting 1788–1970*, 2nd edn, OUP, Melbourne, 1971, chap. 4; Charles S. Blackton, 'Australian Nationality and Nativism: The Australian Natives Association, 1882–1901', *Journal of Modern History*, 30 (1958), pp. 33–46.

2 Lane quoted in Turner, *The Australian Dream*, p. 160; Rev. Andrew Harper, *The Honourable James Balfour, M.L.C. A Memoir*, Critchley Parker, Melbourne, 1918, p. 285.

3 Hughes, interviewed in the *Bulletin*, 10 February 1901, quoted in L.F. Fitzhardinge, *William Morris Hughes: A Political Biography. That Fiery Particle, 1862–1914*, Angus and Robertson, Sydney, 1964, p. 116. See also Patrick Weller (ed.), *Caucus Minutes 1901–49: Minutes of the Meetings of the Federal Parliamentary Labor Party*, MUP, Melbourne, 1975, vol. 1, p. 46.

4 R. Norris, *The Emergent Commonwealth. Australian Federation: Expectations and Fulfilment 1899–1910*, MUP, Melbourne, 1975, p. 89.

5 *CPD*, vol. 4, pp. 4624–33, 4807 (6, 12 September 1901).

6 A.T. Yarwood, 'The "White Australia" Policy: Some Administrative Problems, 1901–20', *AJPH*, 7 (1961), pp. 245–60; Yarwood, *Asian Migration to Australia. The Background to Exclusion, 1896–1923*, MUP, Melbourne, 1964, pp. 101–2.

7 The contest can be traced through the 'Despatch from the Secretary of State', *CPP*, 1905, vol. 2, pp. 1319–39; 'Report of the Royal Commis-

sion on the Navigation Bill', *CPP*, 1906, vol. 3; and 'Report of the Navigation Conference', *CPP*, 1907–08, vol. 3, where Lyne's rebuff occurs on p. 252.

8 B.K. de Garis, 'The Colonial Office and the Commonwealth Constitution Bill', in A.W. Martin (ed.), *Essays in Australian Federation*, MUP, Melbourne, 1969, pp. 94–121; Alfred Deakin, *The Federal Story*, MUP, Melbourne, 1963, chaps 20–22; A.J. Hannan, *The Life of Chief Justice Way*, Angus and Robertson, Sydney, 1960, pp. 172–91; J.A. La Nauze, *The Making of the Australian Constitution*, MUP, Melbourne, 1972, chap. 16.

9 Richard Jebb, *Studies in Colonial Nationalism*, Edward Arnold, London, 1905, p. 156.

10 Ben H. Morgan, *The Trade and Industry of Australasia*, Eyre and Spottiswoode, London, 1908, is a survey undertaken on behalf of British manufacturers.

11 Taken from Neville Meaney, *The Search for Security in the Pacific 1901–14*, SUP, Sydney, 1976, pp. 278, 284. Some columns do not add precisely because of rounding.

12 Alfred Buchanan, *The Real Australia*, Fisher Unwin, London, 1907, pp. 28, 31; C.P. Trevelyan, *Letters from North America and the Pacific 1898*, Chatto and Windus, London, 1968, p. 205; Alex Castles, *An Australian Legal History*, Law Book Company, Sydney, 1982, p. 344.

13 Novar Papers, NLA, 696/680; W.F. Connell, 'British Influence on Australian Education in the Twentieth Century', in A.F. Madden and W.H. Morris-Jones (eds), *Australia and Britain: Studies in a Changing Relationship*, Institute of Commonwealth Studies, London, 1980, pp. 162–79; S.G. Firth, 'Social Values in the New South Wales Primary School, 1880–1914', in R.J. Selleck (ed.), *Melbourne Studies in Education 1970*, MUP, Melbourne, 1970, pp. 123–49.

14 J.G. Latham to Ella Latham, NLA 1009/20/614; Lieutenant G.W. Gorman, quoted in Patsy Adam-Smith, 'All Those Empty Pages', *La Trobe Library Journal*, 4, 14 (October 1974), p. 47; Casey to Niall, 7 December 1900, Goldsbrough Mort Papers, 1, 289; Atlee Hunt quoted in Donald C. Gordon, *The Dominion Partnership in Imperial Defence, 1870–1914*, Johns Hopkins, Baltimore, 1965, p. 209; Roger B. Joyce, *Samuel Walker Griffith*, UQP, St Lucia, 1984, pp. 324–5; Dame Mabel Brookes, *Crowded Galleries*, Heinemann, London, 1956, p. 75; John Barnes, 'Henry Lawson in London', *Quadrant*, 144 (July 1979), pp. 24–35; James R. Tyrrell, *Old Books, Old Friends, Old Sydney*, Angus and Robertson, Sydney, 1952, p. 124. Desmond Zwar, *In Search of Keith Murdoch*, Macmillan, London, 1980, p. 13; H.M. Moran, *Viewless Winds*, Peter Davies, London, 1939, p. 69. See also Gavin Souter, *Lion and Kangaroo: The Initiation of Australia 1901–19*, Collins, Sydney, 1976, pp. 117–27.

15 John Foster Fraser, *Australia: The Making of a Nation*, Cassell, London, 1910, pp. xiii–xiv; Adelaide Lubbock, *People in Glass Houses: Growing Up at Government House*, Nelson, Melbourne, 1977, p. 52; C.E. Jacomb, *'God's Own Country': An Appreciation of Australia*, Max Goschen, London, 1914, p. 233; Will J. Sowden, *An Australian Native's Standpoint*, Macmillan, London, 1913, pp. 4–5. See the comparative study of Patrick Dunae, *Gentlemen Emigrants: From the British Public School to the Canadian Frontier*, Douglas and McIntyre, Vancouver, 1981.

16 *VPD*, vol. 92, p. 1765 (10 October 1899) quoted in Barbara R. Penny, 'Australia's Reaction to the Boer War—A Study in Colonial Imperial-

ism', *Journal of British Studies*, 7 (1967–68), p. 111; Forrest in *WAPD*, vol. 15, p. 1558, quoted in L.M. Field, *The Forgotten War: Australian Involvement in the South African War of 1899–1902*, MUP, Melbourne, 1979, p. 25; Alfred Deakin, *Federated Australia: Selections from Letters to the Morning Post, 1900–10*, ed. J.A. La Nauze, MUP, Melbourne, 1968, p. 34.

17 Field, *The Forgotten War*; C.N. Connolly, 'Manufacturing "Spontaneity": the Australian Offer of Troops for the Boer War', and 'Class, Birthplace and Loyalty: Australian Attitudes to the Boer War', *Historical Studies*, 18 (1978–79), pp. 106–17 and 210–32.

18 Quoted in Maurice French, 'The Ambiguities of Empire Day in New South Wales, 1901–21: Imperial Consensus or National Division?', *AJPH*, 24 (1978), p. 61.

19 F.B. Boyce, *Fourscore Years and Seven*, Angus and Robertson, Sydney, 1934, pp. 115–16; see also Maurice French, 'One People, One Destiny—A Question of Loyalty: the Origins of Empire Day in New South Wales, 1900–1905', *JRAHS*, 61 (1975), pp. 236–48.

20 French, 'The Ambiguities of Empire Day', and Stewart Firth and Jeanette Hoorn, 'From Empire Day to Cracker Night', in Peter Spearritt and David Walker (eds), *Australian Popular Culture*, Allen and Unwin, Sydney, 1979, pp. 17–38.

21 *The First Fifty Years of the Rhodes Trust and the Rhodes Scholarships 1903–53*, Basil Blackwell, Oxford, 1955, pp. 4–20.

22 J.C.V. Behan, 'The Record of the Rhodes Scholars', *The Australian Rhodes Review*, 5 (1940), pp. 11–32; Max Beerbohm, *Zuleika Dobson, or an Oxford Love Story* (1911), Heinemann, London, 1964 edn, p. 86; W.R. Crocker, *Australian Ambassador*, MUP, Melbourne, 1971, p. 123; G.V. Portus, *Happy Highways*, MUP, Melbourne, 1953, chap. 9.

23 Latham Papers, NLA 1009/12/7. The successful candidate was Sir John Behan, for whom see *ADB*, vol. 7, pp. 247–8.

24 Walter Nimocks, *Milner's Young Men: the 'Kindergarten' in Edwardian Imperial Affairs*, Duke University Press, Durham, N.C., 1968; J. Kendle, 'The Round Table Movement', *NZHJ*, 1 (1967), pp. 33–50.

25 Harrison Moore, typescript 'Notes for a History of Round Table', and Round Table minutes in Harrison Moore Papers, University of Melbourne Archives, 11/6/2; John Kendle, *The Round Table Movement and Imperial Union*, University of Toronto Press, Toronto, 1975, pp. 94–9; Leonie Foster, 'High Hopes: the Men and the Motives of the Australian Round Table', PhD thesis, LaTrobe University, 1985.

26 R.B. Walker, 'Some Social and Political Aspects of German Settlement in Australia to 1914', *JRAHS*, 61 (1975–76), pp. 26–40. See W.D. Borrie, *Italians and Germans in Australia: A Study of Assimilation*, Cheshire, Melbourne, 1954.

27 *Commonwealth Year Book* 1901–07, p. 168. These statistics are for place of birth and thus suffer from the limitation discussed in the text.

28 R. Arnold, 'Some Australasian Aspects of New Zealand Life, 1880–1914', *NZJH*, 4 (1970), pp. 54–76. See also R.P. Davis, 'New Zealand Liberalism and Tasmanian Labor, 1891–1916', *Labour History*, 21 (November 1971), pp. 24–35.

29 L.G. Churchward, *Australia and America 1788–1972: An Alternative History*, APCOL, Sydney, 1979, chap. 5; Gordon Greenwood, 'Australia and the United States', in Richard Preston (ed.), *Contemporary Australia: Studies in History, Politics and Economics*, Duke University Press, Durham, N.C., 1969, pp. 417–51.

30 Roger C. Thompson, *Australian Imperialism in the Pacific: The Expan-

sionist Era 1820–1920, MUP, Melbourne, 1980, p. 164; K. Buckley and K. Klugman, *The History of Burns Philp: The Australian Company in the South Pacific*, Burns Philp, Sydney, 1981, p. 259.

31 Buckley and Klugman, *The History of Burns Philp*, p. 198.

32 J.A. La Nauze, *Alfred Deakin: A Biography*, MUP, Melbourne, 1965, vol. 2, p. 489. See Meaney, *The Search for Security*, pp. 94–107.

33 Norris, *The Emergent Commonwealth*, p. 132; *CPD*, vol. 14, p. 1798 (7 July 1903).

34 Quoted in Charles Grimshaw, 'Australian Nationalism and the Imperial Connection, 1900–14', *AJPH*, 3 (1957–58), p. 177.

35 Meaney, *The Search for Security*, pp. 153–73; Ruth Megaw, 'Australia and the Great White Fleet, 1908', *JRAHS*, 56 (1970), pp. 121–33; Donald C. Gordon, 'The Admiralty and the Dominions' Navies, 1902–14', *Journal of Modern History*, 33 (1961), pp. 407–22.

36 *CPD*, vol. 32, p. 2588 (9 August 1906).

37 Quoted in Barbara Penny, 'The Australian Debate on the Boer War', *Historical Studies*, 14 (1969–71), p. 544.

38 Henry Lawson, 'Australia's Peril', in *For Australia and Other Poems*, Melbourne Standard Publishing Company, Melbourne, 1913, pp. 73–6.

39 J.E. Kendle, *The Colonial and Imperial Conferences 1887–1911*, Longmans, London, 1967, p. 225; Ian Hancock, 'The 1911 Imperial Conference', *Historical Studies*, 12 (1965–67), pp. 356–72; Meaney, *The Search for Security*, pp. 226–7.

40 Gordon Greenwood and Charles Grimshaw (eds), *Documents on Australian International Affairs 1901–18*, Nelson, Melbourne, 1977, pp. 561–7.

CHAPTER 7: WAR

1 C.E.W. Bean, *Two Men I Knew*, Angus and Robertson, Sydney, 1957, pp. 21–2, 33–4; Bill Gammage, *The Broken Years: Australian Soldiers in the Great War*, Penguin, Ringwood, 1975, pp. 6–8; E.C. Buley, *The Glorious Deeds of the Australasians in the Great War*, Melrose, London, 1915, p. 16.

2 A.B. Facey, *A Fortunate Life*, Fremantle Arts Centre Press, Fremantle, 1981, p. 240; C.E.W. Bean, *The Story of Anzac from the Outbreak of War to the End of the First Phase of the Gallipoli Campaign*, Official History of Australia in the War of 1914–18, vol. 1, Angus and Robertson, Sydney, 1921, chap. 3; Ernest Scott, *Australia During the War*, Official History, vol. 11, Angus and Robertson, Sydney, 1936, chap. 5.

3 Scott, *Australia During the War*, pp. 35–7; C.E.W. Bean, *Anzac to Amiens*, Australian War Memorial, Canberra, 1946, chaps 3, 6.

4 Colin Forster, 'Australian Manufacturing and the War of 1914–18', *Economic Record*, 29 (1953), pp. 211–30; *Commonwealth Yearbook 1901–15*, p. 1000.

5 *CPP*, 1914–17, vol. 5, p. 165; L.F. Fitzhardinge, *The Little Digger 1914–52: William Morris Hughes, a Political Biography*, Angus and Robertson, Sydney, 1979, pp. 18–26; W.S. Robinson, *If I Remember Rightly*, ed. Geoffrey Blainey, Cheshire, Melbourne, 1967, pp. 81–2; Frank Carrigan, 'The Imperial Struggle for Control in the Broken Hill Base-Metal Industry', in E.F. Wheelwright and Ken Buckley (eds), *Essays in the Political Economy of Australian Capitalism*, vol. 5, ANZ, Sydney, 1983, pp. 164–86.

6 Kevin Fewster (ed.), *Gallipoli Correspondent: The Frontline Diary of*

C.E.W. Bean, Allen and Unwin, Sydney, 1983, p. 39; Bean, *The Story of Anzac*, vol. 1, p. 130.

7 Gammage, *The Broken Years*, p. 44.

8 The day's events occupy more than two hundred pages of Bean's *Story of Anzac*, vol. 1, chaps 12–20, and they are, even by the standards of that monumental work, a remarkable feat of sustained historical narrative. See also Sir Ian Hamilton, *Gallipoli Diary*, Edward Arnold, London, 1920, vol. 1, p. 143.

9 Ellis Ashmead-Bartlett in the Melbourne *Argus*, 8 May 1915, and Kevin Fewster, 'Ellis Ashmead-Bartlett and the Making of the Anzac Legend', *JAS*, 10 (June 1982), pp. 17–30; John Masefield, *Gallipoli*, Heinemann, London, 1916, p. 19; Compton Mackenzie, *Gallipoli Memories*, Cassell, London, 1929, p. 81; Bean's images are in Fewster (ed.), *Gallipoli Correspondent*, pp. 86, 129, and there is further material in Dudley McCarthy, *Gallipoli to the Somme: The Story of C.E.W. Bean*, John Ferguson, Sydney, 1983. See generally K.S. Inglis, 'The Anzac Tradition', *Meanjin*, 24 (1965), pp. 25–44.

10 Geoffrey Serle, *John Monash: a Biography*, MUP, Melbourne, 1982, p. 240; Hamilton, *Gallipoli Diary*, vol. 2, p. 8; Fewster (ed.), *Gallipoli Correspondent*, p. 157; Gammage, *The Broken Years*, p. 97; Facey, *A Fortunate Life*, p. 260.

11 *CPD*, vol. 75, pp. 146–9 (14 October 1914), vol. 76, p. 2677 (28 April 1915); H.V. Evatt, *William Holman: Australian Labour Leader*, Angus and Robertson, Sydney, 1940, p. 273.

12 Scott, *Australia During the War*, p. 871; Michael McKernan, 'Sport, War and Society: Australia 1914–18', in Richard Cashman and Chris McKernan (eds), *Sport in History*, UQP, St Lucia, 1979, pp. 1–16; Walter Phillips, '"Six O'Clock Swill": the Introduction of early Closing of Hotel Bars in Australia', *Historical Studies*, 19 (1980–81), pp. 250–66.

13 Carmel Shute, 'Heroines and Heroes: Sexual Mythology in Australia, 1914–18', *Hecate*, 1, 1 (January 1975), pp. 7–22; Scott, *Australia During the War*, chap. 21.

14 *Round Table*, 21 (December 1915), p. 158.

15 *CPD*, vol. 79, p. 7711 (9 May 1916).

16 N.G. Butlin, *Australian Domestic Product: Investment and Foreign Borrowing 1861–1938/9*, CUP, Cambridge, 1962, p. 461; Ian McLean and Jonathan Pincus, *Living Standards in Australia 1890–1940: Evidence and Conjectures*, Working Papers in Economic History, ANU, Canberra, 1982, p. 30.

17 Scott, *Australia During the War*, chap. 14.

18 Quoted in Jenny Tilby Stock, 'South Australia's "German" Vote in World War One', *AJPH*, 28 (1982), p. 251.

19 Scott, *Australia During the War*, pp. 105–8; Michael McKernan, *The Australian People and the Great War*, Nelson, Melbourne, 1980, chap. 7; Serle, p. 204; Dan Coward, 'The Impact of War on New South Wales: Some Aspects of Social and Political History 1914–17', PhD thesis, ANU, 1974, p. 135.

20 Brian Kennedy, *Silver, Sin and Sixpenny Ale: A Social History of Broken Hill 1883–1921*, MUP, Melbourne, 1978, pp. 130–1; Anthony Splivalo, *The Home Fires*, Fremantle Arts Centre Press, Fremantle, 1980, pp. 51–4, 87.

21 *ADB*, vol. 9, pp. 253, 280, 636; 'Nomenclature Committee on Enemy Place Names', *SAPP*, 1916, vol. 3; Scott, *Australia During the War*, p. 154.

22 Bean, *Anzac to Amiens*, pp. 249, 251–2, 330–4; Suzanne Welborn, *Lords of Death*, Fremantle Arts Centre Press, Fremantle, 1982, p. 113; Scott, *Australia During the War*, p. 871.
23 W.F. Mandle argues in *Going it Alone: Australia's National Identity in the Twentieth Century*, Allen Lane, Ringwood, 1978, pp. 3–23, for the decisive significance of Pozières in fixing Australian attitudes to the war.
24 Sydney *Daily Telegraph*, 11 September 1915; Melbourne *Age*, 11 September, 23 November 1915; Latham Papers, NLA 1009, series 17.
25 Novar Papers, NLA 696/699. On Munro Ferguson see Christopher Cunneen, *Kings' Men: Australia's Governors-General from Hopetoun to Isaacs*, Allen and Unwin, Sydney, 1983, chap. 4.
26 *ADB*, vol. 8, p. 623; Sir Robert Randolph Garran, *Prosper the Commonwealth*, Angus and Robertson, Sydney, 1958, pp. 221–3; Coward, 'The Impact of War', pp. 99–100.
27 Hughes described himself thus during a visit to his old school; Scott, *Australia During the War*, p. 321.
28 Fitzhardinge, *The Little Digger*, pp. 179–91; Patrick Weller (ed.), *Caucus Minutes 1901–49: Minutes of the Meetings of the Federal Parliamentary Labor Party*, vol. 1, MUP, Melbourne, 1975, pp. 434–5; Governor-General to the Secretary of State for the Colonies, 30 August 1916, Novar Papers, NLA 696/820; E.J. Holloway, 'From Labour Council to Privy Council', NLA 2098, pp. 28–30; J.T. Lang, *I Remember*, 2nd edn, McNamara's Books, Sydney, 1980, p. 66.
29 *Australian Worker*, 11 November 1915, quoted in L.L. Robson, *Australia and the Great War*, Macmillan, Melbourne, 1969, p. 56; Ian Turner, *Industrial Labor and Politics: The Dynamics of the Labour Movement in Eastern Australia 1900–21*, ANU Press, Canberra, 1962, pp. 71–96, 252–4.
30 Adelaide Lubbock, *People in Glass Houses: Growing up at Government House*, Nelson, Melbourne, 1977, p. 110.
31 Marilyn Lake, *A Divided Society: Tasmania during World War I*, MUP, Melbourne, 1975, p. 60; G.P. Shaw, 'Patriotism versus Socialism: Queensland's Private War, 1916', *AJPH*, 19 (1973), p. 172.
32 The trade union resolution is reprinted in J.M. Main (ed.), *Conscription: The Australian Debate*, Cassell, Melbourne, 1970, pp. 37–8. The classic exposition of these attitudes is V.G. Childe, *How Labour Governs: A Study of Workers' Representation in Australia*, Labour Publishing Company, London, 1923, which can be supplemented by Ian Bedford, 'The Industrial Workers of the World in Australia', *Labour History*, 13 (November 1967), pp. 40–6, and P.J. Rushton, 'Revolutionary Ideology of the Industrial Workers of the World in Australia', *Historical Studies*, 15 (1971–73), pp. 424–46.
33 On the conscription campaign there is Leslie C. Jauncey, *The Story of Conscription in Australia*, Allen and Unwin, London, 1935, chaps 6–8 and L.L. Robson, *The First AIF: A Study of Its Recruitment 1914–18*, MUP, Melbourne, 1970, chap. 6; Gavin Souter, *Company of Heralds*, MUP, Melbourne, 1981, p. 118, quotes the agreement with the press.
34 Catholic spokesmen quoted in Michael McKernan, 'Catholics, conscription and Archbishop Mannix', *Historical Studies*, 17 (1976–77), pp. 299–314; Frank Cotton, *The Curse of Cairo*, The Worker, Sydney, 1916; Carmel Shute, 'Blood Votes and the "Bestial Boche": A Case Study in Propaganda', *Hecate*, 2, 2 (July 1976), pp. 6–22.
35 Somerville to Pearce, 20 October 1916, Battye Acc. 460; Ian Moles, *A Majority of One: Tom Aikens and Independent Politics in Townsville*,

UQP, St Lucia, 1979, p. 29.

36 Fitzhardinge, *The Little Digger*, pp. 209–13; Robson, *The First AIF*, pp. 118–20; Glenn Withers, 'The 1916–17 Conscription Referenda: a Cliometric Re-appraisal', *Historical Studies*, 20 (1982–83), pp. 36–47.

37 The effect of the conscription issue on the state Labor branches is discussed by the contributors to D.J. Murphy (ed.), *Labor in Politics: The State Labor Parties in Australia 1880–1920*, UQP, St Lucia, 1975, events in the federal caucus in Weller (ed.), *Caucus Minutes*, vol. 1, pp. 438–42, and subsequently in Fitzhardinge, *The Little Digger*, pp. 226–49; the expulsion of the conscriptionists is reported in *The Report of the Proceedings of the Special Commonwealth Conference of the Australian Labor Party*, Melbourne, 1917, pp. 4–16.

38 J. Hume Cook, 'Recollections and Reflections', NLA 601, p. 216; the bribery allegations appear in *CPD*, vol. 81, pp. 108–48 et seq. (2 March 1917).

CHAPTER 8: A NATION DIVIDED

1 Munro Ferguson quoted in L.F. Fitzhardinge, *The Little Digger 1914–52: William Morris Hughes, a Political Biography*, Angus and Robertson, Sydney, 1979, pp. 269–70.

2 Hughes in J.M. Main (ed.), *Conscription: the Australian Debate*, Cassell, Melbourne, 1970, p. 98; *Australian Worker*, 12 April 1917.

3 Ian Turner, *Sydney's Burning*, Alpha Books, Sydney, 1969 edn; Verity Burgmann, 'The Iron Heel: The Suppression of the IWW during World War I', in Sydney Labour History Group (ed.), *What Rough Beast? The State and Social Order in Australian History*, Allen and Unwin, Sydney, 1982, pp. 171–91; Frank Cain, 'The Industrial Workers of the World: Aspects of its Suppression in Australia, 1916–19', *Labour History*, 42 (May 1982), pp. 54–62. See generally, Frank Cain, *The Origins of Political Surveillance in Australia*, Angus and Robertson, Sydney, 1983, chaps 1–5.

4 Ian Turner, *Industrial Labour and Politics: The Dynamics of the Labour Movement in Eastern Australia 1900–21*, ANU Press, Canberra, 1962, chap. 6; Dan Coward, 'Crime and Punishment: the Great Strike in New South Wales, August to October 1917', in John Iremonger et al. (eds), *Strikes: Studies in Twentieth Century Australian History*, Angus and Robertson, Sydney, 1973, pp. 51–80.

5 Pankurst in *Dawn*, September 1917, and at her trial, quoted by Judith Smart, 'Feminists, Food and the Fair Prince: The Cost of Living Demonstrations in Melbourne, August–September 1917', *Labour History*, no. 50 (May 1986).

6 C.E.W. Bean, *Anzac to Amiens*, Australian War Memorial, Canberra, 1946, p. 376; Ernest Scott, *Australia During the War*, Official History, vol. 11, Angus and Robertson, Sydney, 1936, p. 872; Fitzhardinge, *The Little Digger*, p. 285.

7 Leslie C. Jauncey, *The Story of Conscription in Australia*, Allen and Unwin, London, 1935, p. 291.

8 ibid., p. 275. Michael McKernan, 'Catholics, Conscription and Archbishop Mannix', *Historical Studies*, 17 (1976–77), pp. 299–314; D.J. Murphy, 'Religion, Race and Conscription in World War I', *AJPH*, 20 (1974), pp. 135–53; Patrick O'Farrell, *The Catholic Church and Community in Australia: A History*, Nelson, Melbourne, 1977, p. 232.

9 Fitzhardinge, *The Little Digger*, pp. 289–96; D.J. Murphy, *T.J. Ryan*,

A Political Biography, UQP, St Lucia, 1975, chap. 13; Cook, 'Recollections and Reflections', NLA 601, pp. 230–5.

10 Keith Murdoch to Hughes, 24 November 1917, Murdoch Papers, NLA 2823; H.W. Allen to J.G. Latham, 21 November 1917, Latham Papers, NLA 1009/1/330.

11 Stephen Murray-Smith, 'On the Conscription Trail: the Second Referendum Seen from Beside W.M. Hughes', *Labour History*, 33 (November 1977), p. 102; Fitzhardinge, *The Little Digger*, pp. 299–306; Scott, *Australia During the War*, pp. 431–7.

12 Murray Perks, 'Labour and the Governor-General's Recruiting Conference, Melbourne, April 1918', *Labour History*, 34 (May 1978), pp. 28–44.

13 Scott, *Australia During the War*, p. 450; Bean, *Anzac to Amiens*, p. 415; Peter Firkins, *The Australians in Nine Wars*, Pan Books, Sydney, 1982 edn, pp. 146, 172; Geoffrey Serle, *John Monash*, MUP, Melbourne, 1984, p. 380.

14 A.J. Hill, *Chauvel of the Light Horse*, MUP, Melbourne, 1978, especially pp. 9, 192; H.S. Gullett, *The Australian Imperial Force in Sinai and Palestine*, Official History, vol. 7, Angus and Robertson, Sydney, 1923, pp. 787–9; Suzanne Brugger, *Australians and Egypt, 1914–19*, MUP, Melbourne, 1980, chaps 5–8.

15 Bill Gammage, *The Broken Years: Australian Soldiers in the Great War*, Penguin, Ringwood, 1975, p. 236; Gordon Greenwood and Charles Grimshaw (eds), *Documents on Australian International Affairs 1901–18*, Nelson, Melbourne, 1980, pp. 134, 143–4; A.G. Butler, *The Australian Army Medical Services in the War of 1914–18*, Official History, vol. 3, Australian War Memorial, Canberra, 1943, chap. 3.

16 Michael McKernan, 'Clergy in Khaki: the Chaplain in the Australian Imperial Force, 1914–18', *JRAHS*, 64 (1978), p. 160; Serle, *Monash*, pp. 336–7.

17 J.G. Latham to Hilda Latham, 19 May 1918, Latham Papers, NLA 1009/10/204; Fitzhardinge, *The Little Digger*, p. 396; W.J. Hudson, *Billy Hughes in Paris: The Birth of Australian Diplomacy*, Nelson, Melbourne, 1978, chaps 3–6; *CPD*, vol. 89, p. 12169 (10 Sept. 1919).

18 Fitzhardinge, *The Little Digger*, chaps 14 and 15; Hudson, *Billy Hughes in Paris*, chaps 3–5; Peter Spartarlis, *The Diplomatic Battles of Billy Hughes*, Hale and Iremonger, Sydney, 1983, chaps 7 and 8.

19 K. Buckley and K. Klugman, *'The Australian Presence in the Pacific': Burns Philp 1914–46*, Allen and Unwin, Sydney, 1983, p. 7; the principal steps in these tortuous negotiations appear in Neville Meaney (ed.), *Australia and the World: A Documentary History from the 1870s to the 1970s*, Longman Cheshire, Melbourne, 1985, chap. 5.

20 Hudson, *Billy Hughes in Paris*, chap. 2; C.D. Rowley, *The Australians in German New Guinea 1914–21*, MUP, Melbourne, 1958, p. 287.

21 *CPD*, vol. 94, p. 7266 (7 April 1921), quoted in J.R. Poynter, 'The Yo-Yo Variations: Initiative and Dependence in Australia's External Relations, 1918–23', *Historical Studies*, 14 (1969–71), p. 242.

22 Munro Ferguson to the Secretary of State for the Colonies, 22 August 1916, 27 December 1917, 25 June 1918; Novar Papers, NLA 696/987, 1055, 1066. See generally Christopher Cunneen, *Kings' Men: Australia's Governors-General from Hopetoun to Isaacs*, Allen and Unwin, Sydney, 1983, chap. 4 and McKernan, *The Australian People and the Great War*, chap. 6.

23 F.W. Eggleton to T.H. Laby, 4 December 1918; Harrison Moore Col-

lection, University of Melbourne Archives 11/6/5/14; John Kendle, *The Round Table Movement and Imperial Union*, University of Toronto Press, Toronto, 1975, pp. 261–2.

24 Greenwood and Grimshaw (eds), *Documents on Australian International Affairs*, pp. 738–9; Desmond Zwar, *In Search of Keith Murdoch*, Macmillan, London, 1980, p. 54; Fitzhardinge, *The Little Digger*, p. 359; Poynter, 'The Yo-Yo Variations', p. 232.

25 Turner, *Industrial Labour and Politics*, chap. 8; Brian Kennedy, *Silver, Sin and Sixpenny Ale: A Social History of Broken Hill 1883–1921*, MUP, Melbourne, 1978, chap. 11; Richard Morris, 'Mr Justice Higgins Scuppered: The 1919 Seamen's Strike', *Labour History*, 37 (November 1979), pp. 52–62.

26 Fitzhardinge, *The Little Digger*, pp. 443–5; Higgins, *A New Province for Law and Order*, 2nd edn, Dawsons, London, 1968, pp. 172–6. See generally Humphrey McQueen, 'Higgins and Arbitration', in E.L. Wheelwright and Ken Buckley (eds), *Essays in the Political Economy of Australian Capitalism*, vol. 5, ANZ, Sydney, 1983, pp. 145–63; and John Rickard, *H.B. Higgins: The Rebel as Judge*, Allen and Unwin, Sydney, 1984, chap. 10.

27 *The Times*, 15 June 1920.

28 Ian Bedford, 'The One Big Union, 1918–23', in Ian Bedford and Ross Curnow, *Initiative and Organization*, Cheshire, Melbourne, 1963, pp. 5–45.

29 Jensen, 'The Darwin Rebellion', *Labour History*, 11 (November 1966), pp. 3–13; Allen Powell, *Far Country: A Short History of the Northern Territory*, MUP, Melbourne, 1982, pp. 156–9.

30 Terrence Cutler, 'Sunday, Bloody Sunday: The Townsville Meatworkers' Strike of 1918–19', in Iremonger et al. (eds), *Strikes*, pp. 81–102; Murphy, *Ryan*, pp. 434–7.

31 P. Hopper, 'The 1919 Fremantle Lumpers Strike', BA thesis, UWA, 1975; B.K. de Garis, 'An Incident at Fremantle', *Labour History*, 10 (May 1966), pp. 32–7.

32 Scott, *Australia During the War*, chaps 24 and 25.

33 P. Adam Smith, 'All Those Empty Pages', p. 52; Marilyn Lake, *A Divided Society: Tasmania during World War I*, MUP, Melbourne, 1975, p. 168; Dennis R. Shoesmith, 'Boom Year: A Study of Popular Leisure in Melbourne in 1919', MA thesis, ANU, 1971, pp. 207–8.

34 F.K. Crowley, *Modern Australia in Documents*, vol. 1, Wren, Melbourne, 1973, pp. 319–21; Raymond Evans, 'Loyalty and Disloyalty: Social Conflict on the Queensland Homefront 1914–18', PhD thesis, University of Queensland, 1984, chap. 8; Humphrey McQueen, 'Shoot the Bolshevik! Hang the Profiteer! Reconstructing Australian Capitalism, 1918–21', in E.L. Wheelwright and Ken Buckley (eds), *Essays in the Political Economy of Australian Capitalism*, vol. 2, ANZ, Sydney, 1978; Terry King, 'The Tarring and Feathering of J.K. McDougall: "Dirty Tricks" in the 1919 Federal Election', *Labour History*, 45 (November 1983), pp. 54–67.

35 Rohan Rivett, *Australian Citizen: Herbert Brookes 1867–1963*, MUP, Melbourne, 1965, chap. 5; Loyalist League records in Brookes Papers, NLA 1924/21; O'Farrell, *The Catholic Church and Community in Australia*, pp. 343–4; H.M. Moran, *Viewless Winds*, Peter Davis, London, 1939; pp. 158–9.

36 Humphrey McQueen, 'The "Spanish" Influenza Pandemic in Australia, 1918–1919', in Jill Roe (ed.), *Social Policy in Australia: Some Perspectives 1901–75*, Cassell, Stanmore, 1976, pp. 131–47; O'Farrell, *The*

Catholic Church and Community in Australia, p. 339.

37 Munro Ferguson quoted in Fitzhardinge, *The Little Digger*, p. 421.

38 G.L. Kristianson, *The Politics of Patriotism: The Pressure Group Activities of the Returned Servicemen's League*, ANU Press, Canberra, 1966, chaps 1 and 2; Mary Wilson, 'The Making of Melbourne's Anzac Day', *AJPH*, 20 (1974), pp. 197–209.

39 Kevin Fewster, 'Politics, Pageantry and Purpose: the 1920 Tour of Australia by the Prince of Wales', *Labour History*, 38 (May 1980), pp. 59–66.

40 Gerard Henderson, 'The Deportation of Charles Jerger', *Labour History*, 31 (November 1976), pp. 61–78; Peter L'Estrange, 'Alien and Alienated: the Case of Charles Jerger, 1916–20', BA thesis, University of Melbourne, 1974.

41 *CPD*, vol. 88, p. 10106 (26 June 1919); vol. 94, p. 6385 (11 November 1920).

42 Randolph Bedford, *Nought to Thirty-three*, 2nd edn, MUP, Melbourne, 1976, p. 236.

43 Fitzhardinge, *The Little Digger*, pp. 428–9.

44 Quoted in Baiba Berzins, 'Symbolic Legislation: the Nationalists and Anti-Profiteering in 1919', *Politics*, 6 (1971), p. 49. See further Humphrey McQueen, 'Shoot the Bolshevik! Hang the Profiteer!'

45 S.J. Butlin, *Australia and New Zealand Bank: The Bank of Australasia and the Bank of Australia Limited, 1928–1951*, Longmans, London, 1961, pp. 366–7.

46 *Census of the Commonwealth of Australia*, 1921, pp. 39–40; Helen Hughes, 'Federalism and Industrial Development in Australia', *AJPH*, 10 (1964), pp. 323–40; Helen Hughes, *The Australian Iron and Steel Industry, 1848–1962*, MUP, Melbourne, 1964, chap. 4.

47 B.D. Graham, *The Formation of the Australian Country Parties*, ANU Press, Canberra, 1966, chaps 4 and 5.

48 *CPP*, 1920–21, vol. 4, pp. 529–645; Turner, *Industrial Labour and Politics*, pp. 211–12, 253–4. See Bettina Cass, 'Population Policies and Family Policies: State Construction of Domestic Life', in Cora Baldock and Bettina Cass (eds), *Women, Social Policy and the State*, Allen and Unwin, Sydney, 1983, pp. 164–85; Geoffrey Sawer, *Australian Federal Politics and Law 1901–29*, MUP, Melbourne, 1956, pp. 216–17.

49 *Labour Reports*, 7–12 (1916–21),

50 Turner, *Industrial Labour and Politics*, chap. 9.

51 Graham, *Formation of the Australian Country Parties*, chap. 5.

52 Fitzhardinge, *The Little Digger*, chap. 19; L.F. Crisp, 'New Light on the Trials and Tribulations of W.M. Hughes, 1920–22', *Historical Studies*, 10 (1961–63), pp. 86–91; B.D. Graham, 'The Place of Finance Committees in Non-Labor Politics, 1901–30', *AJPH*, 6 (1971), pp. 41–52; Bernie Schedvin, 'E.G. Theodore and the London Pastoral Lobby', *Politics*, 6 (1971), p. 27.

53 Ken Turner, 'From Liberal to National in New South Wales', *AJPH*, 10 (1964), pp. 205–20; H.V. Evatt, *William Holman: Australian Labour Leader*, Angus and Robertson, Sydney, 1940, chaps 52–55.

CHAPTER 9: AUSTRALIA UNLIMITED

1 E.J. Brady, *Australia Unlimited*, Robertson and Mullens, Melbourne, 1918, pp. 101, 628, 630. Some of the more optimistic estimates of future population are given in G.L. Wood and P.D. Phillips (eds), *The*

Peopling of Australia, Macmillan, Melbourne, 1928, chap. 1.

2 J.W. Gregory, *The Dead Heart of Australia*, John Murray, London, 1909, pp. 275–8; Griffith Taylor, *Australia in its Physiographic and Economic Aspects*, 3rd edn, Clarendon Press, Oxford, 1919; 'Geography and Australian National Problems', *Proceedings of the Sixteenth Meeting of the Australasian Association for the Advancement of Science*, Government Printer, Wellington, 1924, pp. 466, 481; and *Australia*, Methuen, London, 1940, p. 100; *CPD*, vol. 107, pp. 2592–3 (30 July 1924); Somerville correspondence, Battye Library, Acc. 460A. See generally J.M. Powell, 'Taylor, Stefansson and the Arid Centre', *JRAHS*, 66 (1980–81), pp. 163–83.

3 *WAPD*, vol. 70, p. 64 (30 July 1924), quoted in Sean Glynn, *Government Policy and Agricultural Development*, UWA Press, Nedlands, 1975, p. 4; Bruce in *BPP*, cmd 2009 of 1924, p. 73.

4 Lado T. Ruzicka and John Caldwell, *The End of Demographic Transition in Australia*, ANU Press, Canberra, 1977, p. 26; D. Pope, 'Contours of Australian Immigration 1901–30', *AEHR*, 21 (1981), pp. 29–52.

5 R.C. Mills, 'Australian Loan Policy', in Persia Campbell et al. (eds), *Studies in Australian Affairs*, Macmillan, Melbourne, 1928, pp. 95–117.

6 Bruce quoted in W.H. Richmond, 'S.M. Bruce and Australian Economic Policy, 1923–29', *AEHR*, 23 (1983), p. 241.

7 *BPP*, cmd 2009 of 1924, p. 58; *Argus*, 9 April 1925.

8 Conference of Commonwealth and State Ministers, Melbourne, 1921, quoted in Colin Forster, *Industrial Development in Australia 1920–30*, ANU Press, Canberra, 1964, p. 171.

9 N.G. Butlin, *Australian Domestic Product, Investment and Foreign Borrowing 1861–1938/9*, CUP, Cambridge, 1962, pp. 460–1.

10 M. Keating, *The Australian Workforce 1910–11 to 1960–61*, Department of Economic History, RSSS, ANU, Canberra, 1973, pp. 356–7.

11 Craig Munro, *Wild Man of Letters: The Story of P.R. Stephensen*, MUP, Melbourne, 1984, p. 32. See generally Ian M. Drummond, *British Economic Policy and the Empire 1919–39*, Allen and Unwin, London, 1972, chaps 1 and 2.

12 W.K. Hancock, *Survey of British Commonwealth Affairs*, vol. 2, part 1, OUP, London, 1940, pp. 132–53; F.A. Bland, 'Development and Migration', in Campbell et al. (eds), *Studies in Australian Affairs*, pp. 49–77; Glynn, *Government Policy and Agricultural Development*, p. 130.

13 A.J. Reitsma, *Trade Protection in Australia*, UQP, St Lucia, 1960, pp. 44–51; Neville Cain, 'Trade and Economic Structure at the Periphery', in C. Forster (ed.), *Australian Economic Development in the Twentieth Century*, Allen and Unwin, Sydney, 1970, p. 77; *Commonwealth Yearbooks*.

14 The wartime rumour is reported by Cecil Edwards, *Bruce of Melbourne: Man of Two Worlds*, Heinemann, London, 1965, p. 86; W.J. Hudson and Jane North (eds), *My Dear PM: R.G. Casey's Letters to S.M. Bruce 1924–29*, AGPS, Canberra, 1980, p. 333. With similar insight, the Colonial Office advised the press of another governor-designate that 'a reference to his cricket would probably go down well in Australia'; Christopher Cunneen, *Kings' Men: Australia's Governors-General from Hopetoun to Isaacs*, Allen and Unwin, Sydney, 1983, p. 153.

15 W.K. Hancock, *Survey of British Commonwealth Affairs*, vol. 1, OUP, London, 1937, p. 52.

16 Hancock, *Survey*, vol. 1, chap. 5; T.B. Millar, *Australia in Peace and War: External Relations 1788–1977*, ANU Press, Canberra, 1978, chap. 4.
17 *Argus*, 11 February 1927; *Age*, 28–29 March 1928; Peter M. Sales, 'White Australia, Black Americans: A Melbourne Incident, 1928', *AQ*, 46, 4 (December 1974), pp. 74–81.
18 Ruth Megaw, 'Undiplomatic Channels: Australian Representation in the United States, 1918–39; *Historical Studies*, 15 (1971–73), pp. 610–30. See generally Richard White, '"Americanization" and Popular Culture in Australia', *Teaching History*, 12, 2 (August 1978), pp. 3–21, and L.G. Churchward, *Australia and America 1788–1972: An Alternative History*, APCOL, Sydney, 1978, chap. 5.
19 Nancy Keesing (ed.), *The Autobiography of John Shaw Neilson*, NLA, Canberra, 1978, p. 132; Maurice Levine, *Cheetham to Cordova*, Neil Richardson, Manchester, 1984, p. 17.
20 'Report of the Royal Commission on Social and Economic Effect of the Increase in Number of Aliens in North Queensland', *QPP*, 1925, vol. 3; Ross Fitzgerald, *From 1915 to the Early 1980s: A History of Queensland*, UQP, St Lucia, 1984, pp. 68–70.
21 B.R. Davidson, *European Farming in Australia: An Economic History of Australian Farming*, Elsevier, Amsterdam, 1981, p. 266 for prices. In chapter 13 he presents a set of calculations of expenditure and income for different types of farming at different scales of operation.
22 John McEwen, 'Autobiography', *RHSVJ*, 55, 2 (June 1984), p. 27. See generally, Marilyn Lake, 'The Limits of Hope: Soldier Settlement in Victoria, 1915–38', PhD thesis, Monash University, 1984.
23 Edward Shann, 'Group Settlement of Migrants in Western Australia', *Economic Record*, 1 (1925), pp. 73–93; I.L. Hunt, 'Group Settlement in Western Australia: A Criticism', *University Studies in Western Australian History*, 3, 2 (October 1958), pp. 5–42. The settler is quoted in G.C. Bolton, *A Fine Country to Starve In*, UWA Press, Nedlands, 1972, pp. 40–1.
24 Hunt, 'Group Settlement', p. 26; 'Report of Inquiry into Losses Due to Soldier Settlement', *CPP*, 1929, vol. 2, pp. 1901–59.
25 *Commonwealth Yearbooks*.
26 Bolton, *A Fine Country*, p. 43; the Victorian woman is quoted in Marilyn Lake, 'Helpmeet, Slave, Housewife: Women in Rural Families, 1870–1930', in Patricia Grimshaw et al. (eds), *Families in Colonial Australia*, Allen and Unwin, Sydney, 1985, pp. 173–85; Tasmanian man quoted in Quentin Beresford, 'The World War One Settlement Scheme in Tasmania', *THRAPP*, 30 (1983), p. 93.
27 Forster, *Industrial Development*; Helen Hughes, *The Australian Iron and Steel Industry, 1848–1962*, MUP, Melbourne, 1964.
28 Quotation from Forster, *Industrial Development*, p. 17; the incidence of the tariff is noted in W.M. Corden, 'The Tariff', in Alex Hunter (ed.), *The Economics of Australian Industry*, MUP, Melbourne, 1963, p. 186.
29 Stuart Rosewarne, 'Capital Accumulation in Australia and the Export of Capital before World War II', in E.L. Wheelwright and Ken Buckley (eds), *Essays in the Political Economics of Australian Capitalism*, vol. 5, ANZ, Sydney, 1982, pp. 187–218.
30 Peter Cochrane, *Industrialization and Dependence: Australia's Road to Economic Development*, UQP, St Lucia, 1980, chaps 5 and 6; R.W. Connell and T.H. Irving, *Class Structure in Australian History: Docu-*

ments, Narrative and Argument, Longman Cheshire, Melbourne, 1980, pp. 270–4; C.P. Haddon-Cave, 'Trends in the Concentration of Operations of Australian Secondary Industries, 1923–43', *Economic Record*, 21 (1945), pp. 65–78.

31 Quoted in Forster, *Industrial Development*, p. 218; see further J.R. Poynter, *Russell Grimwade*, MUP, Melbourne, 1967, pp. 117–19.

32 Forster, *Industrial Development*, pp. 39–43.

33 Jim Hagan, *The History of the ACTU*, Longman Cheshire, Melbourne, 1981, p. 33; Peter Poynter, 'The Development of the Assembly Line in Australia', *Arena*, 58 (1981), p. 74.

34 C.D. Kemp, *Big Businessmen: Four Biographical Essays*, Institute of Public Affairs, Melbourne, 1964, pp. 13–84 and Gepp Papers, NLA, 2052; Geoffrey Blainey, *The Steel Master: A Life of Essington Lewis*, Macmillan, Melbourne, 1971; *ADB*, vol. 8, pp. 1–3, and *Railways Union Gazette*, 24 August 1927 (I am grateful to Alison Churchward for this reference); R.I. Jennings, *W.A. Webb: South Australian Railways Commissioner*, Nesfield Press, North Plympton, South Australia, 1973, and *SAPD*, 1926, vol. 1, p. 256 (4 August 1926).

35 Raelene Frances, '"No more Amazons": Gender and Work Process in the Victorian Clothing Trades, 1890–1939', *Labour History*, 50 (May 1986); Janet McCalman, *Struggletown: Public and Private Life in Richmond 1900–65*, MUP, Melbourne, 1984, pp. 122–3.

36 Hagan, *The History of the ACTU*, p. 31; W.S. Robinson to W.L. Baillieu, quoted in Frank Strahan, 'The Documental Dropout', *Melbourne Historical Journal*, 9 (1970), p. 50. See generally Meredith Atkinson, *The New Social Order: A Study of Post-War Reconstruction*, WEA, Sydney, 1919; Tim Rowse, *Australian Liberalism and National Character*, Kibble Books, Malmsbury, 1978, chap. 2, and Helen Bourke, 'Industrial Unrest as Social Pathology: the Australian Writings of Elton Mayo', *Historical Studies*, 20 (1982–83), pp. 217–33.

37 Harold Hunt, *The Story of Rotary in Australia 1921–71*, Rotary, Sydney, 1971, chaps 2 and 3; *Round Table*, 47 (June 1922), pp. 693–702.

38 Aspects of the periods' lifestyles are treated in Robin Boyd, *Australia's Home*, 2nd end, Penguin, Ringwood, 1968, chap. 7; Peter Spearritt, *Sydney Since the Twenties*, Hale and Iremonger, Sydney, 1978, chap. 3; Michael Symons, *One Continuous Picnic: A History of Eating in Australia*, Duck Press, Uraidla, South Australia, 1982, chap. 9. Robert Murray, *The Confident Years: Australia in the Twenties*, Allen Lane, Ringwood, 1978, chap. 7 gives a rosy general survey.

39 Arthur O. Richardson, *The New Era in Advertising*, Alan Norman, Sydney, 1927, p. 39.

40 Diane Collins, 'The Movie Octopus', in Peter Spearritt and David Walker (eds), *Australian Popular Culture*, Allen and Unwin, Sydney, 1979, pp. 102–20; Diane Collins, 'The 1920s Picture Palace', in Susan Dermody et al. (eds), *Nellie Melba, Ginger Meggs and Friends: Essays in Australian Cultural History*, Kibble Books, Malmsbury, 1982, pp. 60–75; Ruth Megaw, 'The American Image and Australian Cinema Management', *JRAHS*, 54 (1968–69), pp. 194–204; M.S. Counihan, 'The Construction of Australian Broadcasting: Aspects of Radio in the 1920s', MA thesis, Monash University, 1981.

41 Leslie Johnson, 'Wireless and Women', *Melbourne Working Papers*, University of Melbourne, Faculty of Education, 5 (1984), pp. 74–94; Jenni Cook, 'Homework as Scientific Management', *Time Remembered*, 5 (1982), pp. 84–102. See generally Kerreen Reiger, *The Disen-*

chantment of the Home: Modernizing the Australian Family 1880–1914, OUP, Melbourne, 1985, and Jill Julius Matthews, *Good and Mad Women: The Historical Construction of Femininity in Twentieth Century Australia*, Allen and Unwin, Sydney, 1984, chaps 4 and 5.

42 Beverley Kingston, *My Wife, My Daughter and Poor Mary-Ann: Women and Work in Australia*, Nelson, Melbourne, 1975, chap. 3.

43 Ruzicka and Caldwell, *The End of Demographic Transition*, pp. 81, 103.

44 Spearritt, *Sydney Since the Twenties*, pp. 30–1; F.J. Garlick, 'Melbourne Suburban Expansion in the 1920s', MA thesis, University of Melbourne, 1983; L.G. Hovenden, 'The Motor Car in New South Wales, 1900–37', MA thesis, University of Sydney, 1981.

45 Royal commission on health, transcript of evidence, quoted in Humphrey McQueen, *Social Sketches of Australia, 1888–1975*, Penguin, Ringwood, 1978, p. 110.

46 Wendy Lowenstein, *Weevils in the Flour: An Oral Record of the 1930s Depression in Australia*, Hyland House, Melbourne, 1978, p. 67.

CHAPTER 10: HOLDING THE CENTRE

1 *Argus*, 24 April 1918, and *Australian National Review*, 14 May 1923, quoted in David Potts, 'A Study of Three Nationalists in the Bruce-Page Government of 1923–29', MA thesis, University of Melbourne, 1972, pp. 49, 70.

2 Cecil Edwards, *Bruce of Melbourne: Man of Two Worlds*, Heinemann, London, 1965, pp. 82, 133; *ADB*, vol. 7, pp. 453–60.

3 *CPD*, vol. 114, p. 4801 (3 August 1926).

4 J.R. Williams, 'The Organisation of the Australian National Party', *AQ*, 41, 2, (June 1969) pp. 41–51; Trevor Matthews, 'Business Associations and Politics: Chambers of Manufacturers and Employers' Federations in NSW, Victoria and Commonwealth Politics to 1939', PhD thesis, University of Sydney, 1971, chap. 5; B.D. Graham, 'The Place of Finance Committees in Non-Labor Politics, 1901–30', *AJPH*, 6 (1971), pp. 44–5. The quotation comes from J. Hume Cook, 'Recollections and Reflections', NLA 601, p. 274, and his papers include material on the grievances of the federation.

5 Edwards, *Bruce of Melbourne*, p. 82; Earle Page, *Truant Surgeon: The Inside Story of Forty Years of Australian Political Life*, Angus and Robertson, Sydney, 1963, chaps 12 and 13; B.D. Graham, *The Formation of the Australian Country Parties*, ANU Press, Canberra, 1966, especially pp. 159, 231.

6 *Argus*, 9 April 1925, quoted in F.K. Crowley (ed.), *Modern Australia in Documents*, vol. 1, Wren, Melbourne, 1973, pp. 401–2.

7 Keith Trace, 'Australian Overseas Shipping, 1900–60', PhD thesis, University of Melbourne, 1965, pp. 65–8. See generally, Kevin Burley, *British Shipping and Australia 1900–39*, CUP, Cambridge, 1968.

8 L.F. Giblin, *The Growth of a Central Bank*, MUP, Melbourne, 1951, pp. 14–19; Gepp quoted in C.D. Kemp, *Big Businessmen: Four Biographical Essays*, Institute of Public Affairs, Melbourne, 1964, p. 31.

9 Sir George Currie and John Graham, *The Origins of the CSIRO: Science and the Commonwealth Government, 1901–26*, CSIRO, Melbourne, 1966.

10 Warren Denning, *Caucus Crisis: The Rise and Fall of the Scullin Government*, Cumberland Argus, Parramatta, 1937, p. 43; Frank Anstey, 'The Viceroy', *Overland*, 32 (August 1965), p. 20.

11 Bruce in *Argus*, 8 September 1925; Peter Coleman, *Obscenity, Blasphemy, Sedition: 100 Years of Censorship in Australia*, 2nd edn, Angus and Robertson, Sydney, 1974, chap. 5.
12 Richard White, *Inventing Australia: Images and Identity 1688–1980*, Allen and Unwin, Sydney 1981, chap. 9; Humphrey McQueen, *The Black Swan of Trespass: The Emergence of Modernist Painting in Australia to 1944*, APCOL, Sydney, 1979, pp. 5, 114.
13 Andrew Moore, 'Guns Across the Yarra, Secret Armies and the 1923 Melbourne Police Strike', in Sydney Labour History Group (ed.), *What Rough Beast? The State and Social Order in Australian History*, Allen and Unwin, Sydney, 1982, pp. 220–33; Keith Richmond, 'Reactions to Radicalism: Non-Labour Movements 1920–79', *JAS*, 5 (November 1979), pp. 50–63.
14 Graham, *The Formation of the Australian Country Parties*, p. 232; *Sydney Morning Herald*, 6 October 1925, quoted in Heather Radi, '1920–29', in Frank Crowley (ed.), *A New History of Australia*, Heinemann, Melbourne, 1974, p. 399. See generally David Carment, 'Sir Littleton Groom and the Deportation Crisis of 1925: A Study of Non-Labor Response to Trade Union Militancy', *Labour History* 32 (May 1977), pp. 46–54.
15 *CPD*, vol. 112, pp. 459, 466 (28 January 1926).
16 The contrast is discussed by Joan Rydon, 'The Conservative Electoral Ascendancy between the Wars', in Cameron Hazlehurst (ed.), *Australian Conservatism: Essays on Twentieth Century Political History*, ANU Press, Canberra, 1979, pp. 51–70.
17 A. Barnard and N.G. Butlin, 'Australian Public and Private Capital Formation, 1901–1975', *Economic Record*, 57 (1981), pp. 354–67.
18 Fran Jelley, 'Child Endowment', in Heather Radi and Peter Spearritt (eds), *Jack Lang*, Hale and Iremonger, Sydney, 1977, pp. 88–98.
19 Wendy Brady, '"Serfs of the Sodden Scone": Women Workers in the Western Australian Hotel and Catering Industry', *SWAH*, 7 (December 1983), pp. 33–45.
20 *Royal Commission on National Insurance. Minutes of Evidence: Unemployment, Destitution Allowances*, Government Printer, Melbourne, 1927, pp. 624, 771. See generally Stuart Macintyre, *Winners and Losers: The Pursuit of Social Justice in Australian History*, Allen and Unwin, Sydney, 1985, chap. 4.
21 Bavin to J.G. Latham, 23 June 1925, Latham Papers, NLA 1009/1/1435.
22 *Pastoral Review*, 29 (1919), p. 417; *Brisbane Courier*, 4 October 1920; Irwin Young, *Theodore: His Life and Times*, Alpha Books, Sydney, 1971, pp. 34–40. See generally Bernie Schedvin, 'E.G. Theodore and the London Pastoral Lobby', Politics, 6 (1971), pp. 26–41.
23 Lyons in Michael Denholm, 'The Lyons Tasmanian Labor Government', *THRAPP*, 24 (1977), p. 54; McCormack in E.M. Higgins, 'The Queensland Labor Governments, 1915–29', MA thesis, University of Melbourne, 1954, p. 118.
24 James Hagan, 'Lang and the Unions', in Radi and Spearritt (eds), *Jack Lang*, pp. 38–48.
25 McMinn, *A Constitutional History of Australia*, OUP, Melbourne, 1979, pp. 144–52.
26 *VPD*, vol. 185, p. 1394 (21 July 1931); 'Report of the Royal Commission following Statements in the Press', *CPP*, 1926–28, vol. 4, pp. 1235–45. See Generally Chris McConville, 'John Wren: Machine

Boss, Irish Chieftain or Meddling Millionaire?', *Labour History*, 40 (May 1981), pp. 49–67, and K.H. Kennedy, *The Mungana Affair: State Mining and Political Corruption in the 1920s*; UQP, St Lucia, 1978.

27 D.W. Rawson, 'The Organization of the Australian Labor Party, 1916–41', PhD thesis, University of Melbourne, 1954, chap. 2.

28 There is no adequate study of this major union. Aspects of the developments discussed here are treated in Ross Fitzgerald, *From 1915 to the Early 1980s: A History of Queensland*, UQP, St Lucia, 1984, especially pp. 3–7; Don Hopgood, 'The View from Head Office: The South Australian Political Machine, 1917–30', *Politics*, 6 (1971), pp. 70–8; Stuart Macintyre, *Militant: The Life and Times of Paddy Troy*, Allen and Unwin, Sydney, 1984, chap. 3.

29 Jim Hagan, *The History of the ACTU*, Longman Cheshire, Melbourne, 1981, chap. 1.

30 G.G. Headford, 'The Australian Loan Council—its Origin, Operation and Significance in the Federal Structure', in Colin A. Hughes (ed.), *Readings in Australian Government*, UQP, St Lucia, 1968, pp. 271–82. Cf. R.J. May, *Financing the Small States in Australian Federation*, OUP, Melbourne, 1971, chap. 1.

31 Aaron Wildavsky and Dagmar Cărboch, *Studies in Australian Politics*, Cheshire, Melbourne, 1958, part I; the quotation is from p. 34.

32 Quoted in Edwards, *Bruce of Melbourne*, p. 89.

33 May, *Financing the Small States*, chap. 1.

34 'Tariff Board Annual Report, 1926–27', *CPP*, 1926–28, vol. 4, p. 23; Judge Lukin to J.G. Latham, 1 September 1929, Latham Papers, NLA 1009/39/161.

35 W.H. Parsons MHR, letter to *Age*, 10 December 1927, quoted in Matthews, 'Business Associations and Politics', chap. 5.

36 Douglas Copland, 'The Economic Society—Its Origins and Constitution', *Economic Record*, 1 (1925), p. 142. The role of the profession is discussed in Crauford D.W. Goodwin, *Economic Enquiry in Australia*, Duke University Press, Durham, N.C., 1966, chap. 15; C.B. Schedvin, *Australia and the Great Depression: A Study of Economic Development and Policy in the 1920s and 1930s*, SUP, Sydney, 1970, pp. 218–25; Tim Rowse, *Australian Liberalism and National Character*, Kibble Books, Malmsbury, 1978, pp. 96–104.

37 Douglas Copland (ed.), *The Scholar and the Man*, Cheshire, Melbourne, 1960; Giblin Papers, NLA 366.

38 Herbert Heaton in *Economic Record*, 1 (1925), p. 100; L.F. Giblin, 'Some Costs of Marketing Control', Marketing Supplement, *Economic Record*, 4 (1928), p. 154; J.B. Bridgen et al., *The Australian Tariff*, MUP, Melbourne, 1929, p. 99. The economic debates have been discussed in a series of articles by Neville Cain; particularly relevant is his 'Political Economy and the Tariff: Australia in the 1920s', *Australian Economic Papers*, 12 (1973), pp. 1–20.

39 Shann to C.A.S. Hawker, undated [1929], Hawker Papers, NLA 4848, Series 2, Sub-series IV.

40 G.L. Wood, 'The Immigration Problem in Australia', *Economic Record*, 2 (1926), p. 239; D.B. Copland, 'The Trade Depression in Australia in Relation to Economic Thought', *Proceedings of the Sixteenth Meeting of the Australasian Association for the Advancement of Science*, p. 555.

41 W.K. Hancock, *Australia*, Ernest Benn, London, 1930, pp. 54, 72; see also Rowse, *Australian Liberalism*, chap. 3.

42 Lilias Needham, *Charles Hawker: Soldier-Pastoralist-Statesman*, Griffin Press, Adelaide, 1969; Cook, 'Recollections and Reflections', pp. 374–5; Eggleston set down his misgivings in *State Socialism in Victoria*, P.S. King, London, 1932.
43 Frederick Howard, *Kent Hughes*, Macmillan, Melbourne, 1972, p. 48.
44 G.L. Wood, *Borrowing and Business in Australia*, OUP, London, 1930, p. 208; S.R. Cooke and E.H. Davenport, *Australian Finance*, Pelican Press, London, 1926, pp.3–4, 8.
45 W.J. Hudson and Jane North (eds), *My Dear P.M.: R.G. Casey's Letters to S.M. Bruce, 1924–29*, AGPS, Canberra, 1980, p. 380.
46 'Report of the British Economic Mission to Australia', *CPP*, 1929, vol. 2, pp. 1231–72.
47 Quoted in W.H. Richmond, 'S.M. Bruce and Australian Economic Policy, 1923–29', *AEHR*, 23 (1983), p. 249. See also Bruce, *The Financial and Economic Position of Australia*, Joseph Fisher Lecture, Adelaide, 1927.
48 Hawker to Latham, 22 January 1928, Latham Papers, NLA 1009/28/499.
49 *Labour Reports*, 13–22 (1922–31).
50 Beeby to Latham, 25 October 1928, Latham Papers, NLA 1009/1/1835; Miriam Rechter, 'The Strike of Waterside Workers in Australian Ports, 1928, and the Lockout of Coalminers in the Northern Coalfield of New South Wales', MA thesis, University of Melbourne, 1957, chap. 1; Graeme T. Powell, 'The role of the Commonwealth Government in Industrial Relations, 1923–29', MA thesis, ANU, 1974, chap. 5.
51 Rechter, 'The Strike', chaps 2 and 3; R.N. Wait, 'Reactions to Demonstrations and Riots in Adelaide, 1928 to 1932', MA thesis, University of Adelaide, 1973, chap. 1; Wendy Lowenstein and Tom Hills, *Under the Hook: Melbourne Waterside Workers Remember: 1900–80*, Melbourne Bookworkers, Melbourne, 1982, pp. 57–66. Attorney-General's Maritime File, Latham Papers, NLA 1009/36.
52 *Argus*, 17 January 1929.
53 Alastair Davidson, *The Communist Party of Australia: A Short History*, Hoover Institution Press, Stanford, California, 1969, chap. 3.
54 *Sydney Morning Herald*, 28 March 1929.
55 Miriam Dixson, 'The Timber Strike of 1929', *Historical Studies*, 10 (1961–63), pp. 479–92.
56 F.R.E. Mauldon, *The Economics of Australian Coal*, MUP, Melbourne, 1929.
57 Robin Gollan, *The Coalminers of New South Wales*, MUP, Melbourne, 1963, chap. 9; Edgar Ross, *A History of the Miners' Federation of Australia*, Australasian Coal and Shale Employees' Federation, Sydney, 1970, chap. 14; Miriam Dixson, 'Stubborn Resistance: the Northern New South Wales Miners' Lockout of 1929–30' in John Iremonger et al. (eds), *Strikes: Studies in Twentieth Century Australian History*, Angus and Robertson, Sydney, 1973, pp. 128–42.
58 *ADB*, vol. 7, pp. 441–3.
59 *CPD*, vol. 121, p. 51 (15 August 1929); Sydney *Sun*, 14 August 1929.
60 *Australasian Manufacturer*, 20 April 1928; see generally Peter Cochrane, 'Dissident Capitalists: National Manufacturers in Conservative Politics, 1917–34', in E.L. Wheelwright and Ken Buckley (eds), *Essays in the Political Economy of Australian Capitalism*, vol. 4, ANZ, Sydney, 1980, pp. 122–47.

61 Letters from Bruce to Sir Archdale Parkhill et al., 25 May 1929, Latham Papers, NLA 1009/39/44 et seq.; Hawker to J.G. Duncan Hughes, 1 June 1929, NLA 4847, Box 7, Folder 3.
62 Denning, *Caucus Crisis*, pp. 45–7; see generally Wildavsky and Carboch, *Studies in Australian Politics*, part II.

CHAPTER 11: A CHANCE OF 'A BAD SMASH'

1 Nancy Keesing (ed.), *The Autobiography of John Shaw Neilson*, NLA, Canberra, 1978, pp. 142–6.
2 J.C. Docherty, *Newcastle: The Making of an Australian City*, Hale and Iremonger, Sydney, 1983, p. 72; Wendy Lowenstein, *Weevils in the Flour: An Oral Record of the 1930s Depression in Australia*, Hyland House, Melbourne, 1978, pp. 47–8.
3 *J.B. Were's Weekly Letter*, 31 May 1929, quoted in *The House of Were 1839–1954*, J.B. Were, Melbourne, 1954, p. 262; J.B. Bridgen, *Escape to Prosperity*, Macmillan, Melbourne, 1930, p. 80; *The Australian Traveller*, 11 May 1931.
4 *CPD*, vol. 155, p. 480 (8 December 1937); John Robertson, *J.H. Scullin: A Political Biography*, UWA Press, Nedlands, 1974, p. 185.
5 The best account is C.B. Schedvin, *Australia and the Great Depression: A Study of Economic Development and Policy in the 1920s and 1930s*, SUP, Sydney, 1970. Statistics are taken from N.G. Butlin (ed.), *Selected Comparative Economic Statistics 1900–40: Australia and Britain, Canada, Japan, New Zealand and USA*, Source Papers in Economic History, ANU, Canberra, 1984, and G.L. Wood, *Borrowings and Business in Australia*, OUP, London, 1930, pp. 258–9.
6 Butlin, *Selected Comparative Economic Statistics*, pp. 10–11. Unemployment figures are based on returns furnished by unions and thus do not cover the entire workforce, but their general reliability is suggested by the 1933 census.
7 N.G. Butlin, *Australian Domestic Product, Investment and Foreign Borrowing 1861–1938/9*, CUP, Cambridge, 1962, pp. 460–1.
8 A comparison is suggested in Schedvin, *Australia and the Great Depression*, pp. 43–6.
9 M. Keating, *The Australian Workforce 1910–11 to 1960–61*, Department of Economic History, RSSS, ANU, Canberra, 1973, pp. 356–7.
10 Warren Denning, *Caucus Crisis: The Rise and Fall of the Scullin Government*, Cumberland Argus, Parramatta, 1937, p 62; Peter Cook, 'The Scullin Government, 1929–32', PhD thesis, ANU, 1971, p. 76.
11 Robertson, *J.H. Scullin*, pp. 180–8, 356; Cook, 'The Scullin Government', pp. 133–71.
12 Peter Cook, 'Frank Anstey: Memoirs of the Scullin Labor Government 1929–32', *Historical Studies*, 18 (1978–79), pp. 365–92. The manuscript is in NLA, 4636. L.F. Giblin, *The Growth of a Central Bank*, MUP, Melbourne, 1951, pp. 111–15; R.F. Holder, *Bank of New South Wales: A History*, vol. 2, Angus and Robertson, Sydney, 1970, pp. 650–4; Schedvin, *Australia and the Great Depression*, pp. 172–6.
13 Peter Love, *Labour and the Money Power: Australian Labour Populism 1890–1950*, MUP, Melbourne, 1984, chap. 5. The inconsistencies and inadequacies of Labor doctrine were trenchantly criticized by Lloyd Ross, 'Australian Labour and the Crisis', *Economic Record*, 8 (1932), pp. 204–22, reprinted in L.J. Louis and Ian Turner (eds), *The Depression of the 1930s*, Cassell, Melbourne, 1968, pp. 130–45.

14 Schedvin, *Australia and the Great Depression*, chap. 7; Love, *Labour and the Money Power*, chap. 5; Robertson, *J.H. Scullin*, p. 248; Bruce to Latham, 15 July 1930, Latham Papers, NLA 1009/1/1981.
15 Bethia Foott, *Dismissal of a Premier: The Philip Game Papers*, Morgan, Sydney, 1968, p. 19.
16 Peter Love, 'Niemeyer's Australian Diary', *Historical Studies*, 20 (1982–83), p. 268; Makin quoted in John Lonie, '"Good Labor Men": The Hill Government in South Australia, 1930–33', *Labour History*, 31 (November 1976), p. 20. See generally W.F. Mandle, *Going It Alone: Australia's National Identity in the Twentieth Century*, Allen Lane, Ringwood, 1978, chap. 4.
17 E.O.G. Shann and D.B. Copland (eds), *The Crisis in Australian Finance 1929 to 1931: Documents on Budgetary and Economic Policy*, Angus and Robertson, Sydney, 1931, pp. 18–29; Schedvin, *Australia and the Great Depression*, pp. 181–4; Giblin, *The Growth of a Central Bank*, p. 84.
18 Shann and Copland (eds), *The Crisis in Australian Finance*, pp. 29–34.
19 Love, 'Niemeyer's Australian Diary', p. 273.
20 L.F. Crisp, 'The Appointment of Sir Isaac Isaacs as Governor-General of Australia, 1930', *Historical Studies*, 11 (1963–65), pp. 253–7; Christopher Cunneen, *Kings' Men: Australia's Governors-General from Hopetoun to Isaacs*, Allen and Unwin, Sydney, 1983, pp. 173–82. Lord Stonehaven to Latham, 5 October 1930, Latham Papers, NLA 1007/1/1094.
21 Cook, 'Frank Anstey', p. 379.
22 Shann and Copland (eds), *The Crisis in Australian Finance*, pp. 61–6, 70–1; Lyons in the *Argus*, 12 November 1930; Patrick Weller (ed.), *Caucus Minutes 1901–49: Minutes of the Meetings of the Federal Parliamentary Labor Party*, vol. 2, MUP, Melbourne, 1975, pp. 395–8.
23 Shann and Copland (eds), *The Crisis in Australian Finance*, pp. 102–46; Holder, *Bank of New South Wales*, chap. 34; Schedvin, *Australia and the Great Depression*, pp. 155–68; Robertson, *J.H. Scullin*, p. 298.
24 Giblin to Lyons, 1 September 1930, quoted in Philip R. Hart, 'J.A. Lyons: A Political Biography', PhD thesis, ANU, 1967, p. 67. In a voluminous literature E.C. Dyason, *The Australian Economic Outlook* (Herald and Weekly Times, Melbourne, 1930), L.F. Giblin, *Australia, 1930* (MUP, Melbourne, 1930) and Giblin, *Letters to John Smith* (Herald and Weekly Times, Melbourne, 1930), are noteworthy; other views are to be found in issues of the *Economic Record* during this period and the more important collective statements are reprinted in Shann and Copland (eds), *The Crisis in Australian Finance* and its sequel, *The Battle of the Plans: Documents Relating to the Premiers' Conference, May 25th to June 11th, 1931* (Angus and Robertson, Sydney, 1931). See also Neville Cain, *Recovery Policy in Australia, 1930–33: Certain Native Wisdom*, Working Papers in Economic History, ANU, Canberra, 1982.
25 R.F.I. Smith, 'The Scullin Government and the Wheatgrowers', *Labour History*, 26 (May 1974), pp. 49–64; R.N. Wait, 'Reactions to Demonstrations and Riots in Adelaide, 1928 to 1932', MA thesis, University of Adelaide, 1974, chap. 4.
26 Weller (ed.), *Caucus Minutes*, vol. 2, pp. 410–11; Denning, *Caucus Crisis*, p. 84; P.R. Hart, 'Lyons: Labor Minister—Leader of the UAP', *Labour History*, 17 (1970), pp. 37–51.
27 'Proceedings and Decisions of the Conference of Commonwealth and State Ministers, February 1931', *CPP*, 1929–31, vol. 2; Shann and

Copland (eds), *The Crisis in Australian Finance*, pp. 182–7.
28 Giblin, *The Growth of a Central Bank*, p. 89.
29 Weller (ed.), *Caucus Minutes*, vol. 2, p. 422; Scullin in the *Age*, 16 December 1931. See generally Heather Radi and Peter Spearritt (eds), *Jack Lang*, Hale and Iremonger, Sydney, 1977.
30 John Senyard, 'Mallee Farming Community in the Depression. The Walpepup Shire in Victoria, 1925–35', MA thesis, University of Melbourne, 1974, p. 226; *Age*, 7 March 1931.
31 Michael Cathcart, 'The White Army of 1931: Origins and Legitimations', MA thesis, University of Melbourne, 1985; C.J. Fox, 'Unemployment and the Politics of the Unemployed: Victoria in the Great Depression, 1930–37, PhD thesis, University of Melbourne, 1985. See generally Frank Cain, *The Origins of Political Surveillance in Australia*, Angus and Robertson, Sydney, 1983, chap. 6.
32 Eric Campbell, *The Rallying Point: My Story of the New Guard*, MUP, Melbourne, 1965; Keith Amos, *The New Guard Movement 1931–35*, MUP, Melbourne, 1976.
33 Alastair Davidson, *The Communist Party of Australia: A Short History*, Hoover Institution Press, Stanford, California, chap. 3; Frank Farrell, *International Socialism and Australian Labour*, Hale and Iremonger, Sydney, 1981, chap. 7; L.J. Louis, *Trade Unions and the Depression: A Study of Victoria, 1930–32*, ANU Press, Canberra, 1968.
34 Sir William Harrison Moore, 'Constitutional Development in Australia', *AQ*, 10 (June 1931), p. 19.
35 The general manager of the Bank of New South Wales, quoted in Shann and Copland (eds), *Battle of the Plans*, p. 41.
36 J.A. Lorimer, 'Riverina Movement', *AQ*, 10 (June 1931), p. 59.
37 Trevor Matthews, 'The All For Australia League', *Labour History*, 17 (1970), pp. 136–47; J. MacCarthy, '"All For Australia": Some Right Wing responses to the Depression in New South Wales, 1929–32', *JRAHS*, 57 (1971), pp. 160–71; *Sydney Sun*, 17 September 1931, quoted in Amos, *The New Guard*, p. 4.
38 Sir A. Grenfell Price, 'The Emergency Committee of South Australia and the Origin of the Premiers' Plan', *South Australiana*, 17 (1978), pp. 5–40; Colin Kerr, *Archie: The Biography of Sir Archibald Grenfell Price*, Macmillan, Melbourne, 1983, chap. 8; John Lonie, 'Non-Labor in South Australia', *Journal of the Historical Society of South Australia*, 2 (1976), pp. 30–45, and the same author's 'Conservatism and Class in South Australia during the Depression Years 1929–34', MA thesis, University of Adelaide, 1973.
39 *The House of Were*, p. 299; Geoff Hewitt, 'The All for Australia League in Melbourne', *La Trobe Historical Studies*, 3 (March 1972), pp. 5–15.
40 Hawker to J.G. Duncan Hughes, 10 January 1932, Needham Papers. NLA 4847, Box 7.
41 Latham, 'Notes', Latham Papers, NLA 1009/49/106; Hart, 'J.A. Lyons', chap. 3; D.R. Hall, 'The Historical Development of Australian Political Parties', in W.G.K. Duncan (ed.), *Trends in Australian Politics*, Angus and Robertson, Sydney, 1935, p. 23.
42 Shann and Copland (eds), *Battle of the Plans*, pp. 75–107.
43 Hogan quoted in Hal Colebatch, 'Inflation, Deflation or Common Sense?', *AQ*, 8 (December 1931), p. 31; see generally Rob Watts, 'Aspects of Unemployment Relief in Greater Melbourne', MA thesis, La Trobe University, 1974, chap. 3. Lonie, '"Good Labor Men"', pp. 14–29.
44 'Conference of the Commonwealth and State Ministers', *CPP*, 1929–

31, vol. 2, pp. 177–354; Shann and Copland (eds), *Battle of the Plans*, pp. 126–51.

45 Ricketson in *J.B. Were's Weekly Letter*, 28 May 1931; Menzies quoted in Hart, 'J.A. Lyons', p. 130 and in minutes of a meeting, 3 June, 1931, Harrison Moore Collection 9/3/6, University of Melbourne Archives; Eggleston in *Argus*, 30 May 1931.

46 Peter Cook, 'Labor and the Premiers' Plan', *Labour History*, 17 (1970), pp. 97–110.

47 *CPD*, vol. 132, pp. 1888–1906 (25 November 1931).

48 Robertson, *J.H. Scullin*, p. 363; Hart, 'J.A. Lyons', chap. 4.

49 J.T. Lang, *Why I Fight!*, Labor Daily, Sydney, 1934, pp. 216–17.

50 Brian Head, 'Economic Crisis and Political Legitimacy: the 1931 Federal Election', *JAS*, 3 (June 1978), pp. 14–29.

CHAPTER 12: WINNERS AND LOSERS

1 Australia, *Census*, 1933, vol. 2, p. 1908; *Commonwealth Year Book*, 1935, p. 554. The reliability of trade union returns is discussed by Colin Forster, 'Australian Unemployment, 1900–40', *Economic Record*, 41 (1965), pp. 426–50.

2 Tom Collins (Joseph Furphy), 'Introduction' to *Such Is Life, Bulletin*, Sydney, 1903.

3 These attitudes are discussed in Stuart Macintyre, *Winners and Losers: The Pursuit of Social Justice in Australian History*, Allen and Unwin, Sydney, 1985, chap. 4.

4 The best general survey is F.A. Bland, 'Unemployment Relief in Australia', *International Labour Review*, 30 (1934), pp. 23–57.

5 Events at Cairns are described in B.J. Costar, 'Labor, Politics and Unemployment: Queensland during the Great Depression', PhD thesis, University of Queensland, 1981, pp. 273–6. The principal studies of the unemployed experience are regional in focus: Geoffrey Bolton, *A Fine Country to Starve In*, UWA Press, Nedlands, 1972; Ray Broomhill, *Unemployed Workers: A Social History of the Great Depression in South Australia*, UQP, St Lucia, 1978; C.J. Fox, 'Unemployment and the Politics of the Unemployed'; Phyllis Peter, 'Social Aspects of the Depression in New South Wales, 1930–34', PhD thesis, University of Sydney, 1964; Len Richardson, *The Bitter Years: Wollongong during the Great Depression*, Hale and Iremonger, Sydney, 1984. Queensland and Tasmania await adequate accounts.

6 Muriel Heagney, *Are Women Taking Men's Job?*, Hilton and Veitch, Melbourne, 1935; Margaret Power, 'Women and Economic Crises: the Great Depression and the Present Crisis', in Windschuttle (ed.), *Women, Class and History*, pp. 492–513; Judy Mackinolty, 'Woman's Place ...', in Judy Mackinolty (ed.), *The Wasted Years? Australia's Great Depression*, Allen and Unwin, Sydney, 1981, pp. 94–110; Wendy Lowenstein, *Weevils in the Flour: An Oral Record of the 1930s Depression in Australia*, Hyland House, Melbourne, 1978, p. 153.

7 Bolton, *A Fine Country*, p. 115.

8 H.E. Graves, *Who Rides? Events in the Life of a West Australian Police Officer*, Lovat Dickson, London, 1937, pp. 169–70. There is an extensive body of material on the events in Perth on 6 March 1931 and an inquiry into them, in Battye Library, Acc. 430, File 2426/31.

9 Nadia Wheatley, '"The disinherited of the earth"?', in Mackinolty (ed.), *The Wasted Years?*, pp. 27–41, is a convenient summary of her massive study, 'The Unemployed Who Kicked: A Study of the Poli-

tical Struggles and Organisation of the New South Wales Unemployed in the Great Depression', MA thesis, Macquarie University, 1975.

10 Balmain, *Tocsin*, 1 July 1933, quoted in Wheatley, 'The Unemployed Who Kicked', p. 376.

11 Nadia Wheatley, 'Meeting Them at the Door: Radicalism, Militancy and the Sydney Anti-Eviction Campaign of 1931', in Jill Roe (ed.), *Twentieth Century Sydney: Studies in Urban and Social History*, Hale and Iremonger, Sydney, 1980, pp. 208–30.

12 Richardson, *The Bitter Years*, chap. 4; Sheilah Gray, '"An evil long endured"', in Mackinolty (ed.), *The Wasted Years?*, pp. 58–75; Wheatley, 'The Unemployed Who Kicked', pp. 351–70, 725–839.

13 Nadia Wheatley, 'New South Wales Relief Workers' Struggles, 1933–36', in Jill Roe (ed.), *Social Policy in Australia: Some Perspectives 1901–75*, Cassell, Stanmore, 1976, pp. 192–203; C.J. Fox, 'The Unemployed and the Labour Movement: The West Australia Relief and Sustenance Workers' Union, 1933–34', *SWAH*, 5 (December 1983), pp. 48–61; Shane Carmody, 'Class and Community in the Great Depression: Footscray and Hawthorn 1928–35', BA thesis, University of Melbourne, 1983, p. 61.

14 Dame Mabel Brookes, *Crowded Galleries*, Heinemann, London, 1956, chap. 16, and *Memoirs*, Macmillan, Melbourne, 1974, pp. 113–15; *Argus*, 7 December 1933. See generally Drew Cottle, 'The Rich in the Depression', *Bowyang*, 1, 1 (1979), pp. 2–24, 1, 2 (September–October 1979), pp. 67–102; Robert Pascoe, *Western Australia's Capital Suburb: Peppermint Grove*, OUP, Melbourne, 1983, chap. 3.

15 Geoffrey Blainey, *The Steel Master: A Life of Essington Lewis*, Macmillan, Melbourne, 1971, p. 105; *Charity Organisation Society Annual Report 1930–31*, p. 3; G.F.R. Spenceley, *Charity Relief in Melbourne: The Early Years of the 1930s Depression*, Monash Papers in Economic History, Monash University, 1980, especially p. 16.

16 Peter, 'Social Aspects of the Depression Experience', pp. 227–8.

17 *Australian Statesman*, August 1932, quoted in C.J. Lloyd, 'The Formation and Development of the United Australia Party, 1929–37', PhD thesis, ANU, 1984, p. 440.

18 D.B. Copland, *Australia in the World Crisis 1929–33*, CUP, Cambridge, 1934; L.F. Giblin, 'Australia in the Shadows', *AQ*, 20 (December 1933), p. 5.

19 Giblin to Theodore, 7 November 1932, quoted in Neville Cain, *Recovery Policy in Australia, 1930–33: Certain Native Wisdom*, Working Papers in Economic History, ANU, Canberra, 1982, p. 34.

20 N.G. Butlin, *Australian Domestic Product, Investment and Foreign Borrowing 1861–1938/9*, CUP, Cambridge, 1962, pp. 460–1.

21 M. Keating, *The Australian Workforce 1910–11 to 1960–61*, Department of Economic History, RSSS, ANU, Canberra, 1973, pp. 356–7.

22 The quotation comes from Alan J. Holt, *Wheat Farms of Victoria: A Sociological Survey*, School of Agriculture, University of Melbourne, 1946, p. 154. See generally the 'Royal Commission on the Wheat, Flour and Bread Industries', *CPP*, First Report, 1932–34, vol. 4; Second Report, 1934–37, vol. 4.

23 See generally Edgar Dunsdorfs, *The Australian Wheat Growing Industry 1788–1948*, MUP, Melbourne, 1956, chap. 6.

24 Holt, *Wheat Farms of Victoria*, p. 157.

25 L.F. Giblin, 'Farm Production and the Depression', *Economic Record*, 11 (1935), Recovery Measures Supplement, pp. 40–8; see also K.O.

Campbell, 'Australian Agricultural Production in the Depression: Explanations of its Behaviour', *Economic Record*, 20 (1944), pp. 58–73, and B.R. Davidson, *Australian Agriculture in the Great Depression*, Working Papers in Economic History, ANU, Canberra, 1985.

26 A.W. Dyer, 'Adjusting Primary Producers' Debts: South Australia in the 1930s', *AEHR*, 15 (1975), pp. 162–72; N.G. Butlin, A. Barnard and J.J. Pincus, *Government and Capitalism: Public and Private Choice in Twentieth Century Australia*, Allen and Unwin, Sydney, 1982, chap. 4.

27 W.K. Hancock, *Survey of British Commonwealth Affairs*, vol. 2, part 1, OUP, London, 1940, pp. 211–67; D.B. Copland and C.V. James (eds), *Australian Trade Policy: A Book of Documents 1932–37*, Angus and Robertson, Sydney, 1937, chap. 4; D.C.S. Sissons, 'Manchester v. Japan: The Imperial Background of the Australian Trade Diversion Dispute with Japan, 1936', *Australian Outlook*, 30 (1976), pp. 480–502.

28 Neville Cain and Sean Glynn, *Imperial Relations Under Strain: The British–Australian Debt Contretemps of 1933*, Working Papers in Economic History, ANU, Canberra, 1983; Cecil Edwards, *Bruce of Melbourne: Man of Two Worlds*, Heinemann, London, 1965, p. 220.

29 *The Australian Balance of Payments 1928–29 to 1948–49*, Australian Bureau of Census and Statistics, Canberra, 1950, p. 78.

30 C.B. Schedvin, *Australia in the Great Depression: A Study of Economic Development and Policy in the 1920s and 1930s*, SUP, Sydney, 1970, p. 302.

31 B.D. Haig, 'Manufacturing Output and Productivity 1910 to 1948/49', *AEHR*, 15 (1975), pp. 136–61; Helen Hughes, *The Australian Iron and Steel Industry 1848–1962*, MUP, Melbourne, 1964, chap. 5.

32 Schedvin, *Australia in the Great Depression*, pp. 345–50; Jim Hagan, *The History of the ACTU*, Longman Cheshire, Melbourne, 1981, pp. 139–43.

33 Diane Menghetti, 'The Weil's Disease Strike, 1935', in D.J. Murphy (ed.), *The Big Strikes: Queensland 1889–1965*, UQP, St Lucia, 1983, pp. 202–16; Robert Murray and Kate White, *The Ironworkers: A History of the Federated Ironworkers' Association of Australia*, Hale and Iremonger, Sydney, 1982, p. 70; L.J. Louis, 'Recovery from Depression and the Seamen's Strike of 1935–36', *Labour History* 41 (November 1981), pp. 74–86; Brian Fitzpatrick and Rowan J. Cahill, *The Seamen's Union of Australia 1872–1972*, Seamen's Union, Sydney, 1981, part I, chap. 11. See generally Lloyd Ross, 'Australia', in H.A. Marquand et al. (eds), *Organized Labour in Four Continents*, Longmans, London, 1939, pp. 457–78.

34 Peter Cochrane, 'The Wonthaggi Coal Strike, 1934', *Labour History*, 27 (November 1974), pp. 12–30; Robin Gollan, *The Coalminers of New South Wales*, MUP, Melbourne, 1963, chap. 10; Edgar Ross, *A History of the Miners Federation of Australia*, Australian Coal and Shale Employees' Federation, Sydney, 1970, chap. 14.

35 I have discussed this subject in *Militant: The Life and Times of Paddy Troy*, Allen and Unwin, Sydney, 1984, chap. 3. See generally Robin Gollan, *Revolutionaries and Reformists: Communism and the Australian Labor Movement 1920–50*, ANU Press, 1975, chap. 2.

CHAPTER 13: DEVIL'S DECADE

1 Bruce to Latham, 24 June 1932, Latham Papers, NLA 1009; Lyons quoted in C.J. Lloyd, 'Formation and Development of the United Australia Party 1929–37', PhD thesis, ANU, 1984, p. 353.

2 Lex Watson, 'The United Australia Party and its Sponsors', and Philip Hart, 'The Piper and the Tune', both in Cameron Hazlehurst (ed.), *Australian Conservatism: Essays on Twentieth Century Political History*, ANU Press, Canberra, 1979, pp. 71–149.
3 Earle Page, *Truant Surgeon: The Inside Story of Forty Years of Australian Political Life*, Angus and Robertson, Sydney, 1963, pp. 214–16; Ulrich Ellis, *A History of the Australian Country Party*, MUP, Melbourne, 1963, chap. 16.
4 *CPD*, vol. 138, p. 167 (10 March 1933). Hawker's memorandum of resignation is in the Needham Papers, NLA 4847.
5 R.J. May, *Financing the Small States in Australian Federalism*, OUP, Melbourne, 1971, chaps 2 and 4; R.S. Gilbert, *The Australian Loan Council in Federal Fiscal Adjustments 1890–1965*, ANU Press, Canberra, 1973, chaps 11–14.
6 E.D. Watt, 'Secession in Western Australia', *University of Western Australian History*, 3, 2 (October 1958), pp. 43–86; Geoffrey Bolton, *A Fine Country to Starve In*, UWA Press, Nedlands, 1972, pp. 249–56; Somerville Papers, Battye Library.
7 Lyons quoted in F.C. Green, *Servant of the House*, Heinemann, Melbourne, 1969, p. 116. See generally Lloyd, 'Formation and Development of the United Australia Party'; Philip R. Hart, 'J.A. Lyons: A Political Biography', PhD thesis, ANU, 1967.
8 *Sydney Morning Herald*, 15 March 1939; see generally R.W. Watts, 'The Light on the Hill: the Origins of the Australian Welfare State', PhD thesis, University of Melbourne, 1984, chaps 2 and 3.
9 Menzies diary, 21 March 1935, Menzies Papers, NLA, quoted in *Canberra Times*, 15 July 1982. On the bodyline tour see Brian Stoddart, 'Cricket's Imperial Crisis: the 1932–33 MCC Tour of Australia', in Richard Cashman and Michael McKernan (eds), *Sport in History*, UQP, St Lucia, 1979, pp. 124–47.
10 E.M. Andrews, *Isolation and Appeasement in Australia: Reactions to the European Crises, 1935–39*, ANU Press, Canberra, 1970; J.M. McCarthy, 'Australia and Imperial Defence: Co-operation and Conflict, 1918–1939', *AJPH*, 17 (1971), pp. 19–31.
11 The New South Wales and Victorian governments are discussed in Hazlehurst (ed.), *Australian Conservatism*, pp. 149–91.
12 Forgan Smith quoted in Ross Fitzgerald, *From 1915 to the Early 1890s: A History of Queensland*, UQP, St Lucia, 1984, p. 181. See Brian Carroll, 'William Forgan Smith: Dictator or Democrat?' in D.J. Murphy and R.B. Royce (eds), *Queensland Political Portraits 1859–1952*, UQP, St Lucia, 1978, pp. 397–432; R.F. Pervan, 'Cabinet and Caucus: Labor in Western Australia, 1933–47', *University Studies in History*, 5, 1 (1967), pp. 1–37; Richard Davis, *Eighty Years' Labor: the ALP in Tasmania, 1903–83*, History Department, University of Tasmania, 1983, pp. 31–7.
13 Willcock quoted in Lenore Layman, 'Development Ideology in Western Australia 1933–1965', *Historical Studies*, 20 (1982–83), p. 237; Alan Beever, *The Forty-Hour Week Movement in Australia 1930–48*, Working Papers in Economic History, ANU, Canberra, 1985.
14 Michael Howard, 'The Growth in the Domestic Economic and Social Role of the Commonwealth Government, from the Late 1930s to the Early Post-War Period', PhD thesis, University of Sydney, 1978, chap. 1.
15 T.J. Mitchell, 'J.W. Wainwright: the Industrialisation of South Australia, 1935–40', *AJPH*, 8 (1962), pp. 27–40.

16 Frank Cain, *The Origins of Political Surveillance in Australia*, Angus and Robertson, Sydney, 1983, chap. 7; L.J. Louis, 'Victoria Council Against War and Facism: A Rejoinder', *Labour History*, 44 (May 1983), pp. 39–54; C.J. Fox, 'Unemployment and the Politics of the Unemployed: Victoria in the Great Depression 1930–37', PhD thesis, University of Melbourne, 1985, chap. 9.
17 M.J. Costar, 'Labor, Politics and Unemployment: Queensland during the Great Depression', PhD thesis, University of Queensland, 1981, p. 289.
18 Rolf Gerritsen, 'The 1934 Kalgoorlie Riots: a Western Australian Crowd', *University Studies in History*, 5, 3 (1969), pp. 42–75; Robert Pascoe and Patrick Bertola, 'Italian Miners and Second Generation "Britishers" at Kalgoorlie, Australia', *Social History*, 10 (1985), pp. 9–35.
19 Lawson quoted in Gianfranco Cresciani, *Fascism, Anti-Fascism and Italians in Australia, 1922–45*, ANU Press, Canberra, 1980, p. 139; Menzies quoted in Cameron Hazlehurst, *Menzies Observed*, Allen and Unwin, Sydney, 1979, p. 138.
20 Cresciani, *Fascism, Anti-Fascism and Italians in Australia*; Andrews, *Isolation and Appeasement*, pp. 108, 160–1; Don Watson, *Brian Fitzpatrick: A Radical Life*, Hale and Iremonger, Sydney, 1979, pp. 85–7.
21 Frank Farrell, *International Socialism and Australian Labour: The Left in Australia 1919–39*, Hale and Iremonger, Sydney, 1981, chap. 8; Egon Kisch, *Australian Landfall*, Secker and Warburg, London, 1937; Giblin quoted in Watson, *Brian Fitzpatrick*, p. 54.
22 Jon White, 'The Port Kembla Pig Iron Dispute of 1938', *Labour History*, 37 (November 1979), pp. 63–77.
23 Peter Coleman, *Obscenity, Blasphemy, Sedition: 100 Years of Censorship in Australia*, 2nd edn, Angus and Robertson, Sydney, 1974, chap. 1.
24 James Devaney, *Shaw Neilson*, Angus and Robertson, Sydney, 1944, p. 22.
25 K.S. Inglis, *This is the ABC: The Australian Broadcasting Commission 1932–83*, MUP, Melbourne, 1983, p. 69.
26 Richard Haese, *Rebels and Precursors: The Revolutionary Years of Australian Art*, Allen Lane, Ringwood, 1981, chaps 1 and 2; Humphrey McQueen, *The Black Swan of Trespass: The Emergence of Modernist Painting in Australia to 1944*, APCOL, Sydney, 1979.
27 P.R. Stephensen, *The Foundations of Culture in Australia*, W.J. Miles, Sydney, 1936, p. 17.
28 Watson, *Brian Fitzpatrick*.
29 R.M. Crawford, *An Australian Perspective*, University of Wisconsin Press, Madison, 1960, pp. 68–70.
30 Stephen Alomes, 'Reasonable Men: Middle Class Reformism in Australia 1928–39', PhD thesis, ANU, 1979.
31 *Age*, 15, 20 August 1935; F.O. Barnett, 'I Remember', typescript memoirs, La Trobe Library; Leonie Sandercock, *Cities for Sale*, MUP, Melbourne, 1977, chap. 3.
32 F.A. Bland, 'Recent Housing Legislation in New South Wales', *Economic Record*, 14 (1938), pp. 78–82; Sandercock, *Cities for Sale*, chap. 4.
33 Peter Spearritt, 'Sydney's "Slums": Middle Class Reformers and the Labor Response', *Labour History*, 26 (May 1974), pp. 65–81.
34 Graeme Davison, 'The Melbourne University Social Survey, 1941–43', *Australia 1938–88*, 3 (n.d.), pp. 6–10; N.H. Dick, 'Housing and

Slum Clearance in New South Wales', *AQ*, 28 (December 1935), pp. 81–5.

35 Daisy Bates, *The Passing of the Aborigines*, Murray, London, 1938, p. 237.

36 L.R. Smith, *The Aboriginal Population of Australia*, ANU Press, Canberra, 1980, p. 199; Andrew Markus, 'After the Outward Appearance: Scientists, Administrators and Politicians', in Bill Gammage and Andrew Markus (eds), *All That Dirt: Aborigines 1938*, History Project Incorporated, Sydney, 1982, pp. 83–106; C.D. Rowley, *The Destruction of Aboriginal Society*, Penguin, Ringwood, 1972, chaps 16 and 17.

37 Jack Horner, *Vote Ferguson for Aboriginal Freedom*, ANZ, Sydney, 1974, chaps 3–5.

38 For critical discussion of this view see Ian Hogbin, 'Our Native Policy', *AQ*, 15, 2 (June 1943), pp. 100–8.

39 Edward P. Wolfers, *Race Relations and Colonial Rule in Papua New Guinea*, ANZ, Sydney, 1975, chaps 3–4, 7.

40 The expatriate is quoted in Hank Nelson, *Taim Bilong Masta: The Australian Involvement with Papua New Guinea*, ABC, Sydney, 1982, p. 38. See generally his 'From Kanaka to Fuzzy Wuzzy Angel', in Ann Curthoys and Andrew Markus (eds), *Who are Our Enemies? Racism and the Australian Working Class*, Hale and Iremonger, Sydney, 1978, pp. 172–88.

41 Sir Hubert Murray, *The Response of the Natives of Papua to Western Civilization*, Government Printer, Port Moresby, 1929, p. 13.

42 Peter Cowan, *A Unique Position: A Biography of Edith Dircksey Cowan 1861–1932*, UWA Press, Nedlands, 1978, p. 193. I draw here on the arguments of Kerreen Reiger, *The Disenchantment of the Home: Modernizing the Australian Family 1880–1940*, OUP, Melbourne, 1985.

43 Andrew Wright, 'Women in the Family: Women's Weekly Between the Wars', *Refractory Girl*, 3 (1973), pp. 9–13; Denis O'Brien, *The Weekly*, Penguin, Ringwood, 1982, chaps 1–3.

44 Graham Dabscheck, *Arbitrator at Work: Sir William Raymond Kelly and the Regulation of Australian Industrial Relations*, Allen and Unwin, Sydney, 1983, p. 94.

45 Patricia Ranald, 'Feminism and Class: The United Associations of Women and the Council of Action for Equal Pay', in Margaret Bevege et al. (eds), *Worth Her Salt: Women at Work in Australia*, Hale and Iremonger, Sydney, 1982, pp. 270–85; Peter Sekuless, *Jessie Street*, UQP, St Lucia, 1978, chap. 3.

46 C. Hartley Grattan, 'The Future in Australia', *AQ*, 10, 4 (December 1938), pp. 7–29; see also his *Introducing Australia*, John Day, New York, 1942, especially chaps 1 and 2.

47 *Argus*, 7 April 1932, quoted in F.K. Crowley, *Modern Australia in Documents*, vol. 1, Wren, Melbourne, 1973, p. 513.

48 P.G. Edwards (ed.), *Australia Through American Eyes 1935–45: Observations by American Diplomats*, UQP, St Lucia, 1979, especially pp. 11–13, 25.

49 Ian McLean and Jonathan Pincus, *Living Standards in Australia 1890–1940: Evidence and Conjectures*, Working Papers in Economic History, ANU, Canberra, 1982; W.D. Rubinstein, 'The Distribution of Personal Wealth in Victoria 1860–1974', *AEHR*, 19 (1979), pp. 26–41; and 'The Top Wealth-holders of New South Wales, 1817–1939', *AEHR*, 20 (1980), pp. 136–52.

50 Alan Walker, *Coaltown: A Social History of Cessnock*, MUP, Mel-

bourne, 1945, especially pp. 27–8; Alan J. Holt, *Wheat Farms of Victoria: A Sociological Survey*, School of Agriculture, University of Melbourne, 1946, especially p. 65.

CHAPTER 14: 'FREE OF ANY PANGS'

1 R.G. Neale (ed.), *Documents on Australian Foreign Policy 1937–49*, vol. 2, AGPS, Canberra, 1976, p. 221.
2 ibid., p. 226.
3 ibid., p. 256; Paul Hasluck, *The Government and the People 1939–41*, Australia in the War of 1939–45, series 4, vol. 1, Australian War Memorial, Canberra, 1952, p. 152; J.M. McCarthy, 'Australia: a View from Whitehall 1939–45', *Australian Outlook*, 28 (1974), pp. 318–31.
4 Neville Meaney (ed.), *Australia and the World: A Documentary History* from the 1870s to the 1970s, Longman Cheshire, Melbourne, 1985, p. 462.
5 John Hetherington, *Blamey: Controversial Soldier*, Australian War Memorial, Canberra, 1973, p. 183; see generally D.M. Horner, *High Command: Australia and Allied Strategy 1939–45*, Allen and Unwin, Sydney, 1982, chaps 2, 4–5.
6 Neale (ed.), *Documents*, vol. 2, pp. 98, 441. See P.G. Edwards, 'Menzies and the Imperial Connection, 1939–41', in Cameron Hazlehurst (ed.), *Australian Conservatism: Essays in Twentieth Century Political History*, ANU Press, Canberra, 1979, pp. 193–212.
7 Gavin Long, *To Benghazi*, Australia in the War of 1939–45, series 1, vol. 1, Australian War Memorial, Canberra, 1952, chap. 3.
8 Gianfranco Cresciani, *Fascism, Anti-Fascism and Italians in Australia, 1922–45*, ANU Press, Canberra, 1980, chap. 8; Craig Johnston, 'The "Leading War Party": Communists and World War Two', *Labour History*, 39 (November 1980), pp. 62–77; Bruce Muirden, *The Puzzled Patriots: The Story of the Australia First Movement*, MUP, Melbourne, 1968; Benzion Patkin, *The Dunera Internees*, Cassell, Sydney, 1979; Hasluck, *The Government and the People 1939–41*, pp. 238–42.
9 S.J. Butlin, *War Economy 1939–42*, Australia in the War of 1939–45, series 4, vol. 3, Australian War Memorial, Canberra, 1955, chap. 15.
10 Menzies quoted in Meaney (ed.), *Australia and the World*, p. 457; Blamey in Hetherington, *Blamey*, p. 189.
11 A.P. Elkin, *Our Opinions and the National Effort*, the author, Sydney, 1941, p. 17, quoted in Hasluck, *The Government and the People 1939–41*, p. 381.
12 The atmosphere is best conveyed in the memoirs of the participants: see Arthur Fadden, *They Called Me Artie: The Memoirs of Sir Arthur Fadden*, Jacaranda, Brisbane, 1969; Sir Robert Menzies, *Afternoon Light*, Cassell, Melbourne, 1967, chap. 2; Earle Page, *Truant Surgeon: The Inside Story of Forty Years of Australian Political Life*, Angus and Robertson, Sydney, 1963, chaps 29–31; Percy Spender, *Politics and a Man*, Collins, London, 1972, chaps 4–11.
13 Neale (ed.), *Documents*, vol. 3, pp. 256–7.
14 P.G. Edwards, 'S.M. Bruce, R.G. Menzies and Australia's War Aims and Peace Aims, 1939–40', *Historical Studies*, 17 (1976–77), pp. 1–14.
15 D. Dilks (ed.), *The Diaries of Sir Alexander Cadogan 1938–45*, Cassell, London, 1971, p. 359.
16 Hasluck, *The Government and the People 1939–41*, p. 313.
17 ibid., chap. 12; Geoffrey Sawer, *Australian Federal Politics and Law 1929–49*, MUP, Melbourne, 1953, chaps 4 and 5.

18 Meaney (ed.), *Australia and the World*, p. 473; see generally Roger Bell, *Unequal Allies: Australian–American Relations in the Pacific War*, MUP, Melbourne, 1977, chaps 1 and 2.

19 Carl Bridge contends in 'Australia's First Washington Legation and the Origins of the Pacific War 1940–42', *AJPH*, 28 (1982), pp. 181–9, that Australian representatives were aware of the agreement between the United States and Britain; this is disputed by J.R. Robertson, 'Australia and the "Beat Hitler First" Strategy, 1941–42: A Problem in Wartime Consultation', *Journal of Imperial and Commonwealth History*, 11 (1983), pp. 300–21.

20 Noel Barber, *Sinister Twilight: The Fall and Rise Again of Singapore*, Collins, London, 1968, p. 28; Kate Caffrey, *Out in the Midday Sun: Singapore 1941–45*, André Deutsch, London, 1974, p. 77; Horner, *High Command*, p. 171.

21 Lionel Wigmore, *The Japanese Thrust*, Australia in the War of 1939–45, series 1, vol. 4, Australian War Memorial, Canberra, 1957.

22 Curtin quoted in W.J. Hudson and H.J.W. Stokes (eds), *Documents on Australian Foreign Policy 1937–49*, vol. 5, AGPS, Canberra, 1982, p. 463; Gilmore in Denis O'Brien, *The Weekly*, Penguin, Ringwood, 1982, p. 75. See generally Lloyd Ross, *John Curtin, a Biography*, Macmillan, Melbourne, 1977, chaps 18–19.

23 Vance Palmer, 'Battle', *Meanjin*, 1, 8 (March 1942), pp. 5–6.

SOURCES OF ILLUSTRATIONS

Page

xix Battye Library
3 Archives of Business and Labour, Australian National University
8 State Library of Victoria
11 Battye Library
14 Reproduced with the kind permission of Jocelyn Treasure
20 Reproduced with the kind permission of Jeremy Beckett
26 University of Melbourne Archives
29 Thomas Cherry, *Victorian Agriculture*, D.W. Paterson, Melbourne, 1913
33 *Australia To-Day*, 1 November 1911. Reproduced by permission of University of Melbourne Archives
37 Archives Office of New South Wales
40 Archives Office of New South Wales
62 Ernest Scott, *Historical Memoir of the Melbourne Club*, Speciality Press, Melbourne 1936
83 National Library of Australia
90 National Library of Australia
94 Brisbane *Worker*, 18 March 1911
97 E.H. Sugden and F.W. Eggleston, *George Swinburne*, Angus and Robertson, Sydney, 1931
107 *Australia To-Day* collection, University of Melbourne Archives
114 Charles Rasp Memorial Library, Broken Hill

119 Sydney *Mail*, 6 September 1911
120 *Queenslander*, 17 February 1912
131 Archives Office of New South Wales
132 *Australia To-Day*, 14 December 1907, University of Melbourne Archives
43 Australian War Memorial
44 Robert Bodington collection, University of Melbourne Archives
48 Australian War Memorial
50 Australian War Memorial
53 Swallow and Ariell collection, University of Melbourne Archives
58 Australian War Memorial
60 Australian War Memorial
86 National Library of Australia
88 National Library of Australia
11 Huddart Parker collection, University of Melbourne Archives
23 Cecil Edwards, *Bruce of Melbourne: Man of Two Worlds*, Heinemann, London, 1965
49 *Australian Worker*, 25 September 1929
72 National Library of Australia
35 Reproduced with the kind permission of Footscray Historical Society
2 BHP Archives
8 Melbourne *Herald*, 2 January 1932
2 National Library of Australia
9 *Sydney Morning Herald*. Reproduced by courtesy of John Fairfax and Sons Limited
1 *Sydney Morning Herald*, 12 January 1939. Reproduced by courtesy of John Fairfax and Sons Limited
6 *Victorian Parliamentary Papers*, 1937, vol. 2
9 *Bulletin*, 12 February 1941
4 Australian War Memorial

BIBLIOGRAPHIC ESSAY

Suggestions for further reading must of necessity be selective. The following selection lists some standard works, and readers are reminded of the more extensive range of material cited in the endnotes.

The most recent general history, Frank Crowley (ed.), *A New History of Australia* (Heinemann, Melbourne, 1974) is now somewhat dated,but its relevant chapters—and especially that of Heather Radi on the 1920s—still provide the most authoritative introduction to the period. Crowley also compiled *Modern Australia in Documents* (vol. 1, 1901–39 Wren, Melbourne, 1973), which, with Humphrey McQueen's more adventurous *Social Sketches of Australia, 1888–1975* (Penguin, Ringwood, 1975), contains a useful collection of primary sources. The fifth and forthcoming sixth volumes of C.M.H. Clark's magisterial *History of Australia* (MUP, Melbourne, 1981, 1987) cover the first half of the twentieth century. Gavin Souter, *Lion and Kangaroo: The Initiation of Australia 1901–19* (Collins, Sydney, 1976) deftly evokes the early Commonwealth.

State histories are an uneven genre. Western Australia is served best, with C.T. Stannage (ed.), *A New History of Western Australia* (UWA Press, Nedlands, 1981) supplementing and extending F.K. Crowley, *Australia's Western Third* (Heinemann, Melbourne, 1960). The polemical thrust of Ross Fitzgerald's two Queensland volumes, *From the Dreaming to 1915* and *From 1915 to the Early 1980s* (UQP, St Lucia, 1982, 1984) limits his rich material. Victoria's sesquicentennial series—*Arriving* by Richard Broome, *Settling* by Tony Dingle, and *Making Their Mark* by Susan Priestley (Fairfax, Syme and Weldon, Melbourne, 1984)—are also constrained by their themes and deploy impressive techniques only to rework the pioneer legend. Those seeking a more comprehensive approach will have to choose between Donald Garden's stolid *Victoria: A History* (Nelson, Melbourne, 1984) and Geoffrey Blainey's more impressionistic *Our Side of the Country* (Methuen Hayes, Melbourne, 1984). Another sesquicentenary brings the more orthodox *Flinders History of South Australia* (Wakefield Press, Adelaide, 1986–87) in three volumes: social, political and economic. Lloyd Robson, *A Short History of Tasmania* (OUP, Melbourne, 1985) and

Alan Powell, *Far Country: A Short History of the Northern Territory* (MUP, Melbourne, 1982) are both serviceable. All the states have their own historical journal, although New South Wales assumes national horizons with its *Journal of the Royal Australian Historical Society*; and while the Premier State leads the way in the quantity and quality of its local historical activities, it has not so far felt the need for a state history.

Economic history has been heavily influenced by the quantitative methods of N.G. Butlin, whose *Australian Domestic Product, Investment and Foreign Borrowing 1861–1938/9* (CUP, Cambridge, 1962) remains the starting-point for subsequent work. He was a contributor to Colin Forster (ed.), *Australian Economic Development in the Twentieth Century* (Allen and Unwin, Sydney, 1970), and his *Government and Capitalism: Public and Private Choice in Twentieth Century Australia* (Allen and Unwin, Sydney, 1982), written with Alan Barnard and J.J. Pincus, applies the same methodology to the public sector. W.A. Sinclair, *The Process of Economic Development in Australia* (Cheshire, Melbourne, 1976) and E.A. Boehm, *Twentieth Century Economic Development in Australia*, 2nd edn (Longman Cheshire, Melbourne, 1979) are the standard textbooks, exhibiting the same concern for economic growth. In *Industrialization and Dependence: Australia's Road to Economic Development* (UQP, St Lucia, 1980), Peter Cochrane offered a Marxist analysis of the theme, while the five volumes of *Essays in the Political Economy of Australian Capitalism* edited by E.L. Wheelwright and Ken Buckley (ANZ, Sydney, 1975–83) contain a number of relevant contributions. Colin Forster, *Industrial Development in Australia 1920–30* (ANU Press, Canberra, 1964) and C.B. Schedvin, *Australia and the Great Depression* (SUP, Sydney, 1970) are the only substantial treatments of their respective periods. B.R. Davidson, *European Farming in Australia* (Elsevier, New York, 1981) is the best introduction to the primary industries; Geoffrey Blainey, *The Rush That Never Ended*, 3rd edn (MUP, Melbourne, 1978) is the standard mining history; Helen Hughes treats heavy industry in *The Australian Iron and Steel Industry, 1848–1962* (MUP, Melbourne, 1964), and on finance there is A.R. Hall, *The London Capital Market and Australia 1870–1914* (ANU Press, Canberra, 1963) as well as individual bank histories.

Geoffrey Bolton surveys the history of the Australian environment in *Spoils and Spoilers* (Allen and Unwin, Sydney, 1981) and the work of the demography school at the ANU can be approached via Lado T. Ruzicka and John Caldwell, *The End of Demographic Transition in Australia* (ANU Press, Canberra, 1976).

Modern feminist historical writing began in 1975 with a cluster of major publications: Anne Summers, *Damned Whores and God's Police: The Colonization of Women in Australia* (Penguin, Ringwood); Beverley Kingston, *My Wife, My Daughter and Poor Mary-Ann: Women and Work in Australia* (Nelson, Melbourne); Edna Ryan and Anne Conlon, *Gentle Invaders: Australian Women at Work* (Nelson, Melbourne), and Ann Curthoys et al. (eds), *Women at Work* (Australian Society for the Study of Labour History, Canberra). All retain their relevance. Much subsequent work sought to retrieve neglected aspects of women's experience and public activity; Elizabeth Windschuttle (ed.), *Women, Class and History: Feminist Perspectives on Australia 1788–1978* (Fontana, Melbourne, 1980) and Margaret Bevege et al. (eds), *Worth Her Salt: Women and Work in Australia* (Hale and Iremonger, Sydney, 1982) both derived from the immensely successful Conferences on Women and Labour. More recent work has turned towards the construction of the domestic sphere and a more comprehensive treatment of gender relations: Kerreen Reiger, *The Disenchantment of the Home: Modernizing the Australian Family 1880–1940* (OUP, Melbourne, 1985) and Jill Julius Matthews, *Good and Mad Women: The Historical Construction of Femininity in Twentieth Century Australia* (Allen and Unwin, Sydney, 1984) are outstanding examples. Kay Daniels and Mary Murnane offer a rich collection of primary sources in *Uphill All the Way: A Documentary History of Women in Australia* (UQP, St Lucia, 1980).

A quickening interest in the circumstances of Aborigines began with C.D. Rowley's trilogy; *The Destruction of Aboriginal Society* (Penguin, Ringwood, 1972) covers our period. Richard Broome, *Aboriginal Australians: Black Responses to White Dominance 1788–1980* (Allen and Unwin, Sydney, 1982) is a recent survey, while A.T. Yarwood and M.J. Knowling take in other minorities in *Race Relations in Australia: A History* (Methuen, Melbourne, 1982). Such syntheses can only be provisional since this literature, as with that concerning gender, is changing rapidly. The journals *Aboriginal History* and *Hecate* help the interested reader to keep abreast of new work in the two areas.

While there is no lack of material bearing on class relations, much of it is frustratingly inchoate. Labour history is an established sub-discipline with its own society, journal (*Labour History*) and an impressive quantity of detailed studies. On the encouragement of that journal, practitioners have moved from their initial concentration on the institutions of the labour movement to examination of work processes and various aspects of working-class life; yet the new literature has not so far generated a coherent reinterpretation. R.A. Gollan, *Radical and Working Class Politics: A Study of Eastern Australia, 1850–1910* (MUP, Melbourne, 1960) remains the clearest introduction to the period and Ian Turner, *Labour and Politics: The Dynamics of the Labour Movement in Eastern Australia 1900–21* (ANU Press, Canberra, 1965) the most persuasive account of the early history of the labour movement. The capitalist class has not received the same attention. Few business historians bring out the full potential of their subject and even the better biographers (Geoffrey Blainey's *The Steel Master: A Life of Essington Lewis* (Macmillan, Melbourne, 1971) has been described as 'perhaps the most illuminating single book on Australia's recent history') treat class obliquely. Those described here as 'the anxious class'—the small property-owners, shopkeepers, lesser professionals and suburban salariat—remain, as Menzies described them in 1942, the forgotten people. Brian Fitzpatrick's attempt to comprehend the historical dynamics of class relations, notably in *The British Empire in Australia* (MUP, Melbourne, 1941) stood alone until R.W. Connell and T.H. Irving ventured their *Class Structure in Australian History* (Longman Cheshire, Melbourne, 1980). None of its critics have chanced their own arm. On a more restricted scale, John Rickard offers a perceptive view of *Class and Politics: New South Wales, Victoria and the Early Commonwealth 1890–1910* (ANU Press, Canberra, 1976), and Tim Rowse provides an illuminating interpretation of the class dynamics of political ideology in *Australian Liberalism and National Character* (Kibble Books, Malmsbury, 1978).

Geoffrey Sawer's *Australian Federal Politics and Law 1901–29* and *1929–49* (MUP, Melbourne, 1956, 1963) together with Colin Hughes and B.D. Graham, *A Handbook of Australian Government and Politics 1890–1964* (ANU Press, Canberra, 1964) lie at the elbow of the student of political history. But after P. Loveday et al. (eds), *The Emergence of the Australian Party System* (Hale and Iremonger, Sydney, 1977), there is no systematic treatment of politics at the Commonwealth, state or local levels. The reader is thrown back on specialist studies and biographies. Cameron Hazlehurst (ed.), *Australian Conservatism: Essays on Twentieth Century Political History* (ANU Press, Canberra, 1979) has to substitute for comprehensive accounts of the Nationalist and United Australia parties; and D.J. Murphy (ed.), *Labor in Politics: The State Labor Parties in Australia 1880–1920* (UQP, St Lucia, 1975) provides some context for L.F. Crisp, *The Australian Federal Labor Party 1901–51* (Longmans, London, 1955). The Country Party is better served by B.D. Graham, *The Formation of the Australian Country Parties* (ANU Press, Canberra, 1966). Among the biographies, J.A. La Nauze, *Alfred Deakin: A Biography*, 2 vols (MUP, Melbourne, 1965), L.F. Fitzhardinge, *William Morris Hughes: A Political Biography*, 2 vols (Angus and Robertson, Sydney, 1964, 1979) and J.R. Robertson, *J.H. Scullin, A Political Biography* (UWA Press, Nedlands, 1974) are especially useful. The regular commentary on Australian events provided by the conservative *Round Table* is worth consultation.

T.B. Millar, *Australia in Peace and War: External Relations 1788–1977* (ANU Press, Canberra, 1978) is the best introduction to foreign affairs, and Neville Meaney (ed.), *Australia and the World: A Documentary History from the 1870s to the 1970s* (Longman Cheshire, Melbourne, 1985) is a wide-ranging collection. Hank Nelson introduces our principal ex-colony in *Taim Bilong Masta: The Australian Involvement with Papua New Guinea* (ABC, Sydney, 1982). The official histories of the First and Second World Wars, edited by C.E.W. Bean and Gavin Long respectively, remain authoritative. Of recent works, Michael McKernan, *The Australian People and the Great War* (Nelson, Melbourne, 1980) gives general coverage, and Geoffrey Serle, *John Monash: A Biography* (MUP, Melbourne, 1982) is a fine study of the most senior Australian commander. McKernan returns to the fray with *All In! Australia during the Second World War* (Nelson, Melbourne, 1983), though John Robertson, *Australia Goes to War 1939–45* (Doubleday, Sydney, 1984) is more thorough on political as well as military aspects. W.K. Hancock's masterly *Survey of British Commonwealth Affairs*, 2 vols (OUP, London, 1937–42) situates Australia within its imperial context and Donald Denoon's *Settler Capitalism: The Dynamics of Dependent Development in the Southern Hemisphere* (Clarendon Press, Oxford, 1983) is an important exercise in comparative history.

T.H. Kewley's monumental institutional history of *Social Security in Australia 1900–72* (SUP, Sydney, 1973) can be supplemented with the essays edited by Jill Roe, *Social Policy in Australia: Some Perspectives* (Cassell, Stanmore, 1975) and Richard Kennedy (ed.), *Australian Welfare History: Critical Essays* (Macmillan, Melbourne, 1982), as well as Brian Dickey, *No Charity There! A Short History of Social Welfare in Australia* (Nelson, Melbourne, 1980). Alan Barcan, *A History of Australian Education* (OUP, Melbourne, 1980) is stodgy fare but there is no other comprehensive treatment of the subject. L.J. Louis and Ian Turner brought together material on public policy in *The Depression of the 1930s* (Cassell, Melbourne, 1968), and Judy Mackinolty (ed.), *The Wasted Years? Australia's Great Depression* (Allen and Unwin, Sydney, 1981) includes an extensive bibliography.

The editors of the *Journal of Religious History* have complained that their subject receives insufficient attention from Australian historians. The complaint is justified but a large portion of the blame attaches to practitioners who have failed to show how their subject bears on the society at large. This criticism cannot be levelled at Patrick O'Farrell's stimulating account of *The Catholic Church and Community in Australia* (Nelson, Melbourne, 1977), nor at J.D. Bollen, *Protestantism and Social Reform in New South Wales 1890–1910* (MUP, Melbourne, 1972) and Richard Broome, *Treasure in*

Earthen Vessels: Protestant Christianity in New South Wales Society 1900–14 (UQP, St Lucia, 1980)—but the terminal dates of the last two books are significant. The churches between the wars are in need of attention.

Sports historians, on the other hand, have been only too willing to press the larger significance of their enthusiasms. The essays collected in Richard Cashman and Michael McKernan (eds), *Sport in History* (UQP, St Lucia, 1979), and *Sport, Money, Morality and the Media* (New South Wales University Press, Kensington, 1981) range widely, so widely that the games themselves are in danger of disappearing under the epiphenomenal overburden. A similarly portentous tone lingers about some of the contributions to Peter Spearritt and David Walker (eds), *Australian Popular Culture* (Allen and Unwin, Sydney, 1979), and Susan Dermody et al. (eds), *Nellie Melba, Ginger Meggs and Friends: Essays in Australian Cultural History* (Kibble Books, Malmsbury, 1982).

Cultural history deserves greater attention than I have given it. Geoffrey Serle's survey, *From Deserts the Prophets Come: The Creative Spirit in Australia* (Heinemann, Melbourne, 1973) takes in the main forms of artistic expression. The two original volumes of H.M. Green, *A History of Australian Literature*, have been revised by Dorothy Green (Angus and Robertson, Sydney, 1984), and Drusilla Modjeska establishes the crucial significance of the inter-war female novelists in *Exiles At Home: Australian Women Writers 1925–45* (Angus and Robertson, Sydney, 1981). Bernard Smith is also the unrivalled authority on *Australian Painting 1788–1970* (OUP, Melbourne, 1971), though Humphrey McQueen gives a stimulating account of the impact of modernism in *The Black Swan of Trespass* (APCOL, Sydney, 1979). While Richard White, *Inventing Australia: Images and Identity 1688–1980* (Allen and Unwin, 1980) is an illuminating study of the national identity, a full history of Australian nationalism remains to be written.

The student of the period should not overlook memoirs and autobiographies. The better ones are richly evocative, combining a retrospective awareness of the particularities of their childhood with a post-Freudian awareness of its larger significance. Hal Porter, *The Watcher on the Cast-Iron Balcony* (Faber, London, 1963), George Johnston, *My Brother Jack* (Collins, London, 1964), Donald Horne, *The Education of Young Donald* (Angus and Robertson, Sydney, 1967), Amirah Inglis, *Amirah: An un-Australian Childhood* (Heinemann, Melbourne, 1983) and Bernard Smith, *The Boy Adeodatus* (Allen Lane, Ringwood, 1984) are noteworthy examples.

Finally, those seeking further guidance will turn to D.H. Borchardt, *Australian Bibliography: A Guide to the Printed Sources of Information*, 3rd edn (Pergamon, Sydney, 1976) and Henry Mayer et al., *A Research Guide to Australian Politics and Cognate Subjects* (Cheshire, Melbourne, 1976), along with the supplementary *Second Guide* that followed in 1984. These will be complemented by *Australians: a Guide to Australian Historical Sources* compiled by D.H. Borchardt and Victor Crittenden, to be published as a volume of the bicentennial series *Australians 1788–1988: A Historical Library*. Kay Daniels et al. (eds), *Women in Australia: an Annotated Guide to the Records*, 2 vols (AGPS, Canberra, 1977) ranges widely. Journal articles, an important but elusive form of historical writing, are made accessible by Terry Hogan et al., *Index to Journal Articles on Australian History* (University of New England Publishing Unit, Armidale, n.d. [1976]), which covers the periodical literature up to the end of 1973. An index with the same title for the years 1974–78 was prepared by Victor Crittenden et al. under the banner of the bicentennial history project, and has been followed by annual supplements (History Project Incorporated, Kensington). Among works of reference, the *Australian Dictionary of Biography* can be singled out.

INDEX